Environmental Statistics

Environmental Statistics
Methods and Applications

VIC BARNETT
Nottingham Trent University, UK

John Wiley & Sons, Ltd

Telephone (+44) 1243 779777

Email (for orders and customer service enquiries): cs-books@wiley.co.uk
Visit our Home Page www.wileyeurope.com or www.wiley.com

Reprinted November 2004, February 2006

This publication is designed to provide accurate and authoritative information in regard to
the subject matter covered. It is sold on the understanding that the Publisher is not engaged
in rendering professional services. If professional advice or other expert assistance is
required, the services of a competent professional should be sought.

Other Wiley Editorial Offices

John Wiley & Sons Inc., 111 River Street, Hoboken, NJ 07030, USA

Jossey-Bass, 989 Market Street, San Francisco, CA 94103-1741, USA

Wiley-VCH Verlag GmbH, Boschstr. 12, D-69469 Weinheim, Germany

John Wiley & Sons Australia Ltd, 33 Park Road, Milton, Queensland 4064, Australia

John Wiley & Sons (Asia) Pte Ltd, 2 Clementi Loop #02-01, Jin Xing Distripark,
Singapore 129809

John Wiley & Sons (Canada) Ltd, 22 Worcester Road, Etobicoke, Ontario M9W 1L1

Wiley also publishes its books in a variety of electronic formats. Some content that
appears in print may not be available in electronic books.

Library of Congress Cataloging-in-Publication Data

Barnett, Vic.
    Environmental statistics / Vic Barnett.
      p. cm.–(Wiley series in probability and statistics)
    Includes bibliographical references and index.
    ISBN 0–471–48971–9 (acid-free paper)
      1. Mathematical statistics. 2. Environmental sciences–Statistical methods. I. Title. II.
Series.

    QA276.B28484 2004
    519.5–dc22   2003057617

British Library Cataloguing in Publication Data

A catalogue record for this book is available from the British Library

ISBN-10: 0-471-48971-9 (hbk)
ISBN-13: 978–0-471-48971-9 (hbk)

Typeset in 10/12pt Times by Kolam Information Services Pvt. Ltd, Pondicherry, India.

To Hannah

Contents

Preface

As we enter the new millennium we are ever more conscious of the environmental problems we face, whether these relate to depletion of rivers and oceans, despoliation of forests, pollution of land, poor air quality, environmental health issues, and so on. Most countries have active programmes to conserve their environment and to improve the quality of life in the face of the effects of environmental hazard, including the potential dangers of the 'greenhouse effect' and of 'global warming'. At the most basic level it is necessary to monitor what is going on, collecting data in a scientific (i.e. statistical) way to describe the changing scene. More importantly, it is crucial to formally describe the environmental circumstances we face with sound and validated, mathematically structured models and to analyse and interpret the data we can obtain on any environmental problems of interest and concern.

For more than a decade now there has been a clearly identified theme of environmental statistics; many conferences and publications stress this emphasis, and thousands of researchers and developers would describe their professional field as environmental statistics.

University courses are emphasizing the importance of applying scientific method to the study of environmental problems, and environmental statistics features as an important component of undergraduate and postgraduate study of statistics. This book has developed out of a final-year honours course in the Mathematics and Statistics programme at the University of Nottingham, UK. It is hoped that it might serve others as the basis for studying a range of emphases, models and methods relevant to statistical investigation of environmental issues. However, the book has stretched beyond its origins as course material and is designed to provide a broad methodological coverage of most of the areas of statistical inquiry that are likely to be encountered in environmental study. The choice of topics and their ordering differs from most statistical texts; the combination of methods for extremes, for collecting and assembling data, for examining relationship, for dealing with standards and handling temporal and spatial variation provides, in my experience, a relevant base for environmental investigation.

Vic Barnett
April 2003

CHAPTER 1

Introduction

It's all around us.

In this book we will be examining how statistical principles and methods can be used to study environmental problems. Our concern will be directed to:

- probabilistic, stochastic and statistical models;
- data collection, monitoring and representation;
- drawing inferences about important characteristics of the problem;
- using statistical methods to analyse data and to aid policy and action.

The principles and methods will be applicable to the complete range of environmental issues (including pollution, conservation, management and control, standards, sampling and monitoring) across all fields of interest and concern (including air and water quality, forestry, radiation, climate, food, noise, soil condition, fisheries, and environmental standards). Correspondingly, the probabilistic and statistical tools will have to be wide-ranging. *Inter alia*, we will consider extreme processes, stimulus response methodology, linear and generalized linear models, sampling principles and methods, time series, spatial models and methods, and, where appropriate within these themes, give attention to appropriate multivariate techniques and to design considerations including, for example, designed experiments.

Any models or methods applicable to situations involving uncertainty and variability will be relevant in one guise or another to the study and interpretation of environmental problems and will thus be part of the armoury of *environmental statistics* or *environmetrics*. Environmental statistics is a vast subject. In an article in the journal *Environmetrics*, Hunter (1994) remarked: 'Measuring the environment is an awesome challenge, there are so many things to measure, and at so many times and places'. But, however awesome, it must be faced! The recently published four-volume *Encyclopaedia of Environmetrics* (El-Shaarawi and Piegorsch, 2002) bears witness to the vast coverage of our theme and to its widespread following.

Environmental Statistics   V. Barnett
© 2004 John Wiley & Sons, Ltd   ISBN: 0-471-48971-9 (HB)

1.1 TOMORROW IS TOO LATE!

As we enter the new millennium the world is in crisis – in so many respects we are placing our environment at risk and not reacting urgently enough to reverse the effects. Harrison (1992) gives some graphic illustrations (see also Barnett, 1997):

- The average European deposits in a lifetime a monument of waste amounting to about 1000 times body weight; the average North American achieves four times this.
- Sea-floor sediment deposits around the UK average 2000 items of plastic debris per square metre.
- Over their lifetime, each person in the Western world is responsible for carbon dioxide emissions with carbon content on average 3500 times the person's body weight.

The problems of acid rain, accumulation of greenhouse gases, climate change, deforestation, disposal of nuclear waste products, nitrate leaching, particulate emissions from diesel fuel, polluted streams and rivers, etc., have long been crying out for attention. Ecological concerns and commercial imperatives sometimes clash when we try to deal with the serious environmental issues. Different countries show different degrees of resolve to bring matters under control; carbon emission is a case in point, with acclaimed wide differences of attitude and practice between, for example, the United States and the European Union. Environmental scientists, and specialists from a wide range of disciplines, are immersed in efforts to try to understand and resolve the many environmental problems we face.

Playing a major role in these matters are the statisticians, who are uniquely placed to represent the issues of uncertainty and variation inevitably found in all environmental issues. This is vital to the formulation of models and to the development of specific statistical methods for understanding and handling such problems.

1.2 ENVIRONMENTAL STATISTICS

Environmental statistics is a branch of statistics which has developed rapidly over the past 10–15 years, in response to an increasing concern among individuals, organizations and governments for protecting the environment. It differs from other applications topics (e.g. industrial statistics, medical statistics) in the very wide range of emphases, models and methods needed to encompass such broad fields as conservation, pollution evaluation and control, monitoring of ecosystems, management of resources, climate change, the greenhouse effect, forests, fisheries, agriculture and food. It is also placing demands on the statisticians to develop new approaches (e.g. to spatial-temporal modelling) or new methods (e.g. for sampling when observations are expensive or elusive or when we have specific information to take into account) as well as to adapt the whole range of existing statistical methodology to the challenges of the new environmental fields of application.

Environmental statistics is indeed becoming a major, high-profile, identified theme in most of the countries where statistical analysis and research are constantly advancing our understanding of the world we live in. Its growing prominence is evident in a wide range of relevant emphases throughout the world.

Almost all major international statistical or statistically related conferences now inevitably include sessions on environmental statistics. The International Environmetrics Society (TIES), which originated in Canada, has, over more than a decade, held in excess of ten international conferences on environmental statistics and has promoted the new journal *Environmetrics* (published by John Wiley & Sons, Ltd). The SPRUCE organization, established in 1990, is concerned with *Statistics in Public Resources and Utilities, and in Care of the Environment* and has also held major international conferences. Four resulting volumes have appeared under the title *Statistics for the Environment* (Barnett and Turkman, 1993, 1994, 1997; Barnett *et al.*, 1999). A further volume on *Quantitative Methods for Current Environmental Issues* covers the joint SPRUCE–TIES conference in Sheffield, UK, held in September 2000 (Anderson *et al.*, 2002).

Other expressions of concern for environmental statistics are found in the growing involvement of national statistical societies, such as the Royal Statistical Society in the UK and the American Statistical Association, in featuring the subject in their journals and in their organizational structure. Specific organizations such as the Center for Statistical Ecology and Environmental Statistics at Penn State University, US, and the broader-based US Environmental Protection Agency (USEPA) are expanding their work in environmental statistics. Other nations also express commitment to the quantitative study of the environment through bodies concerned with environmental protection, with environmental change networks and with governmental controls and standards on environmental emissions and effects. Many universities throughout the world are identifying environmental statistics within their portfolios of applications in statistical research, education and training.

Of course, concern for quantitative study of environmental issues is not a new thrust. This is evidenced by the many individuals and organizations that have for a long time been involved in all (including the statistical) aspects of monitoring, investigating and proposing policy in this area. These include health and safety organizations; standards bodies; research institutes; water and river authorities; meteorological organizations; fisheries protection agencies; risk, pollution, regulation and control concerns, and so on.

Such bodies are demanding more and more provision of sound statistical data, knowledge and methods at all levels (from basic data collection and sampling to specific methodological and analytic procedures). The statistician is of course ideally placed to represent the issues of uncertainty and variation inevitably found in all environmental problems. An interesting case in point was in relation to the representation of uncertainty and variation in the setting of environmental pollution standards in the 1997 UK Royal Commission on Environmental Pollution study (see Royal Commission on Environmental Pollution, 1998; Barnett and O'Hagan, 1997).

Environmental statistics is thus taking its place besides other directed specialities: medical statistics, econometrics, industrial statistics, psychometrics, etc. It is identifying clear fields of application, such as pollution, utilities, quality of life, radiation hazard, climate change, resource management, and standards. All areas of statistical modelling and methodology arise in environmental studies, but particular challenges exist in certain areas such as official statistics, spatial and temporal modelling and sampling. Environmentally concerned statisticians must be pleased to note the growing public and political acceptance of their role in the environmental debate.

Many areas of statistical methodology and modelling find application in environmental problems. Some notable emphases and needs are as follows:

- The study of extremes, outliers and robust inference methods is relevant to so many fields of inquiry, none more so than environmental issues.
- Particular modern sampling methods have special relevance and potential in many fields of environmental study; they are important in monitoring and in standard-setting. For example, ranked-set sampling aims for high-efficiency inference, where observational data are expensive, by exploiting associated (concomitant, often 'expert-opinion') information to spread sample coverage. Composite sampling seeks to identify rare conditions and form related inferences again where sampling is costly and where sensitivity issues arise, whilst adaptive sampling for elusive outcomes and rare events modifies the sampling scheme progressively as the sample is collected.
- Other topics such as size-biasing, transect sampling and capture–recapture also find wide application in environmental studies.
- Linear and generalized linear models play a central role in statistical investigations across all areas of environmental application. Further developments are needed, particularly in relation to multivariate correlated data, random dependent variables, extreme values, outliers, complex error structure, etc., with special relevance to environmental, economic and social issues.
- Risk evaluation and uncertainty analysis are modern thrusts which still need more careful definition and fuller investigation to formalise their pivotal roles and to elucidate distinctions with conventional statistical concepts and methods (see Barnett and O'Hagan, 1997; Barnett, 2002d). Applications of special relevance occur across the environmental spectrum.
- Temporal and spatial models have clear and ubiquitous relevance; all processes vary in time and space. Time-series methods have been widely applied and developed for environmental problems but more research is needed on non-stationary and multivariate structures, on outliers and on non-parametric approaches. Spatial methods need further major research development, including concern for the highly correlated and multivariate

base of most applications fields. Conjoint spatial-temporal models and methods are not well developed and are ripe for major advances in research.

In this book we will seek to review and represent the wide range of applications and of statistical topics in environmental statistics. This is facilitated by dividing the coverage into a number of thematic parts as follows:

Part I Extremal stresses: extremes, outliers, robustness (Chapters 2 and 3);
Part II Collecting environmental data: sampling and monitoring (Chapters 4, 5 and 6);
Part III Examining environmental effects: stimulus–response relationships (Chapters 7 and 8);
Part IV Standards and regulations (Chapter 9);
Part V A many-dimensional environment: spatial and temporal processes (Chapters 10 and 11).

Each part is, as indicated, divided into separate chapters covering different appropriate aspects of the respective theme.

1.3 SOME EXAMPLES

We will start our study of environmental statistics by considering briefly some practical examples, from different fields, which also illustrate various models and methods which will be developed more formally and in more detail as the book progresses.

1.3.1 'Getting it all together'

Collecting data in an effective and efficient manner is of central importance in studying environmental problems. Often we need to identify those members of a population who possess some rare characteristic or condition or to estimate the proportion of such members in the population. Sometimes the condition is of a 'sensitive' form, and individuals may be loath to reveal it. Alternatively, it may be costly or difficult to assess each member separately.

One possibility might be to obtain material or information from a large group of individuals, to mix it all together and to make a single assessment for the group as a whole. This assessment will reveal the condition if any one of the group has the condition. If it does not show up in our single test we know that all members are free of the condition. A single test may clear 1000 individuals!

This is the principle behind what is known as *composite sampling* (or composite testing). It is also known as *aggregate sampling* or in some contexts as *grab sampling*. For example, we might use a dredger to grab a large sample of soil deposit from a river bed and conduct a single test to show there is no contamination.

Of course, our composite sample might show the condition to be present. We then know nothing about which, or how many, individuals are affected. But that is another matter to which we will return later. If the condition is rare, the single composite sample will often be 'clear' and we will be able to clear all members of the sample with a single test.

Such an approach is not new – early examples of such group testing were concerned with the prevalence of insects carrying a plant virus (Watson, 1936) and of testing US servicemen for syphilis in the Second World War (Dorfman, 1943). An informative elementary review of composite sampling applications is given by USEPA (1997).

So how does this method operate? Typically a (usually large) number of random samples of individuals are chosen from the population. The material collected from each member of a sample is pooled, and a single test carried out to see if the condition is present or absent; for example, blood samples of patients might be mixed together and tested for the presence of the HIV virus.

Thus, suppose that our observational samples are of sizes $n_1, n_2, n_3, \ldots$ (often all n_i are equal) and that our corresponding composite test outcomes are 0, 0, 1, ...(where 0 means negative and 1 means positive). From these data, we can develop an estimator $\tilde{p}$ of the crucial parameter

$$p = P \text{ (individual has the condition)}$$

which expresses the rate of incidence of the condition in the population at large. Further, we will be able to derive the statistical properties of the estimator (i.e. to examine whether it is biased, to determine what is its variance or mean square error, etc.).

An interesting situation arises when, rather than estimating p, our interest is in identifying which specific members of a sample actually have the condition. If the test does not show the condition, all members are free of it. But if it is present, we must then do more testing to find out which members have the condition. Different strategies are possible. Suppose we start with n individuals. The most obvious possibility is *full retesting* (Figure 1.1). Here we do the overall composite test and if it is positive we then proceed to retest each individual, so

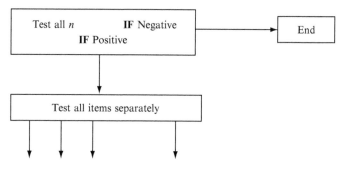

Figure 1.1   Full retesting.

that we either do one test (if the first test result is negative) or $n+1$ tests (otherwise) to uncover the complete situation. If the condition is rare (p is small) the expected number of tests will be just a little larger than 1, compared with the n needed if we just tested all individuals separately at the outset. For example, if $n = 20$ and $p = 0.0005$ the expected number of tests turns out to be just 1.2!

An alternative approach (illustrated in Figure 1.2) is *group retesting*. Here the sample group of size n is divided into k subgroups of sizes $n_1, n_2, \ldots, n_k$ if the first overall composite test is positive, and each of the subgroups is treated as a second-stage composite sample. Each subgroup is then tested as for full retesting and the process terminates.

Different strategies for choice of the number k and sizes $n_1, n_2, \ldots, n_k$ of subgroups have been considered, yielding crucial differences in efficiency of identification of the 'positives' in the overall sample. It is interesting to examine this by trying out different choices of k and $n_1, n_2, \ldots, n_k$.

A special, useful modification of group retesting is *cascading*, where we adopt a hierarchical approach, dividing each positive group or subgroup into precisely two parts and continuing testing until all positives have been identified. Figure 1.3 shows how this might operate.

These three strategies provide much scope for trying to find effective means of identifying the positive individuals – and also for some interesting combinatorial probability theory calculations.

Composite sampling can also be used to estimate characteristics of quantitative variables in appropriate circumstances, such as the mean density of polluting organisms in a water supply, as we shall find in our later, more detailed discussion of the approach (Section 5.3). This intriguing method of composite sampling is just one example of the many modern sampling methods that are being used to obtain data on environmental matters.

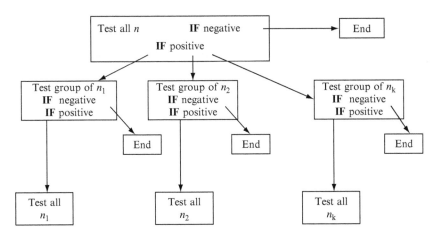

Figure 1.2 Group retesting.

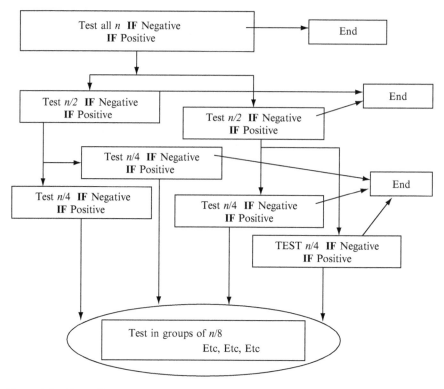

Figure 1.3 Cascading.

1.3.2 'In time and space'

Environmental effects involve uncertainty and need to be represented in terms of random variables. They frequently also exhibit systematic effects related to concomitant variables. For example, the extent of nitrate leaching through fertilized agricultural land may well depend on the concentration of recently applied nitrogenous fertilizer. Thus we will need to study relationships between variables. These may be expressed in simple linear regression model terms, or may require more complicated models such as the generalized linear model or speific non-linear models.

As an example, we might consider how the mid-day temperature on a specific day, at different sites in the UK, varies with the northings and eastings of the sites, which specify where they are located. Perhaps we might feel that the temperature will tend to be lower the more northerly the site (and perhaps the more easterly). This could be expressed in terms of a regression model with two regressor variables representing northing and easting. But what about altitude? This could also be influential.

In this example we see a variable which varies over space – *spatial models* will need to be a feature of our studies and will go beyond simple regression structures. Again, we have considered just a single day. Clearly the time of the year, even the time of the day (if not just noon) will be crucial to the levels of temperature we might expect to encounter. So there will also be a temporal component, and we will need to model the temperature in terms which allow for the effect of time. *Time-series* methods are designed to do this, and we will examine their principles and procedures. The practical example discussed in Section 1.3.3 below also needs precisely such a spatial and temporal basis for its study and raises the prospect of whether we could contemplate a *joint spatial-temporal model*. This is a largely uncharted area.

Regression models are widely used for spatial variation (as are prediction or mapping procedures such as *kriging* – see Section 11.2), while time-series models are used represent time variation; but how are we to combine the spatial and temporal components?

For the moment we notice that a linear (regression) model for the mid-day temperature which expresses the temperature as a linear combination of space and time variables could in principle provide such a *joint model*, but it would have clear limitations.

Consider another atmospheric variable. It is important for various purposes to monitor the amount of ground-level ozone as an environmental health indicator. But where should we measure it (on the roof of the local railway station?) and when (Tuesday, at noon?). In fact we want to know (and to economically represent) how the ozone level is varying from place to place over a defined region and from time to time over a specified time period, as we did for the temperature.

That is, we need realistic models for $Z(t, \mathbf{s})$ (ozone level at time t and location $\mathbf{s}$) which ideally would reflect levels and intercorrelations from time to time and place to place. We need then to develop statistical methods to fit such models, to test their validity, to estimate parameters, to predict future outcomes, etc.

A basic approach to this might be an augmented regression (linear) model. With data for 36 locations on a 6×6 rectangular grid measured monthly over several years, a model for log ozone level Z might take the form

$$z_{ijk} = \mu_{ijk} + \alpha_i + \gamma_{jk} + \varepsilon_{ijk}$$

for location (j, k) at time i (for a fixed discrete set of times and a prespecified and fixed grid of locations). Here μ_{ijk} is a regression component, with added random components $(\alpha, \gamma, \varepsilon)$ which we would need to test for zero means, independence, stationarity, no cross-correlations etc. The α (and ε) might need to be assumed to have common time-series representations, etc. In turn, this may not suffice. Thus we are effectively 'patching' together the space (regression) and time (time-series) approaches to produce a hybrid spatial/temporal analysis, but many other approaches could be (and have been) used even for this single and relatively straightforward problem. Landau and Barnett (1996) used such methods for interpolating several meteorological variables; see the discussion in the next subsection and Examples 11.5.

1.3.3 'Keep it simple'

Environmental problems are usually highly complex. We understand certain aspects of them, but almost never the whole picture. Thus general scientific considerations may suggest how a system should respond to a specific stimulus; observed data may show how it *did* react to some actual conditions. Examples include changes in river stocks of Chinook salmon over time (Speed, 1993) and, on a larger scale, the growth characteristics of winter wheat, a problem examined statistically by Landau *et al.* (1998).

Models for winter wheat growth could easily involve tens, hundreds or even thousands of parameters, including those representing weather characteristics throughout the growth period as well as the many relevant plant physiological features. Such highly parameterized models are quite intractable – we would be unable to fit parameters or interpret results. Occam's razor says 'in looking for an explanation, start with the simplest prospect'; such *parsimony* is vital in the complex field of environmental relationships.

Consider the wheat problem in more detail. Complicated, elaborate *mechanistic models* have been proposed (AFRCWHEAT2 for the UK, CERES for the USA and SIRIUS for New Zealand). These are based on claimed scientific (plant-growth) knowledge and simulated day-to-day growth and climatological conditions (but no actual data from real life). They attempt to represent the physical environmental system by means of a deterministic model, based on differential equations, which usually does not incorporate probabilistic or variational components, although it may in some applications be subjected to a 'sensitivity analysis'. A review of such modelling considerations in the context of the wheat models is given by Landau *et al.* (1998)

In global warming, the vast global circulation models for predicting climate change play a similar role – again they are essentially mechanistic (deterministic) in form.

Do they work? Some recent results by Landau *et al.* (1998) cast doubt on this, for the winter wheat models at least. A major data-assembly exercise was carried out to compile a database of wheat yields for about 1000 sites over many years throughout the UK where wheat was grown under controlled and well-documented conditions. These data constituted the 'observed yields' which were to be compared, for validation purposes, with the corresponding yields which the wheat models would predict. Crucial inputs for the models were daily maximum and minimum temperatures, rainfall and radiation as well as growth characteristics. These climatological measures were available for all relevant days and for 212 meteorological stations. But these meteorological stations were not, of course, located at the sites at which the wheat yields were measured (the wheat trial sites are shown in Figure 1.4). So it was necessary to carry out a major interpolation exercise for the meteorological variables. This is described by Landau and Barnett (1996) and was highly successful (see Figure 1.5, which shows actual *versus* interpolated minimum temperatures, where the fit accounts for 94% of the variation in observed values).

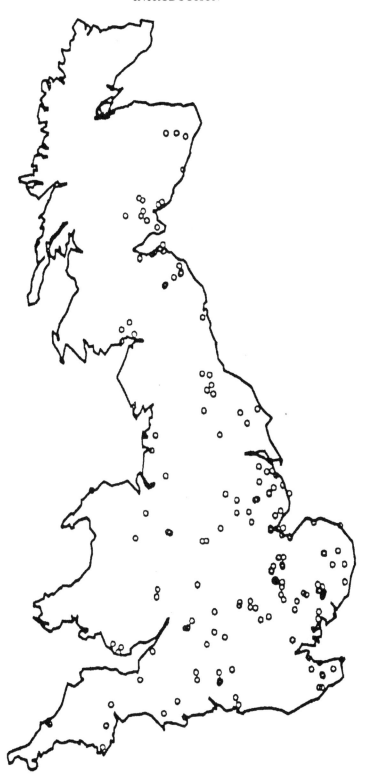

Figure 1.4 Sites of Wheat trials.

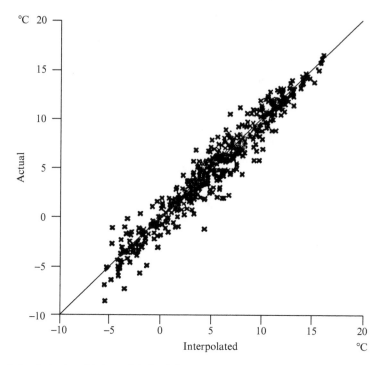

Figure 1.5   Actual and interpolated minimum temperatures.

When all culture and weather information was entered in the wheat models, they produced 'predicted yields' to compare with the 'observed' ones – the results were very surprising, as is shown by the plot (Figure 1.6, from Landau *et al.*, 1998) of actual *versus* predicted yields for the AFRCWHEAT2 model. Such conspicuous lack of association arose with all the models!

At the opposite extreme of sophistication from mechanistic models (see discussion in Barnett *et al.*, 1997), purely statistical (empirical) fits of regression-type models to real data may often predict environmental outcomes rather well but are sometimes criticized for not providing scientific understanding of the process under study. The ideal would presumably be to seek effective *parsimonious data-based mechanistic models*. Landau *et al.* (1998) went on to show how a parsimonious mechanistically motivated regression model predicted wheat yields with correlation in excess of 0.5.

1.3.4 'How much can we take?'

A range of environmental problems centre on the pattern of responses of individuals (or of systems) to different levels of stimulus – of patients to levels of a drug, of citizens to levels of pollution, of physical environment to 'levels' of climate (rain, wind, etc.). In such cases we might be interested in *trends*, in *dose-response relationships*, in *extremes*, and a variety of models and methods will be needed to address such questions as the maximum safe level for particulate

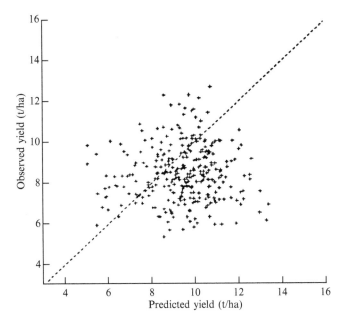

Figure 1.6 Observed versus predicted yields for AFRCWHEAT2 (Landau *et al.*, 1998 with permission from Elsevier).

matter from diesel fuel emissions, or how big a particular dam must be if it is not to overflow more than once in a hundred years?

Consider, as one specific example, a problem in the biomedical field where we are concerned with the control of harmful insects found on food crops by means of an insecticide, and wish to know what 'dose' needs to be used. One model for this is the so-called *logit* model (*probit* models are also used). The logit model expresses the proportion, $p(x)$, of insects killed in terms of the dose, or perhaps the log dose, x, of the applied 'treatment' in the form

$$\ln \frac{p(x)}{1 - p(x)} = \alpha + \beta x.$$

In practice, we observe a binomial random variable $Y \sim B(n(x), p(x))$ at each dose x which represents the number killed out of $n(x)$ observed at that dose level (with possibly different numbers of observations $n(x)$ at different doses). The general shape of this relationship – which is in the broad class of the *generalized linear model* (see Section 7.3) – is ogival (Figure 1.7). Of interest are such matters as the lethal effective doses needed to achieve 50% or 95% elimination of the insects; these are referred to as the LD_{50} and LD_{95}, respectively.

In a development of such interests another relevant topic is *bioassay*. This is concerned with evaluating the relative potency of two forms of stimulus by analysing the responses they produce in biological organisms at different doses. Typical data for what is called a *parallel-line assay* are shown in Table 1.1 (from Tsutakawa, 1982) in the context of examining the relative potencies of two treatment regimes applied at a range of dose levels to laboratory subjects. What

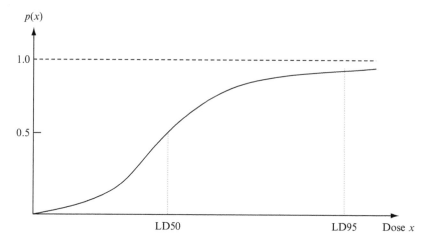

Figure 1.7 A dose–response curve.

would be of interest here is whether the two treatments show similar relationships with dose – apart, perhaps, from a scale factor. We pursue such matters further in Chapter 8.

Table 1.1 Bioassay data on rats: uterine weights (coded).

	Standard treatment			New treatment	
Dose	0.2	0.3	0.4	1.0	2.5
	73	77	118	79	101
	69	93	85	87	86
	71	116	105	71	105
	91	78	76	78	111
	80	87	101	92	102
	110	86		92	107
		101			102
		104			112

Source: Tsutakawa (1982).

1.3.5 'Over the top'

In the design of reservoirs to hold drinking water, it is important to contain and control the water by means of dams which are high and sound enough not to regularly overflow and thus cause loss of resource, and possible serious damage in the outflow area (Figure 1.8). In any year, the major threat of such damage occurs at the annual maximum water level X. If this exceeds the dam height h, we will have overflow and spillage on at least one occasion. It might seem sensible to design the dam to be of *great height*, h_0, with no realistic

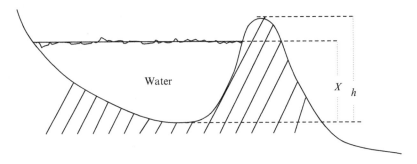

Figure 1.8 A reservoir and dam.

prospect that we will ever have $X > h_0$. But this would be unjustifiably expensive and we must seek a compromise dam height, h_1, where $P(X > h_1) = \gamma$ is designed to be acceptably small. A crucial feature of a dam's design is its *return period*, γ^{-1}. Thus if $\gamma = 0.01$, we would have a return period of 100 years.

Of course, X is a random variable describing the annual maximum water level. It will depend on a variety of features of the prevailing climate and the run-off terrain which drains into the reservoir. We do not know the distribution of X and at the design stage we may not even have any sample data from which to draw references about this distribution. But we do have some structural information about X from which to model it. It is an *annual maximum*. Thus, for example, if Y_i is a random variable describing the highest water level on day i, then $X = \max Y_i = Y_{(n)}$, with $n = 365$. Here, $Y_{(i)}$ is an *order statistic* – the ith largest of a set of identically distributed random variables $Y_i (i = 1, 2, \ldots, n)$. For the dam problem, $Y_{(n)}$ is the potential overall *yearly* maximum, and in a particular year it will take some specific value, $y_{(n)}$. Thus we have an example of a class of environmental problems where we are interested in *ordered random variables* and in particular in the largest, or *upper extreme*, of a set of more basic random variables. Such problems lead to an interest in the theory and methodology of *extremal processes*, a topic which we will examine in some detail in Section 2.3.

The reservoir problem exemplifies some of the features of this interesting area of study. We are unlikely to have much useful information about the characteristics of even our basic random variable Y: the assumed common random variable describing the *daily* maximum level. So what can we hope to say about the random variable of major interest, namely $Y_{(n)}$?

Interestingly, we can say something useful as a result of general *limit-law behaviour* of random variables which comes to our aid. If $Y_1, Y_2, \ldots, Y_n$ are independent and identically distributed random variables then, *whatever their distribution,* it turns out that the distribution of the maximum $Y_{(n)}$ approaches, as $n \to \infty$, just one of three possible forms – the so-called *extreme value distributions*. One of these, the Gumbel extreme-value distribution, has a distribution function (d.f.) in the family

$$F_n(y) = \exp\{-\exp[-(y - \alpha)/\beta]\}.$$

So, in spite of some departures from this structure (the Y_i may not necessarily be identically distributed, n is not infinite but is very large), it may be that we can model X for the dam problem by $F_n(x)$ above reducing our uncertainty to just the values of the two parameters α and β.

Consider another problem with similar features, where we are interested in a random variable Z describing the winter minimum temperature at some location. Here Z is a *minimum* rather than a maximum. So $Y = -Z$ is the *maximum* negative winter temperature, and we might again adopt a model which says that Y has d.f. $\exp\{-\exp[-(y-\alpha)/\beta]\}$.

So

$$P(Z > z) = P(-Y > z) = P(Y < -z) = \exp\{-\exp[(z+\alpha)/\beta]\}.$$

If we had observations $z_1, z_2, \ldots, z_n$, for n years, we could estimate $P(Z > z)$ by the empirical d.f. $(\#z_i > z)/(n+1)$. So if we order our observations as $z_{(1)}, z_{(2)}, \ldots, z_{(n)}$ then we have the approximate relationship

$$\frac{n-i}{n+1} \approx \exp\left[\exp\left(\frac{z_{(i)}+\alpha}{\beta}\right)\right]$$

or

$$-\ln\left[\ln\left(\frac{n+1}{n-1}\right)\right] \approx \frac{z_{(i)}+\alpha}{\beta}.$$

So we could plot a graph of $z_{(i)}$ against $-\ln\ln[(n+1)/(n-i)]$ and expect to find a linear relationship. Such a probability plotting method was illustrated by Barnett and Lewis (1967) in the context of a problem concerned with the deterioration of diesel fuel at low winter temperatures. It was necessary to model the variation in Winter minimum temperatures from year to year at different locations. Figure 1.9 (from Barnett and Lewis, 1967) uses the above plotting technique to confirm the extreme-value distribution for three locations: Kew, Manchester Airport and Plymouth. The linearity of the plots provides compelling empirical evidence of the usefulness of the above extreme-value distribution.

1.4   FUNDAMENTALS

The notation and terminology used throughout the book will conform to the following pattern. A univariate random variable X will take a typical value x. The distribution of X will be represented in terms of a probability density function (p.d.f.) $f_\theta(x)$ (if X is continuous) or a probability function $p_\theta(x)$ (if X is discrete), or in either case by a distribution function $F_\theta(x)$, where θ is a scalar or vector parameter indexing the family of distributions in which that of X resides. Thus for discrete X, for example, we have

$$p_\theta(x) = P(X = x),$$
$$F_\theta(x) = P(X \leq x),$$

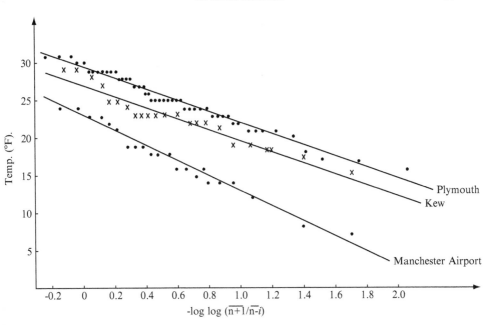

Figure 1.9 Low-temperature plots for typical locations (Barnett and Lewis, 1967).

where $P(\cdot)$ denotes probability. Common families of distribution of X include the discrete *binomial* and *Poisson* distributions and the continuous *normal, exponential* and *gamma* distributions. If X follows these respective distributions, this will be indicated in the following manner:

$$X \sim \text{B}(n,p), \qquad X \sim \text{P}(\mu)$$

and

$$X \sim \text{N}(\mu, \sigma^2), \qquad X \sim \text{Ex}(\lambda), \qquad X \sim \Gamma(r, \lambda),$$

where the arguments $(n,p), \mu, p, (\mu, \sigma^2), \lambda$ and (r, λ) are the symbols used in the distinct cases for the generic family parameter θ.

A random sample $x_1, x_2, \ldots, x_n$ of observations of X has mean

$$\bar{x} = \frac{1}{n}\sum_{i=1}^{n} x_i,$$

which is an unbiased estimator of the population parameter $\mu = \text{E}(X)$ where $\text{E}(\cdot)$ is the expectation operator. The sample variance,

$$s^2 = \frac{1}{n-1}\sum (x_i - \bar{x})^2,$$

is unbiased for the population variance $\sigma^2 = \text{E}[(X - \mu)^2] = \text{Var}(X)$. If we order sample observations as $x_{(1)} \le x_{(2)} \le \ldots \le x_{(n)}$, the $x_{(i)}$ are observations of the *order statistics* $X_{(1)}, X_{(2)}, \ldots, X_{(n)}$, We refer to $x_{(1)}$ and $x_{(n)}$ as the sample *minimum* and sample *maximum* (or lower and upper *extremes*); $x_{(n)} - x_{(1)}$ is the *sample range* and $(x_{(1)} + x_{(n)})/2$ is the *mid-range*. We will denote the sample *median* by m.

A multivariate random variable $\mathbf{X}$ of dimension p has components $X_1, X_2, \ldots, X_p$; a random sample $\mathbf{x}_1, \mathbf{x}_2, \ldots, \mathbf{x}_n$ is an $n \times p$ array

$$
\begin{array}{cccc}
x_{11} & x_{21} & \cdots & x_{p1} \\
x_{12} & x_{22} & \cdots & x_{p2} \\
\vdots & & & \vdots \\
x_{1n} & x_{2n} & \cdots & x_{pn}
\end{array}
$$

The distribution of $\mathbf{X}$ will be described, as appropriate, in terms of joint probability, or probability density, functions $p_\theta(\mathbf{x})$, or $F_\theta(\mathbf{x})$, or by its d.f. $F_\theta(\mathbf{x})$. For a bivariate random variable $\mathbf{X}$ it is often more convenient to represent its components as (X, Y).

It will be assumed that the reader is familiar with basic concepts and methods of probability theory and statistical inference, including correlation, regression, estimation, hypothesis testing, the maximum likelihood method, the maximum likelihood ratio test, etc., at a level covered by standard intermediate texts such as Mood *et al.* (1974) or Garthwaite *et al.* (1995).

1.5 BIBLIOGRAPHY

Throughout our discussion of the statistical models and methods and of practical examples of their use, in the different areas of environmental interest, we will need to consider important recent contributions in the literature. At all stages, references will be given to published papers in professional journals and to books or occasional publications. These will always be of specific relevance to the statistical topic or environmental application being studied or will enable it to be considered in a broader methodological or applications review setting. However, it is useful even at this early stage to refer to a small group of publications of general applicability and relevance.

Firstly, there are a few books which purport to cover the general theme of this book. These include elementary texts or sets of examples whose titles or descriptions indicate coverage of environmental statistics. These include Hewitt (1992), Berthouex and Brown (1994), Cothern and Ross (1994), Ott (1995), Pentecost (1999), Millard and Neerchal (2000) and Townend (2002). They are of varying level and range but do not usually go beyond basic statistical concepts and methods, with corresponding constraints on the types of applications and examples that can be discussed. A set of case studies on environmental statistics are presented in the edited volume by Nychka *et al.* (1998).

As indicated by the breakdown of topics into parts I–V explained in Section 1.2, we will be concerned with distinct areas of statistical principle and method. More detailed treatments are available of many of the areas and include the following:

Part I   Galambos (1987) on extremes; David (1981) on order statistics; Barnett and Lewis (1994) on outliers; and Huber (1981) on robustness.

Part II   Barnett (2002a) on survey sampling; Thompson (2002) and Wheater and Cook (2000) on general sampling methodology; and Thompson and Seber (1996) on adaptive sampling.

Part III   Stapleton (1995) on linear models and regression; and McCullagh and Nelder (1989) on generalized linear models.

Part IV   Barnett and O'Hagan (1997) on standards and regulations.

Part V   Bloomfield (2000) on Fourier analysis of time series; Fuller (1995) on time series in general; Cressie (1993) on spatial models and methods; and Upton and Fingleton (1985, 1989) and Webster and Oliver (2001) on spatial data analysis.

Review papers can provide informative resumes or encapsulations of our theme: see, for example Barnett (1997). The *Encyclopaedia of Statistical Sciences* (Kotz *et al.*, 1982–1988) provides crisp and informative explanations of many of the statistical topics discussed below, whilst the recent *Encyclopaedia of Environmentrics* (El-Shaarawi and Piegorsch, 2002) is to be welcomed for its comprehensive coverage of so much of the field of environmental statistics.

PART I

Extremal stresses: Extremes, outliers, robustness

CHAPTER 2

Ordering and Extremes: Applications, Models, Inference

Hurricane Freda increased in violence as it moved northwards.

We saw in the reservoir example of Section 1.3.5 how important extreme events can be in environmental terms. Extreme values or extreme observations are just examples (the smallest and largest ones) of ordered observations.

We are used to regarding the random sample as the fundamental basis of statistical inference. The idea of ordering the sample values and taking account both of the *value* and the *order* of any observation has a long tradition. Whilst it might seem strange that this should add to our knowledge, the effects of ordering can be impressive in terms of both what aspects of sample behaviour can be usefully employed and the effectiveness and efficiency of resulting inferences.

Thus, for any random sample of observations $x_1, x_2, \ldots, x_n$ of a random variable, X, we might fully order the sample, denoting the ordered sample values as $x_{(1)}, x_{(2)}, \ldots, x_{(n)}$, where $x_{(i)} \leq x_{(j)}$ for $i < j$. Or we might consider various *summary aspects* of the ordered sample. These include the *sample minimum*, $x_{(1)}$, the *sample maximum*, $x_{(n)}$ (e.g. the heaviest frost or the highest sea wave), the *range*, $x_{(n)} - x_{(1)}$ (how widespread are the temperatures that a bridge must withstand) and the *median* (the 'middle observation' in the ordered sample, which can serve as a robust measure of location: see Section 3.12), all of which, depending on context, may serve as useful representations of the ordered sample for inference purposes.

The study of *outliers* in the form of markedly extreme, possibly anomalous, sample behaviour, perhaps caused by contamination of the sample by some occasional unrepresentative values, also depends on first ordering the sample and has played an important role since the earliest days of statistical inquiry – this will be discussed in more detail in Chapter 3. Then again, we will see that linear combinations of all ordered sample values can provide efficient estimators, particularly of location or scale parameters. Barnett (2002c) briefly reviews the different uses of the ordered sample for inference; see also Barnett (1976a, 1988).

Environmental Statistics V. Barnett
© 2004 John Wiley & Sons, Ltd ISBN: 0-471-48971-9 (HB)

So we will start our study of statistical principles and methods in this first part of the book by considering the effects of ordering a random sample, in terms of what this implies for convenient representations of the data, inference procedures for parameter estimation and testing and, in later chapters, for handling outliers and robust procedures for inference.

2.1 ORDERING THE SAMPLE

We start with the random sample $x_1, x_2, \ldots, x_n$ of n observations of a random variable X describing some quantity of, say, environmental interest. If we arrange the sample in increasing order of value as $x_{(1)} \leq x_{(2)} \leq \ldots \leq x_{(n)}$, then these are observations of the *order statistics* $X_{(1)}, X_{(2)}, \ldots, X_{(n)}$ from a potential random sample of size n. Whereas the $x_i (i = 1, 2, \ldots, n)$ are independent observations, the order statistics $X_{(i)}, X_{(j)} (i \neq j)$ are correlated. This often makes them more difficult to handle in terms of their distributional behaviour when we seek to draw inferences about X from the ordered sample. See David (1981) for a detailed general treatment of ordering and order statistics.

At the descriptive level, the *extremes* $x_{(1)}$ and $x_{(n)}$, the *range* $x_{(n)} - x_{(1)}$, the *mid-range* $(x_{(1)} + x_{(n)})/2$ and the *median m* – which is defined as $x_{([n+1]/2)}$ if n is odd, or as $(x_{(n/2)} + x_{([n+2]/2)})/2$ if n is even – have obvious practical appeal and interpretation. In particular, the extremes and the median are frequently employed as basic descriptors in *exploratory data analysis*, illustrating where the sample is located and how widespread it is (see Tukey, 1977). The *studentized range*, $(x_{(n)} - x_{(1)})/S_v$, or *studentized deviation from the mean*, $(x_{(n)} - \bar{x})/S_v$, where S_v^2 is a mean-square estimator of the population variance based on v degrees of freedom, are also frequently employed measures of standardized variation, particularly for use in examining outlying behaviour (see Chapter 3).

Modified order-based constructs such as the *box-and-whisker plot* utilize the ordered sample as a succinct summary of a set of data; see Tukey (1977) or Velleman and Hoaglin (1981) for discussion of such a non-model-based approach.

Example 2.1 A random sample of 12 daily maximum wind speeds (in knots) was taken at a single UK site over a period of a month. The sample values were as follows:

$$19, 14, \ 25, \ 10, \ 11, \ 22, \ 18, \ 17, \ 49, \ 23, \ 31, \ 18.$$

The ordered sample, $x_{(1)}, x_{(2)}, \ldots, x_{(n)}$, can be written

$$10, \ 11, \ 14, \ 17, \ 18, \ 18, \ 19, \ 22, \ 23, \ 25, \ 31, \ 49.$$

So we find a sample minimum value $x_{(1)} = 10$ and a maximum $x_{(12)} = 49$. The maximum seems rather a long way from the rest of the sample; we may want to examine it further as an outlier (see Example 2.8 and Chapter 3). The range is 39 knots, which gives some indication of how widely the wind speeds might vary, whilst the median value of 18.5 knots provides a measure of location. Another such measure is provided by the mid-range which takes the value 29.5 (highly influenced, of course, by the possibly anomalous maximum sample value of 49, but see Example 2.3 below on the potential importance of the mid-range). We might choose to summarize the whole data set by means of a box-and-whisker plot which shows the extremes, the bulk of the data set and the median, as a succinct summary (Figure 2.1)

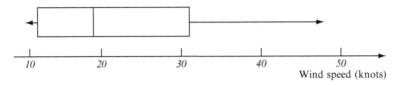

Figure 2.1   Box-and-whisker plot of wind-speed data in Example 2.1.

2.1.1   Order statistics

Suppose X has p.d.f. $f_\theta(x)$ and distribution function $F_\theta(x) = P(X < x)$. Let

$$X_{(1)} < X_{(2)} < \ldots < X_{(n)}$$

be the order statistics. These are of course random variables, so the inequality concept has to be interpreted carefully. In saying, for example, that $X_{(1)} < X_{(2)}$, we are indicating that $X_{(1)}$ and $X_{(2)}$ are the generic random variables corresponding with the smallest and second smallest observations, respectively, in a random sample of size n of the random variable X.

Much effort has gone into examining the distributional behaviour of the ordered sample values. Sometimes this is quite straighforward. Consider the sample maximum. This will have d.f. $F_n(x) = P(X_{(n)} < x)$. If $X_{(n)} < x$ we need to have all $x_i < x$, $i = 1, 2, \ldots, n$.. So we immediately conclude that

$$F_n(x) = \{P(X < x)\}^n = \{F(x)\}^n. \tag{2.1}$$

Example 2.2   Suppose times between $n + 1$ crop infestations are independent and follow an exponential distribution $\text{Ex}(\lambda)$. The maximum time betweeen infestations will be an observation from a distribution with d.f.

$$F_n(x) = (1 - e^{-\lambda x})^n. \tag{2.2}$$

The minimum $x_{(1)}$ arises from a distribution with d.f.

$$F_1(x) = P(X > x)^n = [1 - (1 - e^{-\lambda x})]^n = e^{-n\lambda x},$$

so that $X_{(1)}$ is shown to have an exponential distribution but with parameter $n\lambda$ (i.e. a mean $1/n$ times that of X). In contrast, we see from (2.2) that $X_{(n)}$ has no simple distributional form.

More generally, it is clear that $X_{(r)}$ will have d.f.

$$F_r(x) = P \text{ (at least } r \text{ of } X_i \text{ are less than } x)$$

$$= \sum_{i=r}^{n} \binom{n}{i} F^i(x)[1 - F(x)]^{n-i}. \tag{2.3}$$

The joint distribution of pairs of order statistics can be derived in corresponding terms, and expressions obtained for first- and second-order marginal and joint moments and functions of them. As one example, we can derive an exact form for the d.f. of the range R as

$$F(r) = n \int_{-\infty}^{\infty} \{ F(x + r) - F(x) \}^{n-1} f(x) dx, \tag{2.4}$$

where $f(x)$ is the p.d.f. of X.

The *quantile* essentially extends the notion of the median. Thus the *p-quantile* of a random variable X with d.f. $F(x)$ is ξ_p satisfying $F(\xi_p) = p$. We have a corresponding concept of a sample quantile defined in terms of the observed order statistic value $x_{(np)}$ and confidence intervals for ξ_p can readily be constructed. See David (1981, pp. 15ff.).

David (1981) gives detailed coverage of the properties of, and manipulation of, order statistics and of derived measures such as quantiles, both from the probabilistic and the inferential points of view. In general inference, it is not unusual to encounter order statistics when we seek sufficient statistics or maximum likelihood estimators.

Example 2.3 Suppose that X has a uniform distribution over the interval $(\mu - \sigma, \mu + \sigma)$ and we draw a random sample of size n. Then it turns out that the *minimal sufficient statistic* for (μ, σ) is the pair $(X_{(1)}, X_{(n)})$. Furthermore, the *maximum likelihood estimators* of μ and σ are $(X_{(1)} + X_{(n)})/2$ and $(X_{(n)} - X_{(1)})/2$ – the mid-range and half the range – respectively.

We will find that it is important to be able to characterize order statistics in terms of their means and variances (and covariances), especially when (as in Example 2.3) X has a distribution in a family with d.f. of the form $F[(x - \mu)/\sigma]$. This is so, of course, for many of the well-known and widely used distributions, such as the normal.

If X is symmetric, μ and σ are just the mean and variance of X, respectively; otherwise they are more general measures of location and dispersion. If we transform X to $U = (X - \mu)/\sigma$, then the order statistics $U_{(i)}(i = 1, 2, \ldots, n)$ are known as the *reduced order statistics*. There has been much study of the first- and second-order moments of the $X_{(i)}$ and the $U_{(i)}$ as a basis for inference.

2.2   ORDER-BASED INFERENCE

It is interesting to inquire whether the use of the ordered sample has any advantage from the inference viewpoint. We know that the sample mean

$\bar{x} = \sum_{i=1}^{n} x_i$ is a useful general-purpose estimator of the mean μ of the random variable X. It is the *best linear unbiased estimator* (BLUE) of μ (whenever μ exists). If X is normal $N(\mu, \sigma^2)$, then we can go further and show that $\bar{X}$ is the optimal *minimum-variance bound estimator* or *Cramér–Rao lower bound estimator*. It is also the *maximum likelihood estimator*. Furthermore, it is optimal for other distributions, such as the Poisson and the binomial. The *central limit theorem* confers asymptotic properties of normality and efficiency on $\bar{X}$ at least when μ and σ exist, which transfer useful approximate properties when n is large.

But all is not rosy for $\bar{X}$! It is well known that if X has a Cauchy distribution then $\bar{X}$ is not optimal; it is not even unbiased or consistent (see Barnett, 1999a, p. 146).

One interesting characteristic for present concern, however, is that $\bar{X}$ is a linear combination of the *ordered sample values*, since $\sum_{i=1}^{n} x_i = \sum_{i=1}^{n} x_{(i)}$. So we can conclude that linear functions of the ordered sample values can form not only useful estimators but even optimal ones.

Sometimes they can do better than linear functions of the unordered sample values. For example, $V = 1.7725 \sum_{i=1}^{n} (2i - n - 1)X_{(i)}/[n(n-1)]$ is more easily calculated than the unbiased sample variance estimate S^2, and for normal X it is about 98% efficient relative to S^2 for all sample sizes. No linear estimator based on unordered sample values is of any merit (Downton, 1966; see also Barnett, 1999a, p. 152).

But let us start with an even more startling use of order statistics in estimation.

Example 2.4 Suppose X again has the uniform distribution of Example 2.3 on the interval $(\mu - \sigma, \mu + \sigma)$. This might apply, for suitable σ, to situations where replicate field measurements are being taken of a single environmental quantity but the results are rounded to too great an extent for the required accuracy. Then the matter of interest is estimation of μ. The sample mean $\bar{X}$ is unbiased and consistent and has variance $\sigma^2/(3n)$, where n is the sample size. But oddly enough, in this situation the mid-range $(X_{(1)} + X_{(n)})/2$, which is also unbiased, consistent and the maximum likelihood estimator, can be shown to have variance $2\sigma^2/[(n+1)(n+2)]$. So we have a function of just two order statistics which is a super-efficient estimator against which $\bar{X}$ has *zero asymptotic relative efficiency*.

So clearly estimators based on order statistics can have forms quite different from, and higher efficiency than, those based on the unordered sample under appropriate circumstances.

The organized study of order statistics estimators goes back 50 years or so. Sarhan and Greenberg (1962) report on extensive 'contributions to order statistics' including the first general approach, by Lloyd (1952), to order statistic estimation of location and dispersion parameters. The context is that of the location/scale family with d.f. of the form $F[(x - \mu)/\sigma]$. Using the reduced order statistics $U_{(i)} = (X_{(i)} - \mu)/\sigma$, it is noted that

$$E(X_{(i)}) = \mu + \sigma\alpha_i \qquad (i = 1, 2, \ldots, n) \qquad (2.5)$$

or

$$E(\mathbf{X}) = \mu\mathbf{1} + \sigma\boldsymbol{\alpha} \qquad (2.6)$$

and

$$V(\mathbf{X}) = \mathbf{V}\sigma^2, \qquad (2.7)$$

where $\boldsymbol{\alpha}$ is the n-vector of means $E(U_{(i)})$ and $\mathbf{V}$ is the $n \times n$ matrix of variances and covariances of the $U_{(i)}$. Thus we have a *linear model* (where the means of the random variables of interest are linear functions of the parameters). This will be discussed in detail in Section 7.1, where we will see that the following *extended least-squares* approach is appropriate for estimation of μ and σ.

If we have a random sample of size n then we can express the ith ordered observation $x_{(i)}$ as $X_{(i)} + \varepsilon_i$, where the *residuals* $\varepsilon_i (i = 1, 2, \ldots, n)$ have zero means and variance–covariance matrix $\mathbf{V}\sigma^2$.

So the ordered observation vector $\mathbf{x}$ can be expressed

$$\mathbf{x} = \mathbf{A}\boldsymbol{\theta} + \boldsymbol{\varepsilon},$$

with so-called *design matrix*

$$\mathbf{A}' = \begin{pmatrix} 1 & 1 & \cdots & 1 \\ \alpha_1 & \alpha_2 & \cdots & \alpha_n \end{pmatrix}$$

and where $\boldsymbol{\theta}' = (\mu, \sigma)$. The least-squares approach to this model will yield minimum-variance linear unbiased least-squares estimators

$$\tilde{\boldsymbol{\theta}} = (\mathbf{A}'\mathbf{V}^{-1}\mathbf{A})^{-1}\mathbf{A}'\mathbf{V}^{-1}\mathbf{x}, \qquad (2.8)$$

with variance–covariance matrix $(\mathbf{A}'\mathbf{V}^{-1}\mathbf{A})^{-1}\sigma^2$. These estimators are BLUEs.

Explicit expressions for the individual estimators and their variances and covariances are readily extracted. The estimators take the forms:

$$\tilde{\mu} = -\boldsymbol{\alpha}'\boldsymbol{\Gamma}^{-1}\mathbf{X}, \qquad \tilde{\sigma} = \mathbf{1}'\boldsymbol{\Gamma}^{-1}\mathbf{X}, \qquad (2.9)$$

where

$$\boldsymbol{\Gamma} = \mathbf{V}^{-1}(\mathbf{1}\boldsymbol{\alpha}' - \boldsymbol{\alpha}\mathbf{1}')\mathbf{V}^{-1}/, \qquad \Delta = (\mathbf{1}'\mathbf{V}^{-1}\mathbf{1})(\boldsymbol{\alpha}'\mathbf{V}^{-1}\boldsymbol{\alpha}) - (\mathbf{1}'\mathbf{V}^{-1}\boldsymbol{\alpha})^2. \qquad (2.10)$$

The variances and covariances of these estimates are found to be

$$V(\tilde{\mu}) = (\boldsymbol{\alpha}'\mathbf{V}^{-1}\boldsymbol{\alpha})/\Delta, \qquad V(\tilde{\sigma}) = (\mathbf{1}'\mathbf{V}^{-1}\mathbf{1})\sigma^2/\Delta,$$
$$\text{Cov}\,(\tilde{\mu}, \tilde{\sigma}) = -(\mathbf{1}'\mathbf{V}^{-1}\boldsymbol{\alpha})\sigma^2/\Delta. \qquad (2.11)$$

The use of these estimators depends on being able to readily determine the means, $\boldsymbol{\alpha}$, and the variances and covariances, $\mathbf{V}$, of the reduced order statistics. This can involve a large amount of computation which was challenging in earlier times (see Barnett, 1966, on order statistics estimation for the Cauchy distribution). Modern computers have made such work more accessible, but by no means trivial. For many distributions $\boldsymbol{\alpha}$ and $\mathbf{V}$ are tabulated for $\mathbf{n}$ up to about 20; for example, in Sarhan and Greenberg (1962) we find some tables for the normal distribution as well as for the exponential, rectangular (uniform), extreme value, gamma and other cases. See David (1981) for more recent references to tabulated first- and second-order moments.

In some cases explicit results exist for $\boldsymbol{\alpha}$ and $\mathbf{V}$ and can be used to obtain the BLUEs.

Example 2.5 Suppose that a random sample $x_1, x_2, \ldots, x_n$ is known to arise from a uniform distribution on $(0, \beta)$. The ordered observations $x_{(1)}, x_{(2)}, \ldots, x_{(n)}$ are observed values of order statistics $X_{(1)}, X_{(2)}, \ldots, X_{(n)}$. We note that the p.d.f. of $X_{(i)}$ is

$$f_i(x) = \frac{n!}{(i-1)!(n-i)!} \left(\frac{x}{\alpha}\right)^i \left(1 - \frac{x}{\alpha}\right)^{n-i}.$$

So $E(X_i) = i\beta/(n+1)$. Thus $E(\mathbf{X}) = \mathbf{A}\beta$, where

$$\mathbf{A}' = \left(\frac{1}{n+1}, \frac{2}{n+1}, \ldots, \frac{n}{n+1}\right).$$

We can now show that $\text{Var}(\mathbf{X}) = \mathbf{V}\beta^2$, where

$$\mathbf{V} = \frac{1}{(n+1)^2(n+2)} \begin{bmatrix} n & n-1 & n-2 & \ldots & 1 \\ n-1 & 2(n-1) & 2(n-2) & \ldots & 2 \\ n-2 & 2(n-2) & 3(n-2) & \ldots & 3 \\ \vdots & \vdots & \vdots & & \vdots \\ 1 & 2 & 3 & \ldots & n \end{bmatrix}.$$

Now the least-squares estimator of β is $(\mathbf{A}'\mathbf{V}^{-1}\mathbf{A})^{-1}\mathbf{A}'\mathbf{V}^{-1}\mathbf{x}$, with variance $(\mathbf{A}'\mathbf{V}^{-1}\mathbf{A})^{-1}\beta^2$. In fact,

$$\mathbf{V}^{-1} = (n+1)(n+2) \begin{bmatrix} 2 & -1 & 0 & \ldots & 0 & 0 \\ -1 & 2 & -1 & \ldots & 0 & 0 \\ 0 & -1 & 2 & \ldots & 0 & 0 \\ \vdots & & & & \vdots & \vdots \\ 0 & \ldots & \ldots & & -1 & 2 \end{bmatrix}.$$

So $\mathbf{A}'\mathbf{V}^{-1} = (0, 0, \ldots, 0, (n+1)(n+2))$ and $\mathbf{A}'\mathbf{V}^{-1}\mathbf{A} = n(n+2)$.

Thus $\tilde{\beta} = (n+1)X_{(n)}/n$ with variance $\beta^2/n(n+2)$, which means that we should estimate β from a scaled version of the largest value.

In an effort to reduce the labour of obtaining the BLUE, several approximate forms have been advanced, including the *asymptotically best linear estimators* and the *nearly best linear unbiased estimators* (NBLUES) (Sarhan and Greenberg, 1962, Chapters 4 and 5). In the first case, the coefficients of the $x_{(i)}$ in the linear estimator of either parameter are expressed in the form

$$\gamma_{i,n} = \frac{1}{n} h[i/(n+1)] \tag{2.12}$$

where $h\,(\cdot)$ is chosen to ensure that the estimator of a linear combination $\beta = \gamma\mu + \eta\sigma$ (for arbitrary γ, η) of the basic parameters is 'asymptotically best' in a well-defined sense.

For the NBLUEs, originally proposed by Blom (1958) and David (1981, pp. 146–147), we estimate the quantiles by

$$F(\lambda_i) = P_i = i/(n+1)$$

and the covariance of $x_{(i)}$ and $x_{(j)}$ is approximated by

$$P_i(1 - P_j)/[(n+2)F(\lambda_i)F(\lambda_j)],$$

which is reasonable for large n. So if $\theta_i = F(\lambda_i)$, we have

$$\mathrm{Cov}[\theta_i x_{(i)}, \theta_j x_{(j)}] \approx p_i(i - p_j)/(n+2),$$

and if we put

$$y_{(i)} = \theta_{i+1}x_{(i+1)} - \theta_i x_{(i)} \qquad (i = 1, 2, \ldots, n),$$

with $\theta_0 = \theta_{n+1} = 0$, we obtain interesting distribution-free results, namely that

$$\mathrm{Var}(y_{(i)}) \sim n(n+1)^{-2}(n+2)^{-1}$$
$$\mathrm{Cov}(y_{(i)}, y_{(j)}) \sim -(n+1)^{-2}(n+2)^{-1} \qquad (2.13)$$

So if we are interested in a linear combination β as defined above, it can be shown that use of the weighted differences $y_{(i)}$ reduces the least-squares problem to a more tractable form, resulting in approximations to the BLUEs which do not lose much efficiency. Details can be found in David (1981, pp. 133–135). However, neither of these approaches appears now to be widely used in view of the greater ease of computing BLUEs with modern computers.

A notably simple approach, however, merits consideration. Prompted by the difficulty in calculating the variances and covariances, Gupta (1952) suggested replacing $\mathbf{V}$ by the identity matrix $\mathbf{I}$, yielding revised estimators

$$\mu^* = \sum b_i X_{(i)}, \qquad \sigma^* = \sum c_i X_{(i)}$$

where $c_i = (\alpha_i - \bar{\alpha})/\sum(\alpha_i - \bar{\alpha})^2$ and $b_i = n^{-1} - \bar{\alpha}c_i$. These require us to know just the expected values $\alpha_i(i = 1, 2, \ldots, n)$. Can this be reasonable? It certainly works well for the normal distribution. David (1981, Sections 6.2, 6.3) gives more details and a worked example.

Returning to (2.9) and (2.10), it is interesting to see how these results are affected if $\mathbf{X}$ is symmetric. In this case we find that

$$\boldsymbol{\alpha} = -\mathbf{J}\boldsymbol{\alpha} \text{ and } \mathbf{V} = \mathbf{J}\mathbf{V}\mathbf{J}$$

where $\mathbf{J}$ is the permutation matrix

$$\mathbf{J} = \begin{bmatrix} 0 & & & \cdot{\cdot}\, 1 \\ & & 1 & \cdot{\cdot} \\ & \cdot{\cdot}\, 1 & & \\ 1 & & & 0 \end{bmatrix}.$$

So we find that

$$\mathbf{V}^{-1} = \mathbf{J}\mathbf{V}^{-1}\mathbf{J}, \tag{2.14}$$

with the effect that

$$\mathbf{1}'\mathbf{V}^{-1}\boldsymbol{\alpha} = \mathbf{1}'(\mathbf{J}\mathbf{V}^{-1}\mathbf{J})(-\mathbf{J}\boldsymbol{\alpha}) = -\mathbf{1}'\mathbf{J}\mathbf{V}^{-1}\mathbf{J}^2\boldsymbol{\alpha} = -\mathbf{1}'\mathbf{V}^{-1}\boldsymbol{\alpha}$$

since $\mathbf{1}'\mathbf{J} = \mathbf{1}'$ and $\mathbf{J}^2 = \mathbf{I}$. Thus we conclude that

$$\mathbf{1}'\mathbf{V}^{-1}\boldsymbol{\alpha} = -\mathbf{1}'\mathbf{V}^{-1}\boldsymbol{\alpha} \tag{2.15}$$

and hence that $\mathbf{1}'\mathbf{V}^{-1}\boldsymbol{\alpha} = 0$. This yields simple expressions for the estimators $\tilde{\mu}$ and $\tilde{\sigma}$ as

$$\tilde{\mu} = (\mathbf{1}'\mathbf{V}^{-1}\mathbf{X})/(\mathbf{1}'\mathbf{V}^{-1}\mathbf{1}), \quad \tilde{\sigma} = (\boldsymbol{\alpha}'\mathbf{V}^{-1}\mathbf{X})/(\boldsymbol{\alpha}'\mathbf{V}^{-1}\boldsymbol{\alpha}). \tag{2.16}$$

Further, $\tilde{\mu}$ and $\tilde{\sigma}$ are uncorrelated in view of (2.15) with variances $\sigma^2/(\mathbf{1}'\mathbf{V}^{-1}\mathbf{1})$ and $\sigma^2/(\boldsymbol{\alpha}'\mathbf{V}^{-1}\boldsymbol{\alpha})$, respectively (cf. the general results (2.9) to (2.11)).

The above results for order statistics estimation of μ and σ will be found to have special relevance when applied (in Section 5.4) to ranked-set sampling, as an important component of the modern arsenal of environmental sampling techniques.

Example 2.6 It is interesting to consider some other specific results on order statistics estimators. For the exponential distribution with $F(x) = 1 - \exp[-(x - \mu)/\sigma]$ we find that the BLUEs are

$$\mu^* = (nX_{(1)} - \overline{X})/(n - 1)$$

and

$$\sigma^* = n(\overline{X} - X_{(1)})/(n - 1),$$

with respective variances $\sigma^2/[n(n - 1)]$ and $\sigma^2/(n - 1)$. The BLUE of the mean $(\mu + \sigma)$ is $\overline{X}$ with variance σ^2/n.

For the uniform distribution on $(\mu - \sigma/2, \mu + \sigma/2)$ we conclude that the BLUEs are

$$\mu^* = (X_{(1)} + X_{(n)})/2$$

(the mid-range) and

$$\sigma^* = (n + 1)(X_{(n)} - X_{(1)})/(n - 1)$$

(a function of the sample range), with respective variances $\sigma^2/[2(n + 1)(n + 2)]$ and $2\sigma^2/[(n + 2)(n - 1)]$ (cf. Example 2.5).

It is unusual for us to encounter such simple explicit forms for the BLUEs. Usually we have to construct linear combinations of all the order statistics with weights requiring complex numerical computation. As examples we have from Sarhan and Greenberg (1962, pp. 391–397) the following weights attached to $X_{(1)}, X_{(2)}, \ldots, X_{(n)}$ for estimators μ^* and σ^* for a sample of size $n = 5$, for different symmetric distributions:

	μ^*					σ^*				
1. *U-shaped*	0.558	−0.045	−0.027	−0.045	0.558	−0.489	−0.032	0	−0.032	0.489
2. *Parabolic*	0.386	0.080	0.068	0.080	0.386	−0.411	−0.039	0	0.039	0.411
3. *Triangular*	0.306	0.119	0.150	0.119	0.306	−0.399	−0.064	0	0.064	0.399
4. *Normal*	0.200	0.200	0.200	0.200	0.200	−0.372	−0.135	0	0.135	0.372
5. *Double exponential*	0.017	0.221	0.524	0.221	0.017	−0.326	−0.317	0	0.317	0.326

Note *that the first three have p.d.f.s of the forms*

$$\frac{3(x - \mu)^2}{2\sigma^3} \qquad (\mu - \sigma < x < \mu + \sigma),$$

$$\frac{6(x - \mu - \sigma/2)(\mu + \sigma/2 - x)}{\sigma^3} \qquad (\mu - \sigma/2 < x < \mu + \sigma/2)$$

and

$$4\left(\frac{\sigma}{2} - |x - \mu|\right)/\sigma^2 \qquad (|x - \mu| < \sigma/2),$$

respectively.

2.3   EXTREMES AND EXTREMAL PROCESSES

We have already seen how extreme values, $x_{(1)}$ and $x_{(n)}$, figure in our study of environmental problems. They feature as natural expressions of behavioural outcomes: whether expressing a minimum temperature, or the maximum value in a random sample of n observations of some key pollution variable X, or the range of ozone levels observed at some location. See Example 2.1 on wind speeds.

The extremes also feature in our efforts to obtain estimators of basic parameters in the distribution of the random variable X. Thus we saw in Example 2.5 how the mid-range $(X_{(1)} + X_{(2)})/2$ and the range $X_{(n)} - X_{(1)}$ are the crucial statistics in the BLUEs of the location and scale parameters, respectively, for the modified exponential distribution with p.d.f. $(1/\sigma)\exp[-(x - \mu)/\sigma]$.

The distributional behaviour of $X_{(1)}$ and of $X_{(n)}$ is readily characterized. For example, $X_{(n)}$ has d.f. $F_n(x) = \{F(x)\}^n$, whilst $X_{(1)}$ has d.f. $F_1(x) = \{1 - F(x)\}^n$. Thus in Example 2.2 we found that the distribution of $X_{(1)}$ for an exponential distribution $\mathrm{Ex}(\lambda)$ is also exponential, $\mathrm{Ex}(n\lambda)$, that is, with mean reduced by the factor $1/n$.

Obviously the distributions of $X_{(1)}$ and of $X_{(n)}$ will be stochastically smaller and larger, respectively, than that of X (see Figure 2.2).

The larger the sample size n, the more widely separated will be the distributions of X, $X_{(1)}$ and $X_{(n)}$. One feature of interest was illustrated in the example discussed in Section 1.3.5 on levels of a water supply reservoir. The m-year

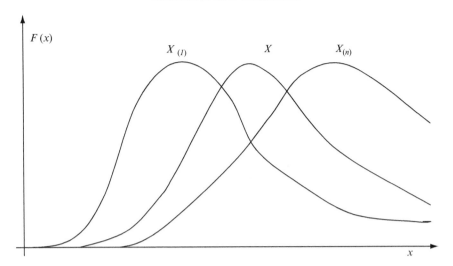

Figure 2.2   The p.d.f.s of X, $X_{(1)}$ and $X_{(n)}$

return period was the level exceeded only once in m years: if the yearly maximum level is $X_{(n)}$ (e.g., with $n = 365$ if we consider the maximum daily level over a year) then the m-year *return period* ξ_m is defined as

$$P\{X_{(n)} > \xi_m\} = 1/m$$

so that ξ_m is the $(m - 1)/m$ quantile of $X_{(n)}$. We will need to return to such concepts later to consider how we might estimate ξ_m.

But the extremes are only limiting cases of the order statistics $X_{(r)}$ ($r = 1, 2, \ldots, n$), and we might ask what, if anything, can be said of the distribution of $X_{(r)}$ as $n \to \infty$. It can be shown that, under quite wide conditions, the distribution of $X_{(r)}$ tends as $n \to \infty$ to a limiting form essentially independent of the form of the underlying distribution F. Specifically, if $n \to \infty$ in such a way that $r/n \to \lambda$, where $0 < \lambda < 1$, then $X_{(r)}$ tends, when suitably standardized, to the normal distribution N(0,1).

What, then, of the sample extremes: for example, the sample minimum, $X_{(1)}$?

Example 2.7   Many natural environmental phenomena can be modelled as a large set of linked units, so that if any one of them fails (or responds to a stimulus) the whole system fails (or responds). Thus with n links, where each link fails (or responds) independently at a level described by a common random variable X, then the distribution of the level at which the whole system fails (or responds) is precisely that of the minimum order statistic $X_{(1)}$ from a sample of large size n. As an example of this phenomenon we clearly have the mechanical failure of linked components such as the links of a chain. In epidemiological terms, we might contemplate a closed community of n individuals being exposed to an infectious organism with an individual 'catching' the related disease at an exposure level X. Once an individual is infected it may be assumed that the disease will spread through the community; the level of exposure at which this happens is governed by the random variable $X_{(1)}$.

Motivated by such considerations, a practical and empirically justified distribution for the failure (or response) level of linked systems is the so-called Weibull distribution with d.f.

$$F(x) = 1 - \exp\{-[(x - \alpha)/\beta]^\gamma\} \quad (x > \alpha > 0, \beta > 0, \gamma > 0). \quad (2.17)$$

David (1981, Chapter 9) traces interest in such 'breaking strength' distributions back nearly 150 years.

So we see that for an empirical model for $X_{(1)}$ when n is large we do *not* encounter the normal distribution (as might be conjectured from the above asymptotic result for $X_{(r)}$) but a distinctly different, well-specified form (2.17). In fact, we have an intriguing dichotomy of asymptotic behaviour for $X_{(r)}$, depending on whether as $n \to \infty$ we assume that

$$r/n \to \lambda, \quad \text{where } 0 < \lambda < 1,$$

or

$$r/n \to \lambda, \quad \text{where } \lambda = 0 \text{ or } \lambda = 1.$$

In the former case, as remarked above, we typically encounter a limiting *normal distribution* – see David (1981, Section 9.2) for more details.

In the latter case, and covering the limiting forms of the distributions of the *extremes*, $X_{(1)}$ and $X_{(n)}$, specifically (or their large-sample approximations – as, for example, the Weibull distribution (2.17) above for $X_{(1)}$) we obtain quite different results, perhaps the most important and intriguing body of work on extremes: namely the *limit laws of extreme values*.

Rather like the central limit theorem for a sample mean, which ensures convergence to normality from almost any distributional starting point, we find that (essentially) whatever the distribution of X, the quantities $X_{(1)}$ and $X_{(n)}$ tend as n increases to approach in distribution one of only three possible forms (which we will refer to as limit laws A, B and C, respectively). For the maximum $X_{(n)}$, for example, we have the *Gumbel*, with d.f.

$$F_A(x) = \exp\{-\exp[-(x - \alpha)/\beta]\} \quad (-\infty < \alpha < x < \infty; \beta > 0),$$

the *Fréchet*, with d.f.

$$F_B(x) = \exp\{-[(x - \alpha)/\beta]^{-\gamma}\} \quad (x > \alpha, \beta > 0, \gamma > 0),$$

and the *Weibull*, with d.f.

$$F_C(x) = \exp\{-[-(x - \alpha)/\beta]^{-\gamma}\} \quad (x > \alpha, \beta > 0, \gamma > 0);$$

see Fisher and Tippett (1928) and Gumbel (1958). The starting point for this work lies in the distant past and is attributed by Lieblein (1954) to W.S. Chaplin in about 1860. David (1981, Chapter 9) gives a clear overview of developments and research continues apace to the present time (see, for example, Anderson, 1984; Gomes, 1994).

Which of these is approached by $X_{(n)}$ (and $X_{(1)}$, which is simply dual to $X_{(n)}$ by a change of sign) is determined by the notion of *zones of attraction* within which all F lead to the same limit law. This is related in part to whether X is bounded below or above, or unbounded.

A key area of research is the rate of convergence to the limit laws as n increases – the question of the so-called *penultimate distributions*. That is to say, we are concerned with how rapidly, and on what possible modelled basis, $X_{(n)}$ approaches a limit law L.

What, in particular, can we say of how the distributions of $X_{(n)}$ stand in relation to each other as n progresses from 40 to 100, 250 or 1000, say? Little, in fact, is known but such knowledge is worth seeking. We shall consider one example of why this is so shortly (see Section 3.6)

Detailed treatments of extreme-value theory are given by Galambos (1987), Leadbetter *et al.* (1983) and Resnick (1987).

Let us examine another aspect of the study of sample extremes.

Example 2.8   Consider again (see Example 2.1) the random sample of 12 daily maximum wind speeds (in knots) from the data of a particular meteorological station in the UK a few years ago:

$$19, 14, 25, 10, 11, 22, 19, 17, 49, 23, 31, 18.$$

We have $x_{(1)} = 10, x_{(n)} = x_{(12)} = 49$. Not only is $x_{(n)}$ (obviously) the largest value – the upper extreme – but it seems extremely extreme!

This is the stimulus behind the study of outliers which we will pursue in Chapter 3. Outliers can be thought of as extreme observations which, by the extent of their extremeness, lead us to question whether they really have arisen from the same distribution as the rest of the data (i.e. from that of X). The alternative prospect, of course, is that the sample is contaminated by observations from some other source. An introductory study of the links between extremes, outliers and contaminants is given by Barnett (1986, 1988) – Barnett and Lewis (1994) provide an encyclopaedic coverage of outlier concepts and methods, demonstrating the great breadth of interest and research the topic now engenders.

Contamination can, of course, take many forms. It may be just a reading or recording error – in which case rejection of the offending observation might be the only possibility (supported by a *test of discordancy*). Alternatively, it might reflect low-incidence mixing of X with another random variable Y whose source and manifestation are uninteresting. If so, a robust inference approach which draws inferences about the distribution of X while accommodating Y in an uninfluential way might be what is needed. Then again, the contaminants may reflect an exciting unanticipated prospect and we would be anxious to identify its origin and probabilistic characteristics if at all possible.

Rejection, accommodation and identification are three of the approaches to outlier study, which must be set in terms of (and made conditional on) some model F for the distribution of X. This is so whether we are examining univariate data, time series, generalized linear model outcomes, multivariate observations, or whatever the basis of our outlier interest within the rich field of methods now available.

But if we are concerned with studying outliers (as we will be in Chapter 3) we are again interested in extremes and their distributional behaviour which in turn is dependent on the form of the underlying family of distribution, F. But we have also seen above that in the limit, and thus effectively for large samples, the extreme has a distribution which is essentially distributionally independent of the family F from which the sample has been chosen – in view of the limit laws.

So whilst the panoply of outlier methods have been developed conditional upon the form of the relevant model, F, (see Barnett and Lewis, 1994) we might in principle contemplate a *new approach* in which we would essentially ignore F and examine outlier behaviour in terms of properties of the extreme-value distribution which is being approached by $X_{(1)}$ or $X_{(n)}$ (or by some appropriate outlier function of them). This is an attractive prospect, a distribution-free and hence highly robust alternative to the usual model-based methods, but one which has not yet borne fruit.

So what is the difficulty? Precisely the following. Although $X_{(n)}$ approaches one of the limit laws A, B or C, we have to deal with finite samples and not enough is known in detail about how quickly and in what manner the forms A or B or C are approached as n progresses from, say, 40 to 100 to 250 etc. The study of convergence to the limit laws and of 'penultimate distributions', as mentioned above, is not yet sufficiently refined for our purpose. See Gomes (1994) for some of the latest developments; see also Anderson (1984) and Barnett (2002c). This matter is further illustrated in the following example.

Example 2.9   Consider Table 2.1, which shows samples of maximum daily, tri-daily, weekly, fortnightly and monthly wind speeds (in knots) at a specific location in the UK. Assuming (reasonably) that these approach the limit law A (they are all maxima) we have plotted, in Figure 2.3, $\ln \ln [(n + 1)/i]$ against $x_{(i)}$ in each case, expecting to find linear relationships. (See also the discussion in Section 1.3.5 above.) We do indeed find rough linearity and with the natural temporal ordering we would expect, reflected in the implied differences in the values of (particularly) α and (possibly) β in the approximating extreme-value distribution in each case.

Davies (1998) also carried out an empirical study of limiting distributions of wind speeds (again from a single UK site) and fitted Weibull distributions to

Table 2.1   Samples of maximum wind speeds (in knots) at a single UK location.

Daily	36	46	13	18	34	19	23	15	18	14	28	10	31	28	22	40	20	23	28
Tri-daily	33	31	21	19	22	28	29	25	36	41	16	24	43	20	38	51	34	20	31
Weekly	40	36	47	21	41	27	34	32	45	42	54	19	30	31	24	31	33	34	36
Fortnightly	35	32	45	37	39	31	34	28	47	58	31	33	51	42	50	47	40	41	52
Monthly	40	44	39	32	48	36	51	40	38	52	62	51	39	50	42	56	29	36	45

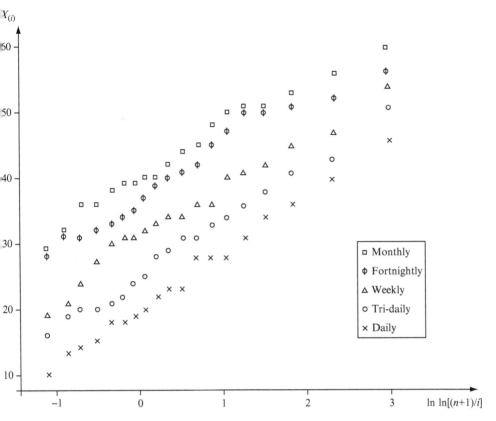

Figure 2.3 Probability plots for windspeed maxima.

daily, weekly, fortnightly, monthly and bimonthly maxima. Figure 2.4 (from Davies, 1998) shows the fitted distributions in which the time periods over which the maxima are taken increase monotonically as we move from the left-hand distribution to the right-hand one.

We need to know much more about how the asymptotic distributions of extremes are approached, that is, about the penultimate distributions. With such knowledge, it might be hoped that we can characterize the proximity to limiting form in any specific case as a function of sample size n and that such knowledge might lead to an essentially new (largely) distribution-free outlier methodology.

2.3.1 Practical study and empirical models; generalized extreme-value distributions

The use of extreme-value distributions in practical problems is widespread in environmental study, particularly for climatological phenomena. The field of hydrology, for example, has seen many applications and developments over a period of more than 70 years. Early applications were concerned with informal

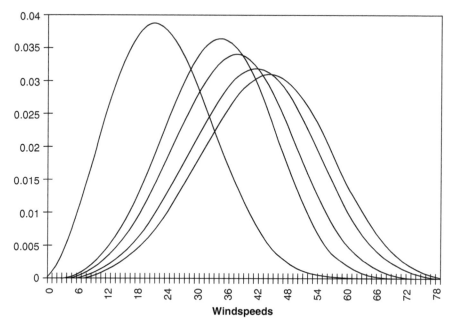

Figure 2.4   Fitted Weibull distributions to maximum wind speeds over different periods of time from 1 day to 2 months (Davies, 1998).

probability plotting methods (we have seen some illustrations in Section 1.3.5 and in Example 2.9 above), whereby a quick assessment could be made of whether a particular limit law seemed to be appropriate. This would be evidenced by the probability plot appearing to be a straight line. Examples include Barnett (1975, 1976b) and Barnett and Lewis (1967); the latter work was described in Section 1.3.5.

Work on extremes is often directly concerned with climate and climate effects. For example, Davison and Smith (1990) examine river flows and wave heights, modelling seasonality and temporal dependence. North (1980) uses time-dependent stochastic models for floods (see also Tadorovic and Zelenhasic, 1970). Tawn and Vassie (1989) and Tawn (1993) study extreme sea levels (possibly linked to global warming), whilst Coles and Wilshaw (1994) are interested in extreme wind speeds and Coles and Tawn (1994) consider extreme effects on structural design.

We find other interesting recent applications of climate-based or meteorologically related extremes in the work of Agnew and Palutikof (1999) on extreme climate impact on the UK retail business sector, Harland *et al.* (1998) on extreme wave forces on offshore structures, Nakai *et al.* (1999) on heatstroke-related deaths in Japan, and Silbergleit (1998) on maximum sunspot numbers. Elsewhere, Scarf and Laycock (1996) examine extremes (largest pits, thinnest walls) in corrosion engineering applications, whilst ApSimon *et al.* (1985) are concerned with long-range transport of radio isotopes.

We referred in Example 2.9 to the use of the Weibull distribution for extremes, as an *empirical* model in large but finite samples and not just as one of the three limit laws. In practical problems the interest will often lie in inference rather than modelling: in asking how we can use available statistical data on extremes to derive *estimates* of, or carry out *significance tests* on, such quantities as some relevant outer quantile γ_p (where $F(\gamma_p) = 1 - p$) with p very close to 0 or 1, or on the associated *return period*, or on what is known as the *extremal index* (effectively the reciprocal of the expected period between specifically defined extreme events). Gomes (1993) highlights the difficulties by explaining that the concern is usually with the need to 'estimate the probability of events rarer than those already observed'. Geladi and Teugels (1996) – who discuss matters as diverse as chemometrics, particle sizes and insurance risk – also refer to the basic problem of estimating characteristics of distribution tails, which behave differently from the mass of the distribution.

The inferential problems are of two interrelated forms. Often the realized sample sizes (e.g. for yearly maxima) are not large enough to justify assuming that one of the extreme-value distributions has set in and, when we take into account serial temporal dependencies, are effectively made even smaller in import. In addition, the small sample sizes are of limited inferential value particularly when we seek to infer aspects of the tail behaviour of the distribution of the extremes – see comments above.

Attempts to overcome these operational difficulties have taken many forms: one is to move away from any direct limit law justification and to employ a workable empirical model (such as the Weibull distribution of Example 2.9), another (to be discussed in the next section) is to shift emphasis from individual observations to those whose values lie beyond some upper or lower threshold values (thereby effectively weakening or removing the serial dependence).

The most widely used empirical model for extremes is the *generalized extreme-value distribution* with d.f.

$$F(x) = \exp\left\{-[1 - k(x - \alpha)/\beta]^{1/k}\right\}, \tag{2.18}$$

for $k(x - \alpha) < \beta$ with $\beta > 0$. This has been in use for practical work for more than 30 years: it was recommended for hydrological applications in the *Flood Studies Report* (National Environmental Research Council, 1975). Sample-based fitting methods were further developed by Prescott and Walden (1980, 1983), Hosking (1985) and Macleod (1989). Asymptotic maximum likelihood results were developed by Smith (1985). Davison and Smith (1990) review this model in the perspective both of the limit laws and of the 'peaks over thresholds' approach.

An important characteristic of (2.18) is that it can be interpreted as equivalent to the classical limit laws (see, for example, Leadbetter *et al.*, 1983). For example, as $k \to 0$ in (2.18), we obtain

$$F(x) = \exp\left\{-\exp[-(x - \alpha)/\beta]\right\} \tag{2.19}$$

which is just the Gumbel (type A) extreme-value distribution discussed above.

Thus we have, in the generalized extreme-value distribution, a model which encompasses the three limit laws and offers a useful basis for seeking to infer extremal characteristics of practical phenomena – notwithstanding the cautionary observations earlier on the difficulties of estimating or testing tail behaviour from the usual modest sized samples.

2.4 PEAKS OVER THRESHOLDS AND THE GENERALIZED PARETO DISTRIBUTION

Frequently, observations of extreme values are taken serially – in close proximity in temporal or spatial terms. For example, our data may be the maximum 24-hour temperatures over successive days at one geographic location. Inevitably, the successive observations are not independent; consider that 'hot spell' in mid-summer or the 'freeze-up' in the winter.

So if we obtain readings (in degrees Celsius) of 28, 27, 26, 27, 29, 30, 29 . . . over a mid-summer week we can hardly take these as independent observations of arbitrary daily maximum temperatures from some single distribution. The daily maximum temperature will vary seasonally and be serially correlated (see Figure 2.5): it may need to be modelled by some stationary time series (if detrended and with seasonal components removed – as described in Chapter 10).

So what are the seven observations worth in inferential terms? Perhaps they merely demonstrate that there can be times when the daily maximum is in the region of 28°C – a single observed feature.

It is such considerations that prompted an alternative representation of extremal processes in terms of 'peaks over thresholds' (POT) models under which a stochastic process is fitted either to the *peaks* or to the *exceedances* over some threshold value, u. Thus for exceedances, we might be interested in the process

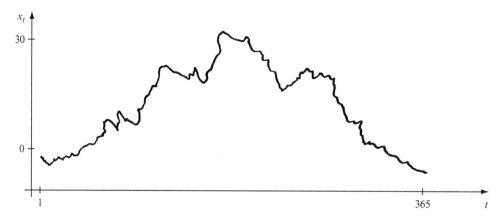

Figure 2.5 A realization of daily maximum temperatures over a year.

$$Y(t) = X(t) - u|_+$$

where $X(t)$ is the basic variational process indexed by t (usually time) and we define $Y(t)$ to be non-zero only if $X(t) > u$. Figure 2.6 illustrates $Y(t)$, shown as the envelopes of the hatched regions. In most applications, the basic process is observed at discrete points (e.g ends of days) and for sufficiently large u the exceedances will typically be single values, effectively independent of each other due to their separation.

Early work on this approach , again rooted in hydrological interests, is to be found in Todorovic and Zelenhasic (1970) and in Todorovic and Rousselle (1971) – assuming an inhomogeneous Poisson process for times of exceedances and independent exponentially distributed exceedances.

Current work assumes a poisson process for times between exceedances but a *generalized Pareto distribution* with d.f.

$$F(y) = 1 - (1 - ky/\sigma)^{1/k} \quad (\sigma > 0) \tag{2.20}$$

for the (independent) exceedances $Y = X - u$ over some high threshold, u. Developments of this model are to be found in Pickands (1975), DuMouchel (1983), Davison (1984) and Smith (1982, 1987). Davison and Smith (1990) give a detailed exposition of the approach with interesting practical applications of its use. Estimation methods are reviewed by Hosking and Wallis (1987).

Equation (2.20) can be shown to arise as the natural asymptotic form by limit law arguments. Its application in practical problems involves the choice of an appropriate threshold value, u, and of methods for effective estimation of k and σ (and of important derived quantities such as the n-year return period). Developments to allow the parameters k and σ to depend on relevant covariates through regression models and examination of the independence assumption for the exceedances have been considered. See Davison and Smith (1990), Gomes (1993), Tawn (1993) for interesting developments and applications.

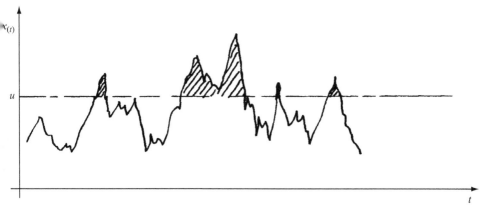

Figure 2.6 The process $Y(t)$.

One interesting feature of the generalized Pareto distribution leads to a useful plotting procedure for initial data scrutiny. We find that, if $k > -1, u > 0$ and $u < \sigma/k$, then

$$E(Y) = (\sigma - ku)/(1 + k) \tag{2.21}$$

(see Yang, 1978; Hall and Wellner, 1981). So if we consider the observed exceedances over u in a set of data and plot the sample mean values $\bar{y}_u$ against u (using what is called a *mean residual life plot*) we should obtain a straight-line plot with intercept $\sigma/(1 + k)$ and slope $k/(1 + k)$ from which informal estimates of σ and of k (and of derived quantities) can be obtained.

Example 2.10   Davison and Smith (1990) discuss a practical problem to do with exceedances of the threshold level $65 \, \text{m}^3 \, \text{s}^{-1}$ by the River Nidd at Hunsingore Weir over the period from 1934 to 1969. There were 154 exceedances. They present a mean residual life plot for the data – see Figure 2.7.

Note, from the form of (2.21), that the plot can be increasing or decreasing, depending on the sign of k. Here it is increasing, but only up to about 110, after which it seems to level off. The erratic behaviour from about 170 onwards is due to the few top order statistics and is not meaningful.

However, the levelling off beyond 110 suggests that we might have not a single generalized Pareto distribution but a mixture of two distributions. The

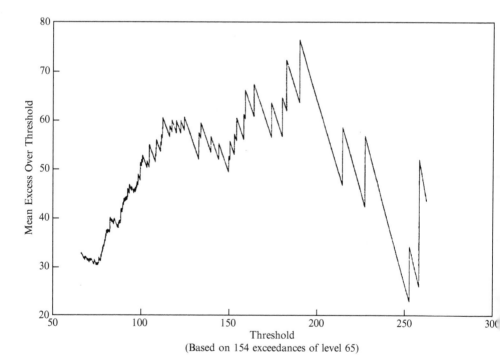

Figure 2.7   *Mean residual life plot for the River Nidd data (Davison and Smith, 1990).*

implications of this for the estimation of parameters and return periods are examined in detail by the authors with discussion of the effects of using different threshold values in the estimation and fitting processes.

We will continue to examine extremes in the next chapter, but with new emphases:

- outliers, as extreme extremes which might manifest contamination in the data set; and
- robust inference, where we seek to infer characteristics of models and estimate relevant parameters with minimal influence from the sample extremes and other model characteristics.

CHAPTER 3

Outliers and Robustness

They may not stick out at the end... but they must stick out somewhere.
(Rohlf, 1975)

3.1 WHAT IS AN OUTLIER?

As we have already discussed, all univariate samples have extremes: the sample *minimum value*, $x_{(1)}$, and sample *maximum value*, $x_{(n)}$, which are observations of the extreme *order statistics*, $X_{(1)}$, and $X_{(n)}$. Such extremes feature widely in the study of environmental issues.

The variable of natural interest is often an extreme: the hottest temperature in the year, the highest ozone level, the weakest link in the chain on a suspension bridge. And if we are interested in the whole range of values of an environmental variable, for example when monitoring pollution levels in an urban area, the extremes will still be important. In particular, if the highest pollution level is especially high it might indicate violation of some *environmental standard* (see Chapter 9). Alternatively, we might believe that it is *too extreme* to be a reflection of the natural random variation and, instead, that it indicates some abnormal event, such as a one-off illegal discharge.

Thus we might be concerned with situations where the extremes are not only the smallest and largest observations but where one or both of them is 'extremely extreme' – apparently inconsistent with the remainder of the sample. Such an observation is known as an *outlier* (or in earlier times a 'spurious', rogue', 'maverick' or 'unrepresentative' observation). We saw this notion of an 'extremely extreme' upper sample value motivated by the study of daily maximum wind speeds in Example 2.8 above. We would declare this to be an *upper outlier*.

For multivariate data the concept of order, extremeness and outlying behaviour is much less readily defined (how do you order a bivariate sample?) although the interest and stimulus of such notions may be no less relevant or important. We will come back to this topic in Section 3.7.

Environmental Statistics   V. Barnett
© 2004 John Wiley & Sons, Ltd   ISBN: 0-471-48971-9 (HB)

3.2 OUTLIER AIMS AND OBJECTIVES

As we have remarked, outliers are commonly encountered in environmental studies and it is important to be able to answer fundamental questions such as the following:

- What does the outlier imply about the generating mechanism of the data and about a reasonable model for it – is there some mixing of distributions, some contamination of the source?
- How can we formally define and examine possible contamination?
- What should we do about outliers or contaminants?

Three possible answers to the last question are that we might

- *reject* them;
- *identify* them for special consideration;
- *accommodate* them by means of robust statistical inference procedures – which are relatively free from the influence of the outliers – in order to study the principal (outlier-free) component of the data generating mechanism.

We will study in more detail the wide range of interests (and associated actions) which underlie the comprehensive machinery of outlier models and methods which has been developed over more than 150 years. The methods allow us to investigate outliers in univariate and multivariate data, in regression and linear models, in generalized linear models, in time series, in contingency tables, in spatial data, in directional data and in many other frameworks. Barnett and Lewis (1994) provide an encyclopaedic coverage of such work; see also Barnett (1978a, 1983b) for motivational introductions to the outlier theme.

Interest in outliers goes back a long way. As long as 150 years ago, Peirce (1852) raised issues of concern (which even at such an early stage involved an application of environmental interest). Peirce remarked:

> In almost every true series of observations, some are found, which differ so much from the others as to indicate some abnormal source of error not contemplated in the theoretical discussions, and the introduction of which into the investigations can only serve . . . to perplex and mislead the inquirer.

Outliers have thus been of interest for a very long time in the historical perspective of statistics – pre-dating by 50 years or so what most would regard as the very origins of the organized study of statistical methods. Inevitably, the early proposals were vague and speculative. Even some present-day attitudes, however, still range in spirit from the earlier extremes (see Barnett and Lewis, 1994, Section 2.1) of never choosing to cast any doubt on the overall integrity of a data set to an almost indiscriminate discarding of any observations which do not seem to fit some convenient pattern (some data-generating model) which the data are expected to follow.

Such an *ad hoc* attitude is unnecessary, however. We now have a comprehensive body of theory and practice to support a rational approach to the handling of outliers (see again Barnett and Lewis, 1994) and research work continues apace on all aspects of the outlier theme.

Let us look in a little more detail at some fundamental issues.

Example 3.1   Peirce (1852) presents data in the form of 30 residuals (about a simple model) of measurements by Professor Coffin and Lieutenant Page in Washington in 1846 of the vertical semi-diameter of the planet Venus. These appear thus:

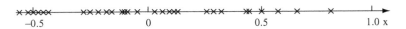

Another data set on the same theme is presented, for data collected by Lieutenant Herndon, also in Washington and also in 1846, in which there are now 15 residuals appearing as follows:

Notice the qualitative difference in the two data sets. In the first we *might* choose to declare $x_{(30)} = 0.83$ an upper outlier. In the second we almost certainly *would* declare $x_{(1)} = -1.40$ a lower outlier; and $x_{(15)} = 1.01$ an upper outlier.

Now consider basic aims. If we want to estimate location and dispersion relatively free of the influence of outliers, we might choose to use estimators which are not greatly affected by the extreme values, such as the sample median, m, and the so-called *median deviation*,

$$S_m = \text{median } \{|x_j - m|\}, \tag{3.1}$$

respectively, which are relatively unaffected by (and thus 'accommodate') any outliers.

But if accommodation is not our aim, we are led to ask a more specific question. Apart from the outlier being an extreme, indeed surprisingly so (which is the very stimulus for its declaration as an outlier), is it also statistically unreasonable when viewed as an extreme value? This is the basis of what is called a *test of discordancy* which is used to statistically support rejection of the outlier – or identification of it for separate investigation. If the outlier is significant on such a test it is termed a *discordant outlier* (at the level of the test).

Consider the Herndon data, which were examined so long ago not only by Peirce (1852) but also later by Chauvenet (1863). The lower outlier, $x_{(1)} = -1.40$, was essentially adjudged to be discordant on their primitive and invalid approaches (although the term 'discordant' was not used at that time).

How do we find it now? Suppose we were to adopt a normal distribution as the basic (non-contamination) model. Different tests of discordancy are possible here (see Barnett and Lewis, 1994), including two whose test statistics are

$$t_1 = (x_{(1)} - \bar{x})/s$$

and

$$t_2 = [x_{(1)} - x_{(2)}]/[x_{(n)} - x_{(1)}],$$

where $\bar{x}$ and s are the sample mean and standard deviation, respectively. We find that $t_1 = -2.58$ and $t_2 = -0.40$, and the 5% points for the appropriate null distributions turn out to be -2.71 and -0.44, respectively (see Tables XIIIa and XIXa in Barnett and Lewis, 1994).

Thus both tests coincide in their conclusion that the lower outlier is not (quite) discordant at the 5% level (in contrast with the early Peirce and Chauvenet conclusions).

Let us move on to more detailed consideration. In doing so we will illustrate ideas in terms of an upper outlier, $x_{(n)}$: the 'extremely extreme' largest sample value. (Analogous ideas will apply to lower outliers or other outlier manifestations, such as groups of outliers.)

We think of outliers as extreme observations which by the extent of their extremeness lead us to question whether they really have arisen from the same distribution as the rest of the data (i.e. from that of the basic random variable, X). The alternative prospect, of course, is that the sample is *contaminated*, that is, that it contains observations from some other source.

An introductory study of the links between *extremes, outliers* and *contaminants* is given by Barnett (1988). In particular, we note the following observations:

- An outlier is an extreme but an extreme need not be an outlier.
- An outlier may or may not be a contaminant (depending on the form of contamination). For example, upward slippage of the mean might be manifest in the appearance of an upper outlier since the larger mean will promote the prospect of larger sample values).
- a contaminant may or may not be an outlier. It could reside in the midst of the data mass and would not show up as an outlier.

Such varying prospects are illustrated in Figure 3.1, where F is the main model for the bulk of the data and G is the contaminating distribution.

Contamination can, of course, take many forms as remarked in Example 2.8. It may be a reading or recording error – in which case *rejection* (or correction) might be the only possibility (supported perhaps by a test of discordancy). Alternatively, it might reflect low-incidence mixing of X with another random variable Y whose source and manifestation are uninteresting. If so, a robust

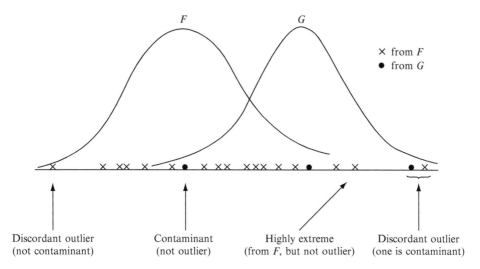

Figure 3.1 Relationships between outliers, extremes and contaminants.

approach which draws inferences about the distribution of X while accommodating Y in an uninfluential way might be required. Then again, the contaminant may reflect an unexpected prospect and we would be interested in *identification* of its origin and probabilistic characteristics if at all possible.

3.3 OUTLIER-GENERATING MODELS

Accommodation, identification and *rejection* – the three basic approaches to outlier handling – must clearly be set in terms of (and made conditional on) some model F for the basic (uncontaminated) distribution of X. This is necessary whether we are examining univariate data, time series, generalized linear model outcomes, multivariate observations, or whatever is our outlier interest within the wide range of methods now available. But if we reject F, what form of model might be relevant to explain any contaminants?

3.3.1 Discordancy and models for outlier generation

Any test of discordancy is a test of an initial (basic, null) hypothesis $H:F$ which declares that the sample arises from a distribution F. A significant result implies rejecting H in favour of an alternative hypothesis $\bar{H}$ which explains the presence of a contaminant likely to be manifest as an outlier (for illustration we continue to restrict attention to upper outliers and to at most one such). Of course, the alternative hypothesis is not needed to carry out a test of discordancy, but it will be if we are to compare and contrast different possible tests (as in Example 3.1 above), or to seek to evaluate their respective powers.

Let us consider the form of a test of discordancy for a single upper outlier. On the null hypothesis, we have a random sample from some distribution F, and we need to examine if the upper outlier, $x_{(n)}$, is a reasonable value for an upper extreme from F or if, on the contrary, it is statistically unreasonably large. (Corresponding principles are invoked if we are considering a lower outlier, $x_{(1)}$, or an outlier pair $(x_{(1)}, x_{(n)})$, etc.) Thus we would need to evaluate $P(X_{(n)} > x_{(n)})$ when the data arise from the model F. If we were to find that $P(X_{(n)} > x_{(n)}) < \alpha$ (for some chosen small value α) we would conclude that the upper outlier is *discordant at level* α and might thus be indicative of contamination.

Example 3.2 The distribution of $X_{(n)}$ is easily determined. Suppose that X has distribution function $F(x)$ and $X_{(n)}$ has distribution function $F_n(x)$. The condition $X_{(n)} < x$ implies that we need all $X_i < x, i = 1, 2, \ldots, n$. So we immediately conclude that

$$F_n(x) = \{P(X < x)\}^n = \{F(x)\}^n.$$

So if X has an exponential distribution with parameter λ, then

$$F_n(x) = (1 - e^{-\lambda x})^n$$

which is quite a complex form.

In contrast, the minimum $x_{(1)}$ arises from a distribution with d.f. $F_1(x)$, where

$$F_1(x) = 1 - P(X > x)^n = 1 - [1 - (1 - e^{-\lambda x})]^n = 1 - e^{-n\lambda x},$$

so that $X_{(1)}$ is shown also to have an exponential distribution but with parameter $n\lambda$ (i.e. with expected value or mean $1/n$ times that of X); see Example 2.2.

Thus a test of discordancy for a lower outlier in an exponential sample is particularly straightforward in form. Suppose λ is (effectively) known. Then we would conclude that a lower outlier $x_{(1)}$ is discordant at level α if

$$1 - \exp[-n\lambda x_{(1)}] < \alpha,$$

that is, if

$$x_{(1)} < \frac{1}{n\lambda} \ln(1 - \alpha) \tag{3.2}$$

Specification of the alternative (outlier-generating) model is important for comparison of different tests of discordancy or for calculating performance characteristics, such as power (see Barnett and Lewis, 1994, pp. 121–125), or if we are interested in the identification aspect of outlier study. We need to consider some of the possible forms which might be adopted for the alternative model. We continue to consider the case of a single upper outlier. The most obvious possibilities are the following:

(i)   *Deterministic alternative.* Suppose that x_i *is known* to be spurious. Then we would have

$$H{:}x_j \in F(j = 1, 2, \ldots, n), \quad \overline{H}{:}x_j \in F(j \neq i).$$

This just says that x_i does not arise from F.

(ii) *Inherent alternative.* Here we would declare

$$H{:}x_j \in F(j = 1, 2, \ldots, n), \quad \overline{H}{:}x_j \in G \neq F, (j = 1, 2, \ldots, n).$$

under which scheme the outlier triggers rejection of the model F for the *whole* sample in favour of another model G for the *whole* sample (e.g. lognormal not normal). Whilst the outlier may be the stimulus for contrasting F and G, the test of discordancy will not, of course, be the only basis for such comparison.

Between these extreme prospects of an obvious specified contaminant, and a replacement homogeneous model, are intermediate possibilities more specifically prompted by the concepts of contamination and of outlying values.

(iii) *Mixture alternative.* Here we contemplate the possibility (under $\overline{H}$) that the sample contains a (small) proportion λ of observations from a distribution G. Then we have:

$$H{:}x_j \in F(j = 1, 2, \ldots, n), \quad \overline{H}{:}x_j \in (1 - \lambda)F + \lambda G(j = 1, 2, \ldots, n). \quad (3.3)$$

For $\overline{H}$ to reflect outliers, G needs to have an appropriate form – for example, larger or smaller location, or greater dispersion, than F. If λ is small enough we might encounter just one, or a few, outliers. For early discussion of the mixture model, see Box and Tiao (1968) and Guttman (1973).

(iv) *Slippage alternative.* This is the most common form of outlier model employed in the literature (going back to Dixon, 1950; Grubbs, 1950; Anscombe, 1960; Ferguson, 1961). Its general form is

$$H{:}x_j \in F(j = 1, 2, \ldots, n), \quad \overline{H}{:}x_j \in F(j \neq i), x_i \in G, \quad (3.4)$$

where, again, G is such that $\overline{H}$ is likely to be reflected by *outliers*. Of course, the slippage model may not be restricted to a single contaminant and may not specify the precise index of the contaminant (in which case we have the *unlabelled* rather than the *labelled* version of the model). Other models are also used – see Barnett (1978a) and Barnett and Lewis (1994).

3.3.2. Tests of discordancy for specific distributions

We saw an illustration of a test of discordancy for an exponential distribution at the beginning of the previous subsection (in Example 3.2).

The literature contains a wealth of different discordancy tests for different univariate situations. There are discordancy tests for the normal distribution (with or without knowledge of the values of the mean and variance), for the exponential, gamma, Pareto, uniform, Gumbel, Fréchet, Weibull and lognormal distributions, as well as for the discrete binomial and Poisson distributions. Barnett and Lewis (1994) present the different forms, review their prospects, describe special features of the individual tests – for example, statistical properties such as being a likelihood ratio test, and providing a degree of protection against masking (see below) – and tabulate relevant critical values. The extent of the state of knowledge is indicated by the fact that over 40 different tests for the normal distribution are described, and almost 20 for the exponential and gamma.

Specific results and recent developments in tests of discordancy for univariate samples include the following.

Normal

We saw above two test statistics t_1 and t_2 for a single lower outlier in a normal sample when the mean and variance are unknown. For a single upper outlier an obvious modification of t_1 yields $(x_{(n)} - \bar{x})/s$. A corresponding test for an upper and lower outlier pair can be based on $(x_{(n)} - x_{(1)})/s$. Other forms have been proposed and investigated for further cases of single or multiple outliers under different sets of assumptions about the state of knowledge of the mean and variance of the normal distribution. A particular example is $(x_{(n)} - \mu)/\sigma$ for an upper outlier from a normal distribution when both the mean, μ, and the variance, σ^2, are known. (See Barnett and Lewis, 1994, Chapter 6, for more details on test forms, properties and tables of critical values.)

Exponential and gamma

Suppose F has probability density function

$$f(x) = [\lambda(\lambda x)^{r-1}e^{-\lambda x}]/\Gamma(r) \quad (x > 0), \tag{3.5}$$

that is X has a gamma distribution so widely relevant to environmental problems (including as special cases the exponential distribution where $r = 1$, and the χ_n^2 distribution: where $\lambda = 1/2$ and $r = n/2$). Some existing tests of discordancy for upper outliers have natural form. Thus we might choose to employ test statistics $x_{(n)}/\sum x_i$ or (specially relevant to an exponential distribution with shifted origin) $[x_{(n)} - x_{(n-1)}]/[x_{(n)} - x_{(1)}]$. The test statistic $x_{(n)}/x_{(1)}$ provides a useful test for an upper and lower outlier pair.

Other distributions

Relationships between distributions often enable outlier tests to be developed by means of transformations. Clearly this is possible in an obvious way for the lognormal distribution, using tests for the normal distribution. For a Pareto distribution with d.f.

$$F(y) = 1 - (a/y)^r \quad (y > a),$$

$X = \ln Y$ has a shifted exponential distribution with origin $\ln a$ and scale parameter $1/r$, and appropriate outlier tests for the exponential distribution, which are independent of the origin, as is that based on $[x_{(n)} - x_{(n-1)}]/$ $[x_{(n)} - x_{(1)}]$, can be applied to the transformed observations $x_i = \ln y_i$ $(i = 1, 2, \dots, n)$. If a is known, it is better to use the transformed variable $X^* = \ln(Y/a)$. For a Weibull distribution an appropriate exponential transformation again enables tests for the exponential distribution to be employed. See Barnett and Lewis (1994, Chapter 6), for further details on discordancy tests for all distributions.

3.4   MULTIPLE OUTLIERS: MASKING AND SWAMPING

Often we may wish to test for *more than one* discordant outlier. For example, we may want to test both $x_{(n-1)}$ and $x_{(n)}$. There are two possible approaches: a *block* test of the pair $(x_{(n-1)}, x_{(n)})$ or *consecutive* tests, first of $x_{(n)}$ and then of $x_{(n-1)}$. Both lead to conceptual difficulties, aside from any distributional intractabilities (which are also likely).

Consider the following sample configuration:

The consecutive test may fail at the first stage when testing $x_{(n)}$ because of the proximity of $x_{(n-1)}$, which masks the effect of $x_{(n)}$. For example, the statistic $[x_{(n)} - x_{(n-1)}]/[x_{(n)} - x_{(1)}]$ will be prone to such *masking* for obvious reasons. On the other hand, we may find that a block test of $(x_{(n-1)}, x_{(n)})$ convincingly declares the *pair* to be discordant.

Consider a slightly different sample :

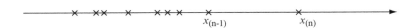

Now a block test of $(x_{(n-1)}, x_{(n)})$ may declare the pair to be discordant, whereas consecutive tests show $x_{(n)}$ to be discordant, but not $x_{(n-1)}$. The marked outlier $x_{(n)}$ has 'carried' the innocuous $x_{(n-1)}$ with it in the block test: this is known as *swamping*.

The dangers of these effects (and masking is a problem even for single outliers, of course) can be minimized by appropriate choice of test statistic, that is, of the form of test of discordancy. Such protection is not so readily available for automated procedures on regularly collected data sets. Different tests will have differing vulnerabilities to masking or swamping and the choice of test needs to take this into account (see Barnett and Lewis, 1994, pp. 109–115.)

Example 3.3   Consider again the discordancy tests in Example 3.1 applied to the lower outlier $x_{(1)} = -1.40$ in the Herndon data. We considered two test statistics,

$$t_1 = (x_{(1)} - \bar{x})/s$$

and

$$t_2 = [x_{(1)} - x_{(2)}]/[x_{(n)} - x_{(1)}],$$

where $\bar{x}$ and s are the sample mean and standard deviation, respectively, and we found that $t_1 = -2.58$ and $t_2 = -0.40$. The 5% points for the appropriate null distributions are -2.71 and -0.44, respectively (see Tables XIIIa and XIXa in Barnett and Lewis, 1994), so that on neither test was the lower outlier quite significant at the 5% level. We might have used other discordancy tests. Consider two more so-called Dixon-type test statistics,

$$t_3 = [x_{(1)} - x_{(2)}]/[x_{(n-1)} - x_{(1)}],$$

$$t_4 = [x_{(1)} - x_{(3)}]/[x_{(n)} - x_{(1)}].$$

These take the values -0.47 and -0.46, respectively and from Tables XIIIc and XIIIe of Barnett and Lewis (1994) we find 5% points -0.38 and -0.43. Thus, in contrast to the results for t_1 and t_2, we now conclude that the lower outlier is discordant at the 5% level.

It is clear that the slight but 'significant' differences in the imports of t_1 and t_3, and t_1 and t_4, are due to the masking effects of the upper outlier, $x_{(n)} = 1.01$, and of the second smallest observation, $x_{(2)} = -0.44$, respectively.

The broader-based and complex issue of testing for *multiple outliers* has been the subject of extensive study in its own right, including proposals for 'inward and outward' procedures, for 'forward selection', for 'backward elimination', for 'sequential tests' and for 'recursive' approaches. Again, details can be found in Barnett and Lewis (1994).

3.5 ACCOMMODATION: OUTLIER-ROBUST METHODS

The gamut of robust methods (though not specific to outliers) can assist in minimizing the influence of contaminants when drawing inferences about the basic distribution, F. Methods range from uses of *trimming* and *Winsorization*, through general estimators of L (linear order statistic), M (maximum likelihood) and R (rank test) type, to proposals even more specific to the outlier problem and the form of the uncontaminated distribution.

Accommodation is a principle of special interest in outlier methodology whenever we wish to carry out statistical inference about the basic distribution, rather than wishing to identify, or test for, possible contaminants. It can be of special relevance for automated data collecting and screening where we suspect outliers might occur.

Accommodation procedures in univariate samples include point and interval estimation, and testing of location and dispersion in the uncontaminated distribution source. Some methods are of general applicability; others are specific to particular distributional forms. Performance can be assessed in various ways; some of these relate especially to outliers and to the contaminating distribution, whilst others are more general. In the latter cases, concepts of *influence* and *sensitivity*, within the broader non-outlier-specific field of *robustness studies*, provide the relevant criteria (see Section 3.12).

A review of robustness studies in general is given in the book by Huber (1981); Barnett and Lewis (1994, Chapters 3 and 5), provide an extensive coverage of accommodation methods for outliers. We will review some of the general principles of robustness in Section 3.12.

3.6 A POSSIBLE NEW APPROACH TO OUTLIERS

It is clear from the discussions above that the methods for analysing outliers depend on assuming a specific form for the null (no-contamination) model. This is readily illustrated. Consider, for example, a discordancy test of a single upper outlier $x_{(n)}$ based on the test statistic $t = (x_{(n)} - x_{(n-1)})/(x_{(n)} - x_{(1)})$ (used in Section 3.3.2). Its null distribution and critical values depend *vitally* on the distribution of F, yet we are often unlikely to have any sound basis for assuming a particular form of F, especially if the only evidence is the single random sample in which we have observed an outlier. This is a fundamental problem in outlier study.

In practice, this dilemma is well recognized and it is usually resolved by a judicious mix of *historical precedent* ('such and such a model has always been used in this problem'), *broad principle* ('randomness in time and space implies a Poisson form'), *association* (linking the current problem to apparently similar types of situation), *some data-fitting* (e.g. probability plotting), and an element of *wishful thinking* (as in all areas of model-based statistics).

Thus it may be that a previous related study, and general scientific or structural features of the practical problem, linked with formal statistical considerations (e.g. the central limit law, characteristics of extremal processes) support a particular form for F, such as a normal distribution or an exponential distribution. We then relate our inferences on outliers to the appropriate null (no-outlier) distributions for that particular chosen F.

But we can look at outliers in a different way. In studying outliers we are, of course, concerned with sample extremes. These in turn must have the distributional behaviour of extremes, which we have seen (in Section 2.3) to be essentially, in the limit, distributionally independent of the family F from which the sample has been chosen – in view of the limit laws. So, in principle, we might question whether we could essentially ignore F and examine outlier behaviour in terms of properties of the extreme-value distribution which is being approached by $x_{(1)}$ or $x_{(n)}$ (or by some other appropriate outlier function of the sample). This would be an attractive prospect: a *distribution-free* and hence highly robust alternative to the usual model-based methods. But is it at all feasible?

Let us consider the basic difficulty in such an approach. Although $x_{(n)}$ approaches one of the limit laws A, B or C (of Section 2.3) as the sample size increases indefinitely, we have to deal with a finite sample and not enough is known in detail about how quickly and in what manner the forms A, B or C are approached as n increases. The study of convergence to the limit laws and of 'penultimate distributions' is not yet refined enough for our purpose (see, for example, Gomes, 1994).

Example 3.4 To consolidate this point, consider again the random sample data of maximum daily, tri-daily, weekly, fortnightly and monthly wind speeds (in knots) at a specific location in the UK, which we exhibited in Table 2.1. In Example 2.9 we observed how all the data sets approach the limit law A of Section 2.2 (as expected). The roughly linear plots of $\ln \ln [(n + 1)/i]$ against $x_{(i)}$ in each case (Figure 2.3) supported this. We also noted the natural temporal ordering we would expect, reflected in the implied differences in the values of (particularly) λ and (possibly) β in the approximating extreme-value distribution in each case.

We also examined Davies' (1998) empirical study of limiting distributions of wind speeds (again from a single UK site) in which Weibull distributions were fitted to daily, weekly, fortnightly, monthly and bimonthly maxima. Figure 2.4 illustrated the fitted distributions, which we noted increased monotonically with the time period over which the maxima are taken.

Such results give general reinforcement to the prospect of such a 'distribution-free' approach. However, it is clear that we need to know much more about how the distributions change in form with changes in the maximizing period. It might be hoped that research will lead to a clearer understanding of how the limit distribution of an extreme is approached in any specific case as a function of sample size n. Such knowledge might lead to a new and (largely) distribution-free outlier methodology.

The foregoing discussion above relates to univariate random samples from specific distributions. Even within the univariate context, some other possibilities arise, for example, outliers in time-series data, in directional data, in regression data and in data from designed experiments (although the latter two areas are in a sense concerned with multivariate data, which we will examine in more detail in the following sections of this chapter). Outliers in all these contexts are considered in detail by Barnett and Lewis (1994).

3.7 MULTIVARIATE OUTLIERS

Environmental problems frequently involve simultaneous consideration of several relevant random variables; thus, for example, health issues may be related to levels of several air pollutants; the structural stability of a bridge may be influenced jointly by wind force, rainfall and temperature; sustainability may be measured by numerous distinct indicators of biological, social and economic condition, and so on. Such problems require use of *multivariate* models and methods.

Often, as we have remarked above, it is the occurrence of extreme values of the random variables which is of principal concern – the maximum wind force or the lowest temperature. We saw that in a univariate random sample the notion of an extreme value, as a smallest or largest observation, is well defined. But the corresponding concept for multivariate data is less accessible. There is no obvious way unambiguously to order multivariate data. Nonetheless, the same intuitive stimulus of extremeness is still present. Observations 'seem' more extreme than others. Barnett (1976a) has discussed the basic principles and problems of 'the ordering of multivariate data'.

Interest in outliers in multivariate data remains the same as for the univariate case. Extreme values can again provide naturally interpretable measures of environmental concern in their own right and if they are not only extreme, but 'surprisingly' extreme or unrepresentative, they may again suggest that some unanticipated influence is present in the data source. So once more we encounter the notion of 'maverick' or 'rogue' extremes which we will term *outliers*.

Thus, as in a univariate sample, an extreme observation may 'stick out' so far from the others that it must be declared an outlier, and an appropriate test of discordancy may demonstrate that it is statistically unreasonable even when viewed as an 'extreme'. Such an outlier is said to be discordant and may lead us to infer that some contamination is present in the data. Discordancy may, as in univariate data, prompt us to *reject* the offending observation from the sample. But rather than rejection we will again wish to consider *accommodation* (outlier-robust) procedures which limit the effects of possible contamination or *identification* of the nature of the contamination as an interest in its own right.

Finally, again as in the univariate case, methods need to be based on appropriate *outlier models* to represent the possible uncontaminated and contaminated forms of the underlying population and its distributional characteristics.

Whilst all approaches (rejection, accommodation, identification), principles, procedures and models are essentially parallel to those for univariate data, there is a crucial and fundamental difference which needs to be resolved. It resides in the extent to which and manner in which the data indicate the presence of outlying observations. For univariate data the outliers are extreme values: they 'stick out at the ends of the sample'. In more structured or higher-dimensional data the stimulus is less obvious. For example, outliers in regression data may be detected by extreme *residuals* which clearly break the pattern of relationship in the data but may not be 'extreme' in any obvious (marginal) sense. As remarked above, for general multivariate data there is no natural concept of an extreme, and preliminary outlier detection procedures will almost always be needed to reveal outlying data points before we can progress to analysing them and taking appropriate action (rejection, accommodation, identification) in respect of them.

3.8   DETECTING MULTIVARIATE OUTLIERS

Outliers in multivariate data are often clearly signalled by observations lying well out on the periphery of the data cloud.

Example 3.5   Consider an environmental example discussed in Barnett and Lewis (1994, pp. 289–290). Rothamsted Experimental Station is one of the oldest agricultural research centres; it was also the scene for much of the early development of statistical methods in the work of R.A. Fisher. One of the so-called 'Classical Experiments' at Rothamsted is Park Grass in which the growth of leas (grasses and associated weeds) has been monitored under a fixed combination of different treatment regimes (on different experimental plots) for about 150 years. Figure 3.2 (from Barnett and Lewis, 1994, Figure 7.3) is a scatter plot of the yields of dry matter (in tonnes per hectare) on two totally untreated plots for the 50 years from 1941 to 1990. Several observations stand out as 'extreme', possibly as outliers (notably those for 1965, 1971, 1969, 1982

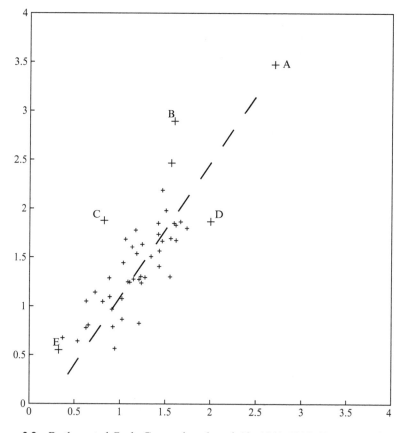

Figure 3.2   Rothamsted Park Grass plots 3 and 12, 1941–1990 (Barnett and Lewis, 1994).

and 1942 which are marked as A, B, C, D and E on Figure 3.2). These clearly lie well out on the data cloud and we should consider the implications of such manifestations. We will return to this example to illustrate various aspects of multivariate outlier study and it will be convenient for some illustrative purposes to assume that the data arise from a bivariate normal distribution.

Let us start with a fundamental question. What can we mean by 'lying well out'? We will need to try to formalize this notion for detecting multivariate outliers, and we will then go on to examine the wide range of procedures available for the statistical analysis of multivariate outliers (a comprehensive review is provided in Chapter 7 of Barnett and Lewis, 1994).

3.8.1 Principles

We start with a basic question: what can we mean, if anything, by 'ordering' multivariate data? There is no natural unambiguous ordering principle for data in more than one dimension; but progress can be made using certain *subordering principles*. Barnett (1976a) describes four types: marginal, reduced (or aggregate), partial and conditional.

For outlier study, *reduced subordering* is essentially the only principle that has been employed (especially if we regard *marginal* subordering as a special case of it). In this approach, we transform any multivariate observation $\mathbf{x}$, of dimension p, to a scalar quantity $R(\mathbf{x})$. We can then order a sample $\mathbf{x}_1, \mathbf{x}_2, \ldots, \mathbf{x}_n$ in terms of the values $R_j = R(\mathbf{x}_j)(j = 1, 2, \ldots, n)$. The observation $\mathbf{x}_i$ which yields the maximum value $R_{(n)}$ is then a candidate for declaration as an outlier – provided its extremeness is surprising relative to some basic model F which we believe to be appropriate for the uncontaminated data.

Further (as for univariate outliers), an outlier $\mathbf{x}_i$ will be declared *discordant* if $R_{(n)}$ is (statistically) unreasonably large in relation to the distribution of $R_{(n)}$ under F. Thus the principle of a *test of discordancy* is just as it was for a univariate outlier.

But special considerations arise in the multivariate case. We can easily lose useful information on multivariate structure by employing reduced (or any other form of sub-)ordering. So we need to choose the reduction measure $R(\mathbf{x})$ appropriately. Subjective choice can be costly; multivariate data do not reveal their structure readily (or reliably) to casual observation.

Barnett (1979) proposed two general principles for the detection of multivariate outliers, as follows.

Principle A.   The most extreme observation is the one, $\mathbf{x}_i$, whose omission from the sample $\mathbf{x}_1, \mathbf{x}_2, \ldots, \mathbf{x}_n$ yields the largest incremental increase in the maximized likelihood under F for the remaining data. If this increase is surprisingly large we declare $\mathbf{x}_i$ to be an outlier.

This principle requires only the basic model F to be specified. If, however, we are prepared to specify an alternative (contamination) model $\bar{F}$ – for example, of slippage type with one contaminant (see above) – we can set up a more sophisticated principle, as follows:

Principle B. The most extreme observation is the one, x_i, whose assignment as the contaminant in the sense of $\bar{F}$ maximizes the difference between the log-likelihoods of the sample under F and $\bar{F}$. If this difference is surprisingly large, we declare x_i to be an outlier.

Such principles have been widely applied; Barnett and Lewis (1994, Section 7.3) discuss applications to multivariate normal, exponential and Pareto models.

Often, however, the reduction metric $R(\mathbf{x})$ is chosen in an *ad hoc* (if intuitively supported) manner. For example, it is common to represent a multivariate observation $\mathbf{x}$ by means of a *distance measure*,

$$R(\mathbf{x}; \mathbf{x}_0, \mathbf{\Gamma}) = (\mathbf{x} - \mathbf{x}_0)'\mathbf{\Gamma}^{-1}(\mathbf{x} - \mathbf{x}_0), \tag{3.6}$$

where $\mathbf{x}_0$ reflects the location of the underlying distribution and $\mathbf{\Gamma}^{-1}$ applies a differential weighting to the components of the multivariate observation inversely related to their scatter or to the population variability. For example, $\mathbf{x}_0$ might be the zero vector, the true mean $\boldsymbol{\mu}$ or the sample mean $\bar{\mathbf{x}}$, and $\mathbf{\Gamma}$ might be the variance–covariance matrix $\mathbf{V}$ or its sample equivalent $\mathbf{S}$, depending on the state of our knowledge about $\boldsymbol{\mu}$ and $\mathbf{V}$.

If the basic model F were multivariate normal, $\mathrm{N}(\boldsymbol{\mu}, \mathbf{V})$, the corresponding form

$$R(\mathbf{x}; \boldsymbol{\mu}, \mathbf{V}) = (\mathbf{x} - \boldsymbol{\mu})'\mathbf{V}^{-1}(\mathbf{x} - \boldsymbol{\mu}) \tag{3.7}$$

has major practical appeal in terms of probability density ellipsoids as well as much broader statistical support, including accord with Principle A. Consider a sample (shown in Figure 3.3; from Barnett and Lewis, 1994, Figure 7.1) from a bivariate normal distribution with two observations, *A* and *B*, marked as possible outliers. The corresponding elliptic probability density contours shown in Figure 3.4 (also from Barnett and Lewis, 1994, Figure 7.1) clearly reinforce any intuitive concern we had for the outlying nature of the observations *A* and *B*. For other distributions, $R(\mathbf{x}; \mathbf{x}_0, \mathbf{\Gamma})$ may or may not be appropriate, as we shall find later, but it is nonetheless widely used.

A Bayesian approach to outlier detection ('detection of spuriosity') arises from the work of Guttman (1973). In this case also the essential concept of extremeness used to detect the outlier is expressible in terms of the distance metric, $R(\mathbf{x}; \boldsymbol{\mu}, \mathbf{V})$.

3.8.2 Informal methods

Due to the greater complexity of issues in the multivariate case, there is a wide range of *informal methods* that are used for the initial detection of outliers – sometimes interpretable in terms of the principles described above. We will examine some of these before going on to consider specific tests of discordancy and accommodation procedures.

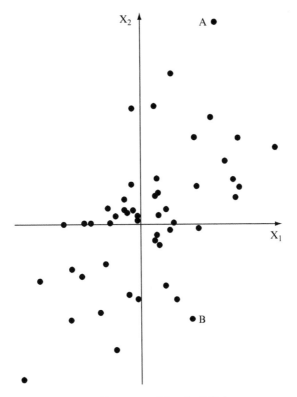

Figure 3.3 A bivariate example (Barnett and Lewis, 1994).

Many informal proposals for detecting multivariate outliers use quantitative or graphical methods. They may be based on derived reduction measures (but with no supporting distribution theory) or, more commonly, they may be offered as intuitively 'reasonable' for picking out those multivariate observations which are suspiciously aberrant from the bulk of the sample. They do not constitute formal tests of discordancy, which we shall consider later.

Initial processing of the data can take various forms. It may involve transformation of the data, study of individual marginal components of the observations, judicious reduction of the multivariate observations to scalar quantities in the form of reduction measures or linear combinations of components, changes in the coordinate bases of the observations, and appropriate methods of graphical representation. These can all help to highlight suspicious observations. If several such procedures are applied simultaneously (or separately) to a set of data they can be highly informative. An observation which stands out clearly on one such representation of the sample, or preferably more, becomes a firm candidate for declaration as an outlier.

An early example of an informal graphical procedure is described by Healy (1968), who proposes plotting the *ordered* $R_j(\mathbf{x}, \mathbf{S})$ against the expected values

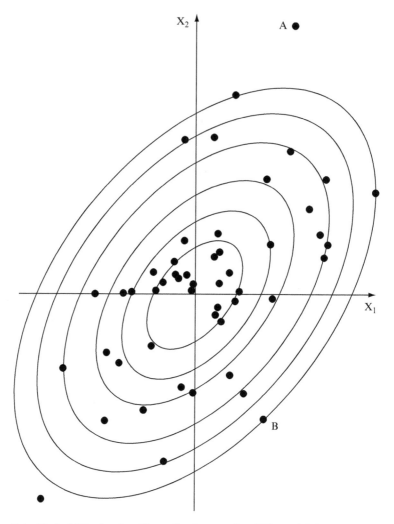

Figure 3.4 Probability density ellipses for the sample of Figure 3.3 (Barnett and Lewis, 1994).

of the order statistics of a sample of size n from a chi-squared distribution on p degrees of freedom.

Then again, the *marginal samples* (that is, the univariate samples of each component value in the multivariate data) can be informative in highlighting outliers. It is perfectly plausible for contamination to occur in one of the marginal variables alone: for example, in the form of a misreading or a recording error. It can even happen that a single marginal variable is intrinsically more liable to contamination. We must be careful, however, not to adopt too simplistic an approach in examining this prospect. This is illustrated by Barnett (1983a). A sample is examined from a bivariate normal distribution where contamination has occurred by slippage of the mean of the first component only for one observation. The detection of such an outlier is by no means simple.

More structured graphical and pictorial methods are often used for multivariate outliers. Rohlf (1975) remarks as follows:

> Despite the apparent complexity of the problem, one can still characterize outliers by the fact that they are somewhat isolated from the main cloud of points. They may not 'stick out on the end' of the distributions as univariate outliers must, but they must 'stick out' somewhere.

With this emphasis it is natural to consider a variety of different ways in which we can merely *look* at the data to see if they seem to contain outliers. Many different forms of pictorial or graphical representation have been proposed, with varying degrees of sophistication. Reviews of such methods of 'informal inference' applied to general problems of analysis of multivariate data, including outlier detection, are given by Gnanadesikan (1977) and in Barnett (1981). (See Barnett and Lewis, 1994, Section 7.4, for further details on published contributions in this spirit.)

A different approach notes that it can be informative to conduct a preliminary *principal component analysis* on the data, and to look at sample values of the projection of the observations on to the principal components of different order. Principal component analysis is part of the arsenal of multivariate analysis or multivariate statistical methods. There is a large bibliography on this important part of statistical methodology: see, for different emphases, Bock (1975), Dillon and Goldstein (1984), Gnanadesikan (1977), Kaufmann and Rousseeuw (1990) and Mardia *et al.* (1995).

Example 3.6 A study was conducted to try to characterize the environmental stresses throughout Poland by examining more than 50 common key indicators in each of the 50 or so voivodeships ('counties'). With $n(> 50)$ observations of a multivariate observation of dimension greater than 50 it is clearly not possible from the basic data to obtain any clear view of what is going on: of what variables are most important or least important, of what they contribute to the overall environmental condition etc. – there's too much wood and too many trees!

Principal component analysis is a multivariate technique for identifying that linear combination (the 'first principal component') of the variables which accounts for the largest amount possible of the overall variation and then (as the 'second principal component') that orthogonal linear combination which accounts for the largest amount possible of the remaining overall variation, and so on. Thus we effect a linear transformation to new variables of highest, second highest etc, import in explaining the variability in the data.

The hope is that after a few principal components (say two or three) we have effectively explained all the variability and that those principal components have a natural interpretation in terms of the original variables. This can often

happen and we obtain a dramatic reduction in the dimensionality and inter-
pretability of the data.

The environmental study of Poland was a case in point where a down-to-earth
and interpretable characterization was obtained in no more than five (compared with originally more than 50) dimensions.

Gnanadesikan and Kettenring (1972) examine in some detail the use of
principal component analysis in the study of multivariate outliers. They show
how the first few principal components are sensitive to outliers inflating variances or covariances (or correlations, if the principal component analysis has
been conducted in terms of the sample correlation matrix, rather than the
sample covariance matrix), whilst the last few are more sensitive to outliers
adding spurious dimensions to the data or obscuring singularities.

Example 3.7 Consider again the Park Grass data of Example 3.1. Figures 3.5
and 3.6 show normal probability plots of the first and second principal components, respectively (from Barnett and Lewis, 1994, Figure 7.5). For the first
principal component (Figure 3.5) we see no strong indication of non-normality.

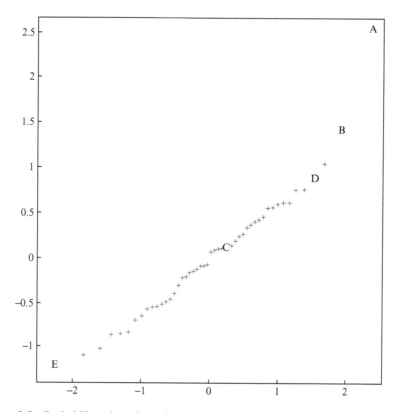

Figure 3.5 Probability plot of the first principal component of the Park Grass data
(Barnett and Lewis, 1994).

Also the previously detected 'extremes', A and B, are clearly confirmed as outliers (extreme and off the straight-line fit). Observation D also shows up as extreme. In Figure 3.6, normality is less evident, and observations B, C and D are clearly distinguished as outliers.

Some modifications of the approach to outlier detection by principal component analysis are suggested by Hawkins (1974) and by Fellegi (1975).

Another way in which informal quantitative and graphical procedures may be used to exhibit outliers is to construct univariate reduction measures. Gnanadesikan and Kettenring (1972) discuss classes of such measures all similar in principle to the 'distance' measures discussed above. Particularly extreme values of such statistics, possibly demonstrated by graphical display, may reveal outliers of different types. For graphical display of outliers, 'gamma-type probability plots' of ordered values, with appropriately estimated shape parameters, are also a useful approximate procedure and have been widely considered.

Outliers may also affect, and be revealed by, the correlation structure in the data. For example, Gnanadesikan and Kettenring (1972) propose examining

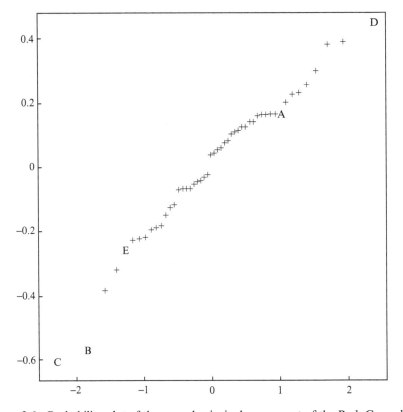

Figure 3.6 Probability plot of the second principal component of the Park Grass data (Barnett and Lewis, 1994).

the product-moment correlation coefficients $r_{-j}(s, t)$ relating to the sth and tth marginal samples after the omission of the single observation x_j. As we vary j we can examine, for any choice of s and t, the way in which the correlation changes – substantial variations reflecting possible outliers. Devlin *et al.* (1975) use the *influence function* to examine how outliers affect correlation estimates in bivariate data (dimension $p = 2$). Influence functions of other statistics (apart from the correlation coefficient) have also been proposed for detecting outliers.

Continuing his general characterization (that multivariate outliers tend to be separated from other observations 'by distinct gaps'), Rohlf (1975) proposes a *gap test* for multivariate outliers based on the *minimum spanning tree* (MST). He argues that a single isolated point will be connected to only one other point in the MST by a relatively large distance, and that at least one edge connection from a cluster of outliers must also be relatively large. An informal gap test for outliers is proposed, based on this principle.

3.9 TESTS OF DISCORDANCY

The concept of discordancy is as relevant to multivariate data as it is to univariate samples, although conceptual and manipulative difficulties have limited the number of formal and specific proposals for discordancy tests. Most results consider the normal distribution, which proves amenable to the construction of tests of discordancy with clear intuitive support and desirable statistical properties.

Suppose $\mathbf{x}_1, \mathbf{x}_2, \ldots, \mathbf{x}_n$ is a sample of n observations from a p-dimensional normal distribution, $N(\boldsymbol{\mu}, \mathbf{V})$. An alternative model which would account for a single contaminant is the *slippage alternative*, obtained as a multivariate version of the univariate models A (slippage of the mean) and B (slippage of the variance) discussed by Ferguson (1961). Using a *two-stage maximum likelihood ratio* principle (i.e. Principle B above) a test of discordancy can be derived. Models A and B have been studied extensively with various assumptions about what parameter values are known (see Barnett and Lewis, 1994, Section 7.3).

As an example, consider model A with $\mathbf{V}$ known. Here we are led to declare as the outlier $\mathbf{x}_{(n)}$ that observation $\mathbf{x}_i$ for which $R_i(\mathbf{x}, \mathbf{V}) = (\mathbf{x}_i - \bar{\mathbf{x}})'\mathbf{V}^{-1}(\mathbf{x}_i - \bar{\mathbf{x}})$ is a maximum, so that implicitly the observations have been ordered in terms of the reduced form of subordering based on the distance measure $R(\mathbf{x};\bar{\mathbf{x}}, \mathbf{V})$. Furthermore, we will declare $\mathbf{x}_{(n)}$ a *discordant* outlier if

$$R_{(n)}(\mathbf{x}, \mathbf{V}) = (\mathbf{x}_{(n)} - \bar{\mathbf{x}})'\mathbf{V}^{-1}(\mathbf{x}_{(n)} - \bar{\mathbf{x}}) = \max \ R_j(\bar{\mathbf{x}}, \mathbf{V}) \qquad (3.8)$$

is significantly large.

The null distribution of $R_{(n)}(\bar{\mathbf{x}}, \mathbf{V})$ is not tractable and certainly not readily determined in exact form, but it has been widely studied. Critical values for discordancy tests for models A and B under various assumption about parameter values being known or unknown are given in Barnett and Lewis (1994, Tables XXX–XXXIV).

For model A with **V** *unknown*, using a similar likelihood approach, we obtain what at first sight seems to be a different principle for the declaration of an outlier and for the assessment of its discordancy. Here we are led implicitly to order the multivariate observations in terms of a reduced form of subordering based on the values of

$$|\mathbf{A}^{(j)}| = \left|\left\{\sum (\mathbf{x}_i - \bar{\mathbf{x}})(\mathbf{x}_i - \bar{\mathbf{x}})'\right\}\right|, \tag{3.9}$$

where the sum is taken over all observations except $\mathbf{x}_j$. The $|\mathbf{A}^{(j)}|$ are ordered and the observation corresponding with the smallest value of $|\mathbf{A}^{(j)}|$ is declared an outlier.

Thus the outlier is that observation whose removal from the sample effects the greatest reduction in the 'internal scatter' of the data set, and it is adjudged discordant if this reduction is sufficiently large. Such a method was proposed by Wilks (1963), but the apparent distinction of principle for declaring an outlier in the case of unknown **V**, compared with the case where **V** is known, is less profound than might appear at first sight. This is because it is possible to re-express the internal scatter in terms of the distance measure $R(\bar{\mathbf{x}}, \mathbf{S})$ and the outlier is *again* found to be that observation whose 'distance' from the body of the data set is a maximum, provided we replace $\boldsymbol{\mu}$ and **V** by $\bar{\mathbf{x}}$ and **S**; see Barnett (1978b).

Example 3.8 Returning once again to the Park Grass data from Rothamsted, we can confirm that the five largest values of $R(\bar{x}, S)$ (in decreasing order) are those for the observations labelled A, B, C, D and E in Figure 3.2. In fact, we have

$$R_{(50)}(\bar{\mathbf{x}}, \mathbf{S}) = 14.07,$$
$$R_{(49)}(\bar{\mathbf{x}}, \mathbf{S}) = 10.87,$$
$$R_{(48)}(\bar{\mathbf{x}}, \mathbf{S}) = 7.53,$$
$$R_{(47)}(\bar{\mathbf{x}}, \mathbf{S}) = 5.49,$$
$$R_{(46)}(\bar{\mathbf{x}}, \mathbf{S}) = 4.55.$$

From Table XXXII in Barnett and Lewis (1994) we find that A is a discordant outlier since the 5% critical value (for $n = 50$ and $p = 2$) is 12.23. More sophisticated studies of the set of outliers A, B, C, D and E, in terms of *block tests* or *sequential tests* (see Barnett and Lewis, 1994, pp. 125–139) also attribute discordancy to B and C.

Frequently, marked *skewness* in a multivariate data set makes it clear that the normal distribution is inappropriate. Thus we might need to consider models expressing skewness. Two such prospects are provided by a *multivariate exponential* model and a *multivariate Pareto* model.

Many forms of *multivariate exponential distribution* have been proposed. One of these, due to Gumbel, has for the bivariate case a p.d.f.

$$f(x_1, x_2) = [(1 + \theta x_1)(1 + \theta x_2) - \theta] \exp(-x_1 x_2 - \theta x_1 x_2). \tag{3.10}$$

Applying a directional form of Principle A, an appropriate reduction measure,

$$R(X) = X_1 + X_2 + \theta X_1 X_2,$$ (3.11)

is obtained. So an upper outlier is detected as that observation (x_{1i}, x_{2i}) which yields the largest value of $R(x)$ over the sample of n observations. It is judged discordant if the corresponding $R_{(n)} = \max R(x_i)$ is sufficiently large. The distribution of $R_{(n)}$ turns out to be tractable and easily handled. For details see Barnett (1979), who also considers a discordancy test for an upper outlier in another skew bivariate distribution: namely, that one of the two *Pareto distributions* considered by Mardia (1962) which has p.d.f.

$$f(x_1, x_2) = a(a+1)(\theta_1\theta_2)^{a+1}(\theta_2 x_1 + \theta_1 x_2 - \theta_1\theta_2)^{-(a+2)},$$ (3.12)

for $x_1 > \theta_1 > 0, x_2 > \theta_2 > 0$ and $a > 0$. The correlation coefficient is $\rho = a^{-1}(a > 2)$. This time the appropriate restricted form of Principle A (assuming θ_1, θ_2 and a known) yields a reduction measure

$$R(X) = \frac{x_1}{\theta_1} + \frac{x_2}{\theta_2} - 1,$$ (3.13)

which is again tractable. It yields the critical value γ_α for a level-α discordancy test, for an upper outlier, determined from

$$\delta\gamma_\alpha^{(a+1)} - (a+1)\gamma_\alpha + a = 0,$$ (3.14)

with $\delta = 1 - (1-\alpha)^{1/n}$.

3.10 ACCOMMODATION

For multivariate data we commonly need to employ statistical methods that are specifically robust against (i.e. which *accommodate*) outliers as manifestations of contamination. Such accommodation procedures exist for estimating parameters (often with specific regard to the multinormal distribution) and for various multivariate procedures (such as principal component and discriminant analysis).

Suppose that under a basic model F, X has mean vector μ and covariance matrix V. Outlier robust estimation of μ and of V (and of the correlation matrix, R) has been widely studied, in terms of both individual components, and the overall mean vector and covariance matrix.

For μ, the starting point is in an obvious generalization of the work of Anscombe (1960). For the normal distribution $N(\mu, V)$, we first order the sample in terms of $R_{(n)}(x; \bar{x}, V)$ (or $R_{(n)}(x; \bar{x}, S)$), depending on our knowledge about V, and if the largest value $R_{(n)}$ is sufficiently large, we omit the observation x yielding $R_{(n)}$ before estimating μ from the residual sample. If $R_{(n)}$ is not sufficiently large, we use the overall sample mean, $\bar{x}$. Such *adaptive trimming* is revised by Golub et al. (1973) who employ a similar approach but based on *Winsorization* or *semi-Winsorization*. This approach can be extended to sequential trimming or Winsorization of several sufficiently 'extreme' values. Guttman (1973) considers the posterior distribution of **a** for a basic model $N(\mu, V)$ and mean-slippage alternative $N(\mu + a, V)$ for at most one of the observations.

Qualitative effects of outliers on estimation of **V**, and corresponding attitudes to robust estimation, are considered by Devlin et al. (1975). Campbell (1980) claims that outliers 'tend to deflate correlations and possibly inflate variance', although this may be rather too simplistic a prescription. A tangible form for outlier-robust *M-estimators* of μ and V, relevant to an elliptically symmetric basic model, is considered by Maronna (1976) and Campbell (1980).

An early accommodation proposal made by Gnanadesikan and Kettenring (1972) takes a different approach. They consider *direct* robust estimation of the matrix **V** in positive definite form, using selective iterative *trimming* of the sample based on values of the measure $R(x; x^*, I)$ where x^* is a robust estimator of μ. The procedure is intuitively attractive but it does not seem to have been widely used.

Outlier-robust estimation of μ and V is also considered by Rousseeuw and van Zomeren (1990). They reject the M-estimators of Campbell (1980) in view of their low breakdown point and suggest alternative robust estimators with a higher breakdown point (see Section 3.12 below).

Gnanadesikan (1977, Section 5.2.3) presents detailed proposals for constructing robust estimators of the individual elements of μ and V as well as for 'multivariate' estimators of μ and V. One difficulty with the former approach is that, if we form estimators of **V** or of **R** from separate robust estimators of their elements, the resulting **V** and **R** may not be positive definite. Devlin et al. (1975) suggest a solution involving 'shrinking' **R** until it is positive definite, and rescaling it if an estimate of **V** is required. (They also review various *ad hoc* estimators of the correlation coefficient, for a bivariate normal distribution, based on partitioning the sample space, on transformations of *Kendall's* τ or on normal scores.)

Another approach to outlier-robust estimation of correlation uses the ideas of *convex hull 'peeling'* or *ellipsoidal peeling* (see Bebbington, 1978; Titterington, 1978), following the suggestion by Barnett (1976) that the most extreme group of observations in a multivariate sample are those lying on the convex hull (with those on the convex hull of the remaining sample as the second most extreme group, and so on). Figure 3.7 (from Barnett and Lewis, 1994, Figure 7.2) shows the successive convex hulls for the bivariate sample presented in Figure 3.3.

Of course, any form of multivariate analysis can be affected by outliers or contamination. Modified forms of multivariate analysis which give protection against outliers include those proposed for *principal component analysis* using M-estimators (Campbell, 1980), for *canonical variate analysis* (Campbell, 1982) and for *discriminant analysis* (Campbell, 1978). In Critchley and Vitiello (1991) we find an extension to the influence of outliers in linear discriminant analysis with particular regard to estimates of misclassification probabilities. Outlier-robust (accommodation) methods have also been proposed for analysis of

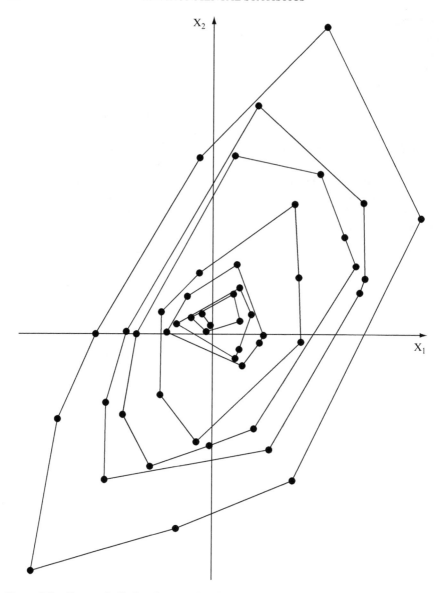

Figure 3.7 Convex hulls for the sample of Figure 3.3 (from Barnett and Lewis, 1994).

covariance (Birch and Myers, 1982), correspondence analysis (Escoffier and Le Roux, 1976) and multidimensional scaling (Spence and Lewandowski, 1989).

3.11 OUTLIERS IN LINEAR MODELS

We have so far considered outliers in unstructured multivariate data. Often several random variables are linked by a model under which one or more of them depends on the others. For example, in *simple linear regression* a random variable Y has a mean which can be expressed in terms of values of a second variable X in the form:

$$E(Y|X = x) = \alpha + \beta x.$$

More generally, we may postulate a broader class of *linear model* where Y has an expected value which is a linear function of the values of several dependant variables, X_i ($i = 1, 2, \ldots, k$). *Multilinear regression* and *polynomial regression* can be expressed in this way, and a special form provides the model basis for the analysis of classical *designed experiments*.

For use of the method of *least squares* we need make no further assumptions. However, if we are prepared to specify the error structure (e.g. with variations about the mean normally distributed), or to allow transformation of the data to achieve the linear model form, we can employ such methods as *maximum likelihood* or the range of techniques applicable to the *generalized linear model*. (Models for relationships between variables are considered in Chapters 7 and 8.)

In all such situations we can again encounter outliers. So here also we need methods to detect them in the first instance, and to handle them by tests of discordancy or accommodation procedures. For the simple linear regression model we have observations (y_i, x_i; $i = 1, 2, \ldots, n$) where $y_i = \alpha + \beta x_i + \varepsilon_i$, the ε_i being *residuals* about the expected values (often assumed to be uncorrelated and with constant variance).

Example 3.9 Barnett (1983b) discusses an environmental problem where there are 12 observations of the tensile strengths, z_i, of separate specimens of rubber after x_i days of exposure to weathering.

z_i	110	81	90	74	20	30	37	22	38	35	18	9
$y_i = \ln z_i$	4.7	4.4	4.5	4.3	3.0	3.4	3.6	3.1	3.6	3.2	2.9	2.2
x_i	4	5	7	9	11	14	17	20	23	26	30	35
e_i	0.85	0.17	0.79	0.59	**-2.5**	-0.91	0.10	-0.68	1.13	0.59	0.51	-0.60

The scatter plot of the $y_i = \ln z_i$ and x_i shown as Figure 3.8 appears to support a simple linear regression model, but the observation at 11 days of weathering (where $\ln z = 3.0$) departs markedly from the linear relationship and we would declare it to be an outlier.

Note, however, that the outlier is not an extreme of either variable: it merely 'stands out' or 'breaks the pattern'. This is typical of the form of outlying behaviour we find in a linear model. We can examine the outlier further by considering the estimated residuals about the linear fit. If we standardize the estimated residuals to allow for their different variances (denoting them by e_i) we find the results shown in the last line of the table above. The estimated standardized residual of -2.5 at 11 days of weathering is highly conspicuous. Reference to Table XXXII in Barnett and Lewis (1994) shows (on a normal

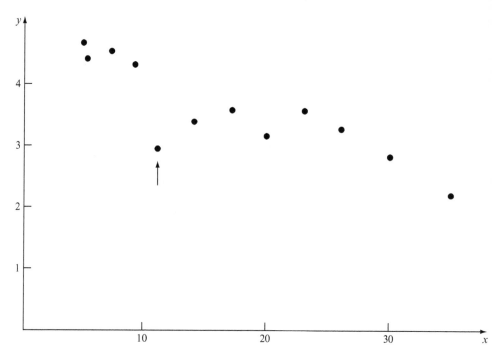

Figure 3.8   Tensile strengths of rubber samples after weathering.

basic model for the uncontaminated data) that this is a discordant outlier (the 5% critical value is −2.43).

Corresponding approaches for detecting and testing discordancy in more complex models, such as the linear model with several dependant variables, designed experiments and the generalized linear model, can be developed analogously using various residual-based methods. Details of such methods and of corresponding accommodation procedures can be found in Chapter 8 of Barnett and Lewis (1994); similar results for *time series* are discussed in Chapter 10 of the same book.

3.12   ROBUSTNESS IN GENERAL

We have considered the accommodation of outliers, in the sense of using methods that are robust against, or relatively unaffected by, outliers or by the contamination of which they are the product. A much broader form of robustness features in statistical methodology, under which we aim to devise methods of general statistical inference which are not strongly influenced by the underlying model, F, for data generation (see Huber, 1981, for a detailed development of this theme). Because accommodation methods for outliers may employ principles and measures from this broader area of study, and because environmental studies may need the wider robustness form, we will briefly comment on some of these principles and measures in their own right, although more detailed study will need to be conducted through appropriate reference texts such as Huber (1981), Hampel *et al.* (1986) and Rousseeuw and Leroy (1987).

The essential concern in robustness is the extent to which an estimator or the outcome of some other inference procedure is affected ('influenced') by any departure from the basic model F under which it is determined. Such considerations include the effects on the bias or precision of estimators, on the power of tests of significance and on more general aspects of the performance of the inference process. Included here is the concept of the *influence* on an estimator, say, of any single observation in the sample. This is particularly pertinent in respect of an outlier as a manifestation of contamination.

Thus we might ask if the effect is proportional to the number of contaminants, how the effect of a single observation relates to its magnitude, if there is a maximum possible effect that can arise from a single observation, if the effect is bounded, and so on.

A powerful armoury of tools has been developed to provide a means of answering such questions. Central to these is what is called the *influence function* or *influence curve* (Hampel, 1968, 1971, 1974). In the context of outliers and contamination (see Barnett and Lewis, 1994, particularly Section 3.1.3), the influence function for an estimator $T(x_1, x_2, \ldots, x_n)$ under the basic distribution F is given by

$$IC_{T,F}(\xi) = \lim_{\lambda \to 0} \{[T\langle(1 - \lambda)F + \lambda G\rangle - T(F)]/\lambda\}, \tag{3.15}$$

where λ is small, G is a model under which a value ξ is inevitably encountered (with probability one) and $T(F)$ is $\lim_{n \to \infty} T(x_1, x_2, \ldots, x_n)$.

Thus if T is the sample mean, $\overline{X}$, we have $IC_{T,F}(\xi) = \xi - \mu$ and the influence of a contaminant on the mean is *unbounded*; in contrast, this is not so for the *median*.

The *sensitivity curve* (Tukey, 1977) provides a finite-sample representation of the influence function. Measures derived from the influence function $IC_{T,F}(\xi)$ provide useful summary representations of influence effects. For example, the *gross-error sensitivity* $\sup_\xi |IC_{T,F}(\xi)|$ expresses the greatest possible influence that a fixed amount of contamination can have on the value of an estimator. If the influence function $IC_{T,F}(\xi)$ vanishes outside some finite interval $|\xi - \mu| \le \rho$, then ρ is the *rejection point* of the estimator; observations outside $|\xi - \mu| \le \rho$ have no influence on T. The trimmed mean provides an example of this. A related concept is that of *breakdown*; the *breakdown point* is the smallest proportion of contamination which can cause the value of the estimator to range to an unlimited extent.

Apart from seeking to characterize the influence effects of contamination or of an inappropriate basic model, we can seek inference procedures which have intrinsic robustness characteristics. For estimation, for example, this can be found in the use of L-estimators, M-estimators and R-estimators; see, for example, Barnett and Lewis (1994, Section 3.2).

PART II

Collecting environmental data: Sampling and monitoring

CHAPTER 4

Finite-Population Sampling

I know a representative sample when I see one!

We are often interested in the characteristics of *finite populations*. For example, in managing a woodland we will want to know how the trees are developing: how many varieties there are, what is the mean girth after 10 years' growth or the estimated gross yield of timber after 20 years. There will be a finite number of trees in the woodland, but we cannot measure or examine all of them – testing is time-consuming and might even be destructive (e.g. cutting a tree down to weigh the useful timber). We will want instead to take a *sample* from the finite population to estimate some population characteristics such as the mean or total of some variable, or the proportion with some qualitative measure (e.g. the proportion of spruce trees).

The statistical methodology for finite-population sampling is what underlies the study of *sample surveys* or *opinion polls*. This differs in some critical ways from that which applies in model-based inference where we assume that we can observe a random variable X which arises in potentially infinite numbers of observations from some family of distributions, perhaps with distribution function $\{P_\theta(x); \theta \in \Omega\}$, where the parameter θ lies in a parameter space Ω. Such a *model-based* approach applies to most of the methodology in this book. But here we take a different view – asking how we make inferences about a population which consists of just N specific values $X_1, X_2, \ldots, X_N$ but where N may be very large and we do not want to, or are not able to, observe every one of the population members. The principles and methods of sample surveys are explained in many books devoted solely to finite populations. See, for example, Barnett (2002a), Cochran (1977), Hedayat and Sinha (1991), Levy and Lemeshow (1991) or Thompson (2002).

Before we examine some finite population methods let us note a couple more practical environmental situations relevant to such an approach. Pollution is one of the major environmental issues of our time. It arises in many forms: as discharge of radioactive substances, in terms of water contamination or from everyday exposure to road traffic exhaust fumes. A large-scale survey was conducted over several metropolitan areas in northern Italy to examine the effects of traffic pollution. It showed that exhaust fumes from heavy road traffic have adverse effects on the respiratory health of school children (Ciccone

Environmental Statistics   V. Barnett
© 2004 John Wiley & Sons, Ltd   ISBN: 0-471-48971-9 (HB)

et al., 1998). The children were seen to have increasing incidence of infections of the lower respiratory tract and of wheezing and bronchitis with increase in exposure to vehicular exhaust fumes. The questionnaire-based survey of about 40 000 children, comprising more than 25% of the reference population, produced a response rate of more than 90%. The results were *stratified* by level of urbanization, that is, obtained and analysed for regions with different proportions of countryside and built environment.

Consider another example. Water or air pollution in socially controlled areas may be measured regularly at a large network of monitoring sites. For example, such regular sampling of bathing water in recreational areas is carried out to check if water quality satisfies required standards. This may well be done by drawing a sample at random from a larger *finite* set of network sites. As an example, weekly readings of mean pollution levels (e.g. of *E. coli*) might be taken in the form of a random sample of, say, ten of the many monitoring sites around a lake. The sample mean level gives an estimate of the population mean pollution level for the lake as a whole, but it is likely to be subject to high sampling variance, which may make it unsuitable for assessing if the standard is being met. Greater sample size would give greater precision but at greater cost. This is the crux of finite-population statistical methodology: how to obtain unbiased estimates of population characteristics of sufficient accuracy for decision-making.

Yet another example of sample survey work with environmental importance is to be found in the theme of 'social noise'. Young people the world over are exposed to very high levels of noise, when they visit nightclubs, movies, discos, etc. All ages have to put up with the 'noise' of their neighbours. Such questions as whether social noise levels have increased or young people are suffering hearing damage are most important in assessing this aspect of 'environmental pollution' (see, for example, Davis *et al.*, 1998).

4.1 A PROBABILISTIC SAMPLING SCHEME

Given a finite population $X_1, X_2, \ldots, X_N$ of size N, where X_i is the value for individual of some variable of interest X, we may want to estimate some population characteristic such as the *population mean* $\overline{X} = \sum_{i=1}^{N} X_i/N$, the *population total* $X_T = N\overline{X}$, or the *population proportion* P of individuals satisfying some condition, such as those for whom $X_i < A$ for some given A. Note that P is also a population *mean* of the derived variable Y, where

$$Y_i = \begin{cases} 1 & \text{if } X_i < A, \\ 0 & \text{if } X_i \geq A. \end{cases}$$

We will need to estimate $\overline{X}, X_T$ or P from a sample $x_1, x_2, \ldots, x_n$ of $n < N$ observed values. For example, the *sample mean* $\bar{x} = n^{-1}\sum_{i=1}^{n} x_i$, the *sample total estimator* $N\bar{x}$, and the *sample proportion* $\{x_i < A\}/n$ are intuitively sensible estimates.

But how are we to choose the sample, and how are we to assess its statistical properties? For example, is an estimator unbiased, and what is its sampling variance? What do we even mean by 'unbiased' or by 'sampling variance' in the finite-population context? We cannot express these in terms of a probabilistic model $P_\theta(x)$ for X. We have just N specific prescribed values $X_1, X_2, \ldots, X_N$.

Primitive approaches such as *accessibility sampling* ('take the most convenient sample') or *judgemental sampling* ('I know how to choose a representative sample') are clearly unsatisfactory. They can (in spite of the claim behind judgemental sampling) be grossly *unrepresentative* and give highly *biased* estimates. But much more important, they have no schematic basis on which to measure long-term behaviour and to assess representativity, precision, bias, etc. In fact, successive accessibility or judgemental samples chosen by the same person are likely to be the same sample each time, and it is meaningless to talk about the sampling variance in such a situation.

It is only by introducing a probabilistic mechanism for choosing the sample that we can usefully define and assess bias, sampling variation, etc. Thus, we need in principle to consider all $\binom{N}{n}$ possible samples s_i of size n (chosen without replacement) which could arise and then choose from these with prescribed probabilities $\{p_i = p(s_i \text{ chosen}); i = 1, 2, \ldots, \binom{N}{n}\}$. In terms of such a *probability sampling scheme* we can define our concepts properly and assess the statistical properties or procedures.

4.2 SIMPLE RANDOM SAMPLING

The simplest probability sampling scheme is one in which each sample has the same probability or occurring. This is called *simple random sampling* and can be shown to be equivalent to picking population members individually and successively at random without replacement. It is the basis of many more-structured sampling schemes (discussed later) such as *stratified simple random sampling* or *one-* or *multi-stage cluster sampling*, or those employing the *probability-proportional-to-size* principle.

Let us show that, as remarked above, such a sample can be obtained sequentially: by drawing members from the population one at a time without replacement, where at each stage every remaining member of the population has the same probability of being chosen. We can show that this produces a simple random sample. Suppose that from this sequential method we obtain n (distinct) population members whose X values are $x_1, x_2, \ldots, x_n$, where x_i refers to the ith chosen member $(i = 1, 2, \ldots, n)$. Such an *ordered* sequence has probability

$$\frac{1}{N} \cdot \frac{1}{N-1} \cdots \frac{1}{N-n+1} = \frac{(N-n)!}{N!}$$

Now any reordering of $x_1, x_2, \ldots, x_n$ yields the same choice of n distinct population members (that is to say, the same sample) and there are in all $n!$

possible reorderings. So the probability of obtaining any sample $x_1, x_2, \ldots, x_n$ of n distinct population members (irrespective of order) is clearly

$$\frac{n!(N-n)!}{N!} = \binom{N}{n}^{-1}.$$

There are $\binom{N}{n}$ such sets (or *samples*) which can arise in precisely this way, so these samples are therefore generated with equal probabilities – that is, by simple random sampling.

We shall need to examine the use of simple random sampling for estimating the three population characteristics described above (the population mean $\overline{X}$, the population total X_T, and the proportion, P, of X-values in the population which satisfy some stated condition) and to see how resulting estimators behave in an aggregate sense – that is, in terms of their sampling distributions. Again, as for traditional estimators of parameters in infinite-population models, the variance of the parent population will be a crucial measure. So we need to define the **variance** of a finite population, which we will express in the form

$$S^2 = \frac{1}{N-1} \sum_{i=1}^{N} (X_i - \overline{X})^2.$$

Note that this *deterministic* form differs from the *probabilistic* average we would use for random variable theory for probability models. The divisor $N-1$ is arbitrary; we could have used N. We just need some convenient measure of the variability of the X-values in the population. S^2 proves to be convenient in that it can lead to simpler algebraic expressions in the later discussion, and produces results sometimes more closely resembling corresponding ones in the infinite-population context.

Note that it is only when we *choose a sample* (e.g. by simple random sampling) that the probability concept arises; it is a property of the probability sampling scheme. Thus for simple random sampling we can introduce the notion of the *expected value* of x_i the ith observation in the sample. We have

$$E[x_i] = \sum_{j=1}^{N} X_j P(x_i = X_j) = \frac{1}{N} \sum_{j=1}^{N} X_j = \overline{X},$$

in which we have put $P(x_i = X_j) = 1/N$ since the number of samples with $x_i = X_j$ is $(N-1)!/(N-n)!$, and each has probability $(N-n)!/N!$. Similar considerations show that

$$E(x_i^2) = \frac{1}{N} \sum_{j=1}^{N} X_j^2,$$

$$E(x_i x_j) = \frac{2}{N(N-1)} \sum_{r<s} X_r X_s \ (i \neq j).$$

So we conclude that the *variance* of x_i and *covariance* of x_i and x_j take the forms

$$\mathrm{Var}(x_i) = \mathrm{E}\big[(x_i - \overline{X})^2\big] = \frac{(N-1)S^2}{N} \qquad (4.1)$$

and

$$\mathrm{Cov}(x_i x_j) = \mathrm{E}[(x_i - \overline{X})(x_j - \overline{X})]$$

$$= \frac{1}{N(N-1)}\left[\left(\sum_{j=1}^{N} X_j\right)^2 - \sum_{j=1}^{N} X_j^2 - N(N-1)\overline{X}^2\right] = -\frac{S^2}{N}, \qquad (4.2)$$

respectively. The form of the covariance implies a small negative correlation between the potential sample observations.

Example 4.1 Suppose we wish to estimate the proportion of young people exposed to 'social noise' in certain venues (e.g. in clubs, pop-concerts, cinemas) who exhibit various forms of hearing impairment. This might take the form of a deterioration in hearing threshold, of tinnitus, etc. We decide to conduct a survey in the form of a simple random sample of size n drawn from a relevant study population of size N. The variable to be measured under appropriate medical protocols is the hearing threshold, in decibels.

How would you attempt to deal with some of the ambiguities and difficulties in this situation:

- What is a 'young person'?
- What is an 'appropriate venue'?
- How can we construct a sampling frame for young persons in appropriate venues?
- How do we define the size N of the study population?
- How do we match the study population (the population we examine) and the target population (the population in which we wish to draw inferences)?
- How (where, when) should we measure hearing threshold?
- What are the cost and access problems in such a survey?

4.2.1 Estimating the mean, $\overline{X}$

An intuitively appealing estimator of $\overline{X}$, based on a simple random sample of size n, is the sample mean,

$$\bar{x} = \frac{1}{n}\sum_{i=1}^{n} x_i.$$

We see immediately that

$$\mathrm{E}(\bar{x}) = \mathrm{E}[(x_1 + x_2 + \ldots + x_n)/n] = n\overline{X}/n = \overline{X},$$

so that $\bar{x}$ is *unbiased*. Also

$$\mathrm{Var}(\bar{x}) = \frac{1}{n^2} \sum_{i=1}^{n} \mathrm{Var}(x_i) + \frac{2}{n^2} \sum_{r<s} \mathrm{Cov}(x_r x_s) = \frac{1}{n^2} [n(N-1)S^2/N - n(n-1)S^2/N]$$

$$= \left(\frac{N-n}{N}\right) \frac{S^2}{n} = (1-f)\frac{S^2}{n}. \tag{4.3}$$

This shows the effect of the population being finite. The sampling variance of $\bar{x}$ is reduced by a factor $f = n/N$, called the *sampling fraction*, compared with what we would have in an infinite population. We refer to this effect as the *finite-population correction*.

A crucial consideration is how well, within the simple random sampling scheme, $\bar{x}$ compares with other possible estimators of $\bar{X}$. We know that it is unbiased, but it also turns out that $\bar{x}$ is the *best linear unbiased estimator* of $\bar{X}$, based on a simple random sample of size n, in the sense of 'having smallest variance'. This is easily demonstrated (see, for example, Barnett, 2002a, p. 36); it does not mean that it is optimal among all estimators, $\tilde{\theta}(x)$, of $\bar{X}$, but it is at least useful to know that in the class of easily calculated *linear unbiased* estimators, the simple form $\bar{x}$ is best.

4.2.2　Estimating the variance, S^2

We may be interested in $\mathrm{Var}(\bar{x})$ for a number of reasons. It is used in three ways in particular:

1. to assess the precision of the estimator $\bar{x}$;
2. to compare $\bar{x}$ with other estimators of $\bar{X}$; and
3. to determine how large a sample is needed to yield a desired precision.

In practice, we will not know the value of S^2, so that to make use of the sampling variance in the ways just described we need to estimate S^2 from sample data. The usual sample estimator of S^2 is the sample analogue (cf. infinite populations)

$$s^2 = \frac{1}{n-1} \sum_{i=1}^{n} (x_i - \bar{x})^2$$

which turns out to be unbiased, since

$$\mathrm{E}(s^2) = \frac{n}{n-1} \left[\sum_{j=1}^{N} X_j^2/N - \mathrm{E}(\bar{x}^2) \right]$$

$$= \frac{n}{n-1} \left[\sum_{j=1}^{N} X_j^2/N - (1-f)S^2/n - \bar{X}^2 \right]$$

$$= \frac{n}{n-1} \left(\frac{N-1}{N} - \frac{1-f}{n} \right) S^2$$

$$= S^2.$$

For purposes 1 and 2 (above), then, we can substitute the unbiased sample estimator, s^2, for the unknown population variance to yield an unbiased estimator of $\text{Var}(\bar{x})$ as $s^2(\bar{x}) = (1 - f)s^2/n$.

Sometimes, estimation of S^2 may be of interest in its own right; again s^2 serves for this purpose.

But problem 3 (above), where we wish to examine what size of sample is needed to achieve a desired precision, is more difficult if (as is usual) S^2 is not known. This is because the sample estimator s^2 is now of no relevance since we do not have a sample from which to determine it! We need to assess how large a sample we require *prior to sampling*. We return to this later (Section 4.2.3).

Invoking a finite-population analogue of the *central limit theorem*, we can often assume that the simple random sample mean is approximately *normally distributed*, that is, $\bar{x} \sim N(\bar{X}, (1 - f)S^2/n)$. This can be used to construct an approximate $100(1 - \alpha)$ % *symmetric two-sided confidence interval* for $\bar{X}$ in the form

$$\bar{x} - z_\alpha S\sqrt{(1 - f)/n} < \bar{X} < \bar{x} + z_\alpha S\sqrt{(1 - f)/n},$$

where z_α is the two-tailed α-point of $N(0, 1)$ satisfying $P(|Z| > z_\alpha) = \alpha$ for $z \sim N(0, 1)$.

But as we noted above, S^2 will not be known in practice and replacing S^2 by the sample estimate, s^2, will be reasonable provided n is sufficiently large. Taking the tone from the infinite-population case, it is not uncommon to seek to allow for not knowing S^2 by using *Student's t distribution* rather than the normal distribution, when n is small (less than about 40 or so). This gives an approximate $100(1 - \alpha)$ % *symmetric two-sided confidence interval for $\bar{X}$* in the form

$$\bar{x} - t_{n-1}(\alpha)s\sqrt{(1 - f)/n} < \bar{X} < \bar{x} + t_{n-1}(\alpha)s\sqrt{(1 - f)/n},$$

where $t_{n-1}(\alpha)$ is the two-tailed α-point of *Student's t* with $n - 1$ degrees of freedom.

There is no detailed formal justification for using the t distribution; it is used by analogy with the infinite-population case. Sample surveys are often taken in very large populations (say $N = 10\,000$, $100\,000$ or more) with substantial sample sizes (say $n = 1000$ or more). So, we will frequently be safe in adopting the above form, replacing S by s, without regard to the fine details of justifying the normal or t distributional form for $\bar{x}$ or serious concern for the sampling fluctuations of s as an estimate of S.

Example 4.2 Consider again Example 4.1. A simple random sample of 345 young adults (aged 18–25, from a target population of 6100 in a particular city) attended a hearing clinic to access the possible effects of 'social noise' as reflected by their hearing threshold levels at 1 kHz sound frequency. There were 190 females and 155 males, with respective mean threshold levels, in decibels, of -4.2 and -5.1, and sample standard deviations of 0.331 and 0.463. Estimate the mean and variance of the hearing threshold for the target population and calculate an approximate 99% confidence interval for the mean.

The sum of all hearing threshold levels is $-(190 \times 4.2 + 155 \times 5.1) = -1588.5$. So our estimate of the mean is -4.604.

The sums of squares for the component samples are $189 \times 0.331^2 + 190 \times (-4.2)^2 = 3372.307$ and $154 \times 0.463^2 + 155 \times (-5.1)^2 = 4064.563$. Thus the sample estimate of the population variance is $(7436.87 - 345 \times (-4.604)^2)/344 = 0.3572$. So we obtain an approximate 99% confidence interval as $-4.604 \pm 2.576 \times 0.5976/18.574$, or

$$-4.687 < \overline{X} < -4.521.$$

4.2.3   Choice of sample size, *n*

Clearly as the sample size increases so does the precision of $\bar{x}$ as an estimator of $\overline{X}$, but it is likely that the sampling costs will also increase perhaps beyond what can be afforded. We really need to be able to state the precision we require, or the maximum cost that we can afford, and then choose the sample size appropriately.

Thus suppose we want to estimate the population mean $\overline{X}$, from a simple random sample mean $\bar{x}$, keeping the probability that the absolute difference between $\overline{X}$ and $\bar{x}$ is greater than some specified value to some acceptable level. Although this involves no direct consideration of costs, if it happens that sampling costs are directly proportional to sample size, then we would achieve our aim for minimum cost.

So let us seek the minimum value of n that ensures that $P(|\overline{X} - \bar{x}| > d) \leq \alpha$ for some prescribed d and (small) α. In this formulation, the survey scientist would need to declare what is required by way of the tolerance d, and the risk α of not obtaining an estimate within such tolerance.

If we standardize the objective and assume an approximate normal distribution for $\bar{x}$, we will need to satisfy

$$\frac{d}{S\sqrt{(1/f)/n}} \geq z_\alpha,$$

or

$$n \geq N\left[1 + N\left(\frac{d}{z_\alpha S}\right)^2\right]^{-1}. \tag{4.4}$$

Equivalently, from the form of $\mathrm{Var}(\bar{x})$, we could write our requirement as

$$\mathrm{Var}(\bar{x}) = (d/z_\alpha)^2 = V$$

(say). The above inequality for n can be written

$$n \geq \frac{S^2}{V(1 + N^{-1}S^2V^{-1})^{-1}} \tag{4.5}$$

so that as a first approximation to the required sample size, we could take $n_0 = S^2/V$. This will be an overassessment, but it will be reasonable unless the provisional sampling fraction, n_0/N, is substantial. If it is, then we would need to reduce our estimate of the required sample size to $n_0(1 + n_0/N)^{-1}$.

This all presupposes, however, that S^2 is known, which is unlikely to be true. So we have to face up to the fact that we must seek to estimate the required sample size without knowing S^2. There are various ways in which we might attempt to do this.

Use of pilot studies
Often we may explore the ground before carrying out a major survey by conducting one or more *pilot studies*. These can help in many ways, including examining different possible *sampling frames* (sets of sampling units from which to choose the sample) and any implicit practical difficulties which may be encountered in the sampling procedure. When a pilot study takes the form of a simple random sample, its results may give some indication of the value of S^2. We could use this provisional estimate of S^2 in the choice of the sample size of the main survey. But we must be careful. Sometimes a pilot study is limited to some resticted part of the population. If so, the estimate of S^2 which it yields can be quite biased.

Use of previous surveys
Frequently other surveys have been conducted elsewhere and at different times, and have studied *similar* characteristics in *similar* populations to the one of current interest (e.g. an environmental health survey last year and for the neighbouring region). This is particularly common in educational, environmental, medical, or sociological investigations. Often the measure of variability from the earlier or nearby survey can be used to estimate S^2 for the present population, so as to enable us to estimate the sample size required to meet any prescription of precision in the current work. But we need to be careful in extrapolating from one survey situation to another.

Using a preliminary sample
This can be the most reliable approach, but it may be too costly or time-consuming. It operates as follows. A preliminary simple random sample of size n_1 is taken. This is used to provide an initial estimate s_1^2 of the variance, S^2. We would ensure that n_1 is unlikely to be a large enough sample to achieve the required precision. We would use s_1^2 to estimate the required sample size n, and then augment the sample with a further simple random sample of size $n - n_1$ in an attempt to obtain the required precision.

4.2.4 Estimating the population total, X_T

Often we are interested in estimating the population total $X_T = N\overline{X}$, rather than the population mean $\overline{X}$. For example, in an annual survey of infected agricultural land (on some appropriate environmental definition) for farms in the Northeast region, we are likely to want to estimate the region's total annual infected land stock rather than the average number of infected hectares per farm. In view of the simple relationship between X_T and $\overline{X}$, no major problems arise; we can immediately extend the results we have obtained concerning the estimation of $\overline{X}$. Thus the simple random sample estimator of X_T will be $x_T = N\bar{x}$ and the earlier results confirm that x_T is unbiased for X_T. Furthermore, X_T will be the minimum-variance linear unbiased estimator of X_T based on a simple random sample of size n.

With similar conditions concerning the sample size, n, and value of the sampling fraction, f, we can again use a normal approximation for the sampling distribution of x_T to construct confidence intervals for X_T, or to choose a sample size to meet specified requirements concerning the precision of estimation of X_T. The detailed forms follow immediately from our results for estimating $\overline{X}$ since X_T is just $N\overline{X}$.

Example 4.3 Consider again the situation concerning hearing thresholds described in Examples 4.1 and 4.2. What sample size would be needed to be 99% sure of our estimate of hearing threshold being within 0.1 decibels of the true value?

The difficulty is that we do not know and cannot estimate beforehand the variance of the population. Suppose that in a corresponding earlier survey the population variance was 0.38. We might adopt this for the present purpose. We would then need

$$n > n_0 = 0.38 \times 2.576^2/0.1^2 = 252.16$$

But $n_0/N = 0.0413$, so that we obtain a minor reduction to $252.16/1.0413 = 242.15$ in the more accurate form; that is, 242 observations will be needed to meet the accuracy requirement.

4.2.5 Estimating a proportion, P

Consider an environmental process in which effluent is produced and discharged each day into a designated set of river sites. If the amount of pollution, Y, is not contained below a required level on any day at any site, damage can arise. To estimate the *proportion, P*, of sites/days over a particular year when damage might occur, a simple random sample of size n sites is taken and a count is made of the number, r, which have Y in excess of the required level. The population of Y values is not in itself of interest; we would merely like to know the proportion P of such values which are at damage levels.

Rather than studying a *quantitative* measure, Y, in relation to whether it satisfies some size criterion, we may be concerned directly with some qualitative attribute or characteristic; for example, with the proportion P of the inhabitants of some town who burn coal fires. Again, a simple random sample of size n can be used to estimate P. If r out of the n sampled inhabitants burn coal fires, a reasonable intuitive approach would be to estimate P by the simple random *sample proportion, $p = r/n$*.

As for the population total, we can again readily modify the results for estimation of the *population mean* to describe the properties of the estimator p. Suppose P is the proportion of members of a finite population of size N who possess some characteristic A, and a variable X_i is defined as follows:

$$X_i = \begin{cases} 1 & \text{if the member possesses characteristic A,} \\ 0 & \text{otherwise.} \end{cases}$$

Then $X_T = \sum_{i=1}^{N} X_i = R$ is the number of population members possessing the characteristic A. Consequently, $\overline{X} = N^{-1} \sum_{i=1}^{N} X_i = R/N = P$, so that the proportion P is merely the *population mean* for the X-values. Likewise, the sample proportion p is just the mean $\bar{x}$ for the *sample* of X values.

So the performance of p as an estimator of P can be obtained from the earlier results on use of the simple random sample mean to estimate the corresponding population mean. The only difference arises from the simple structure of the population of X-values, where only the values 0 or 1 can occur. This implies that the population mean $\overline{X}$ (or P) and the corresponding population variance, S^2, are related, with

$$S^2 = \frac{1}{N-1} \sum_{i=1}^{N} (X_i - P)^2 = \frac{NP(1-P)}{N-1} \tag{4.6}$$

Thus, if $Q = 1 - P$, the earlier results show that p is unbiased for P and that

$$\text{Var}(p) = (1-f)S^2/n = \frac{N-n}{N-1} \frac{PQ}{n}. \tag{4.7}$$

Of course P will be unknown, and so S^2 will also not be known. An unbiased estimator of S^2 can be obtained as

$$s^2 = \frac{1}{(n-1)} \sum_{i=1}^{n} (x_i - \bar{x})^2 = \frac{npq}{n-1}, \tag{4.8}$$

where $q = 1 - p$, so that $s^2(p) = (1-f)pq/(n-1)$ will be unbiased for Var (p). If the sampling fraction f is negligible, we have the simple form $s^2(p) = pq/(n-1)$.

Confidence intervals

As before, we can obtain approximate confidence intervals for P, at different degrees of approximation, depending on what distributional assumptions we make about R.

When n is small relative to NP and NQ, then R is essentially binomial, $B(n,P)$.

At a further stage of approximation we can use the *normal approximation to the binomial distribution*. So for the sample proportion p, we have $p \sim N(P, (1-f)PQ/n)$ as a basis for constructing approximate confidence intervals for P. This yields an approximate $100(1-\alpha)\%$ *two-sided confidence interval for P* in the form of the region between the two roots of

$$P^2[1 + z_\alpha^2(1 - f)/n] - P[2p + z_\alpha^2(1 - f)/n] + p^2 = 0.$$

If we substitute for $\text{Var}(p)$ its unbiased sample estimate, this leads to an approximate $100(1 - \alpha)\%$ confidence interval with limits

$$p \pm [z_a\sqrt{(1 - f)pq/(n - 1)}.].$$

Choice of sample size for prescribed accuracy

Finally, we must consider how to choose the sample size in estimating a proportion with required precision. We start by noting that $\text{Var}(P)$ will be a maximum when $P = Q = \frac{1}{2}$. This means that we can estimate P least accurately when it is in the region of $\frac{1}{2}$. What happens if $P \neq \frac{1}{2}$? The *standard error* of p is a multiple of $\sqrt{PQ}$. For $\frac{1}{4} < P < \frac{3}{4}$, $\sqrt{PQ}$ only varies over the range (0.433, 0.500), so over this region there will be little change in the accuracy of the estimator p.

When we consider what sample size is needed for prescribed accuracy requirements, we find that this depends on whether we are concerned with achieving, with prescribed probability, a specified *absolute accuracy*, or *relative accuracy*, for the estimator. In the two cases we would need, respectively, to satisfy

$$P(|p - P| > d) \leq \alpha \qquad \text{(for absolute accuracy } d\text{)}$$

or

$$P(|p - P| > \xi P) \leq \alpha \qquad \text{(for relative accuracy } \xi P\text{)}.$$

Using the normal approximation and ignoring the finite-population correction, we would need, respectively, to choose n to satisfy

$$\text{S.E.}(p) = \sqrt{(PQ/n)} < d/z_\alpha$$

or

$$\text{S.E.}(p)/P = \sqrt{Q/nP} \leq \xi/z_\alpha.$$

For required absolute accuracy we thus need $n > PQz_\alpha^2/d^2$. But PQ has a maximum value of $\frac{1}{4}$ when $P = \frac{1}{2}$. Thus choosing $n = z_\alpha^2/4d^2$ will suffice whatever the value of P, and this will not be too extravagant a choice over a quite wide range of values of P, say $0.30 < P < 0.70$. No similar basic facility is available if we want to ensure a certain relative accuracy.

Returning to the case of specified *absolute* accuracy, we can examine the effect of retaining the finite-population correction and of incorporating the precise form for $\text{Var}(p)$. This will now (using the appropriate normal approximation form) lead to a sample size satisfying

$$n \geq N\left[1 + \frac{(N - 1)}{PQ}\left(\frac{d}{z_\alpha}\right)^2\right]^{-1} \qquad (4.9)$$

or, putting $(d/z_\alpha)^2 = V,$

$$n \geq \frac{PQ}{V}\left[1 + \frac{1}{N}\left(\frac{PQ}{V} - 1\right)\right]^{-1}. \tag{4.10}$$

So as a first approximation to the required minimum sample size, we have $n_0 = PQ/V$, which is just what was obtained above when we ignored the finite-population correction. However, if n_0/N is not negligible, then we need a more exact specification: $n = n_0[1 + (n_0 - 1)/N]^{-1}$.

Now consider again the *relative* accuracy case. We would then need, as a first approximation, $n = n_0 = (Q/P)(z_\alpha/\xi)^2$, as obtained for the required sample size when ignoring the finite-population correction, or more accurately $n = n_0[1 + (n_0 - 1)/N]^{-1}$.

Of course, in practice P will not be known precisely. Again the three methods of Section 4.2.3 can be used to provide an 'advance estimate' of P to assist in assessment of the required sample size.

Example 4.4 Suppose that regionally based social services workers have been surveyed and found (as in many employment sectors) to take days off work which are not related to sickness or holiday entitlement. A fairly liberal view is taken of this, recognizing that stress factors in the work may well lead to employees taking the odd days off as 'casual holidays'. Suppose that up to 3 days in 6 months is regarded as reasonable and that we want to use the survey data to estimate the proportion of workers taking more than 3 days off in a six-month period of study.

Of 2800 workers, we sampled 332 and found that 56 took more than 3 days of 'casual holiday'. Thus we can estimate the population proportion P by $p = 56/332 = 0.169$. An approximate 95% confidence interval for P is obtained from solving the appropriate quadratic equation as $0.1359 < P < 0.2081$.

4.3 RATIOS AND RATIO ESTIMATORS

We introduced the probability sampling scheme of simple random sampling for estimation of a *single* population characteristic. Often, responses to a survey will consist of completed questionnaires concerning many different factors. Thus we will encounter multivariate (rather than univariate) data. We will not pursue the multivariate case in detail, but one simple extension of the univariate situation needs to be examined. This is the case where we simultaneously observe two variables, X and Y. Two matters are of particular interest, having separate aims but involving related statistical considerations: how we might estimate the *ratio* of two population characteristics, for example $\bar{X}/\bar{Y}$; and how, from observing X and Y simultaneously, we can sometimes make use of any association between the variables to improve the efficiency of estimation of the characteristics of one of them, for example X_T or $\bar{X}$. Motivation for this second consideration is easily observed. Suppose X is the level of one urban pollutant and Y that of another at the same site. It is likely that X and Y will be highly correlated. If, furthermore, our sample-based estimate of the mean

X-level is too high (or too low), then that for the mean Y-level will also tend to be too high (too low). Such a mutual link between the errors in the estimates can be exploited to improve the estimation of the mean X-level by using what are called *ratio estimators* and *regression estimators*. To explore this we need to start with the problem of how to estimate a ratio.

4.3.1 The estimation of a ratio

We will consider the quantity

$$R = \frac{X_T}{Y_T} = \frac{\bar{X}}{\bar{Y}}, \tag{4.11}$$

which is known as the *population ratio*.

Ratios are of interest for many reasons. For example, we may wish to estimate the proportion of local area social reform expenditure in some region that is allocated to environmental issues. So we could sample local area returns in the region and record their total spend on social reform issues and their corresponding spend on environmental matters. If these are X_i and Y_i for the different local areas in the region, it is precisely $R = X_T/Y_T$ that we wish to estimate.

Alternatively, suppose we want to estimate the average number of cars per person, for adult persons living in a particular geographic region. We might envisage taking a simple random sample of adult individuals, noting how many cars each has (mainly 0 or 1) and using the sample mean in each case to estimate the corresponding population mean of interest. But it might be easier to use larger sampling units, say households, rather than individuals If so, we become concerned with ratios rather than means, since the average number of cars is just the ratio of the total number of cars X_T to the total adult population size Y_T, with both characteristics estimated from the sample of households.

So, suppose we want to estimate the population ratio $R = X_T/Y_T$, on the basis of a simple random sample $(x_1, y_1), \ldots, (x_n, y_n)$ of the bivariate population measures (X_i, Y_i) $(i = 1, \ldots, N)$. How are we to estimate a ratio and what properties do we find for the intuitively attractive estimators?

Two distinct prospects arise: we could use the individual sampled ratios x_i/y_i as the basic observations; or we could work in terms of summary sample features for the two variables, such as $\bar{x}$ and $\bar{y}$. These would lead to use of what are called the *sample average ratio* and the *ratio of the sample averages*, respectively; specifically, these are

$$r_1 = \frac{1}{n}\sum_{l=1}^{n}\left(\frac{x_i}{y_i}\right) \tag{4.12}$$

and

$$r_2 = \frac{\bar{x}}{\bar{y}} = \frac{x_T}{y_T}. \tag{4.13}$$

Let us compare and contrast these estimators, starting with r_1 which has obvious intuitive appeal. However, r_1 is not widely used as an estimator of the population ratio R since it turns out to be biased and the bias and mean square error can be large relative to those of other estimators, particularly r_2.

Consider the population of values $R_i = X_i/Y_i$. This has population mean $\bar{R}$ and variance S_R^2, say. Since r_1 is a sample mean from a simple random sample, it has expected value $\bar{R}$ and variance $(1 - n/N)S_R^2/n$. The bias is easily confirmed to be

$$\text{bias}(r_1) = \bar{R} - R = -\frac{1}{Y_T}\sum_1^N R_i(Y_i - \bar{Y}),$$

and the mean square error

$$\text{M.S.E.}(r_1) = \left(1 - \frac{n}{N}\right)\frac{S_R^2}{n} + (N - 1)^2\frac{S_{RY}^2}{Y_T^2},$$

where $S_{RY} = \sum_1^N R_i(Y_i - \bar{Y})/(N - 1)$ is the covariance between R and X.

We have the usual unbiased variance estimator $\sum_1^n (r_i - \bar{r})^2/(n - 1)$ available for S_R^2, and a corresponding unbiased estimator of the covariance S_{RY} in the form

$$\frac{1}{n - 1}\sum_{i=1}^n r_i(y_i - \bar{y}) = \frac{n(\bar{x} - \bar{r}\bar{y})}{n - 1}.$$

Thus we can obtain unbiased estimates of the bias and the mean square error provided Y_T is known (we will encounter this need again below). In this case we could in fact correct r_1 for bias by using a modified estimator,

$$r_1' = r_1 + \frac{(N - 1)n(\bar{x} - r_1\bar{y})}{(n - 1)Y_T}, \tag{4.14}$$

which is known as the *Hartley–Ross estimator*. (see Barnett, 2002a, pp. 71–72, for more detail).

Let us now consider r_2. This estimator finds much wider application although it is still biased (and has a skew distribution) in small samples. However the bias and mean square error tend to be lower than those for r_1. In large samples the bias becomes negligible and the distribution of r_2 tends to normality, thus enabling inferences to be drawn based on a normal distribution with appropriate variance, Var (r_2).

Let us try to determine some properties of r_2. We have

$$r_2 - R = \frac{\bar{x} - R\bar{y}}{\bar{y}},$$

and expanding about the population mean $\bar{Y}$ yields an approximation to the bias in the form

$$E(r_2) - R \approx \frac{(1 - n/N)}{n\bar{Y}^2}(RS_Y^2 - \rho_{YX}S_XS_Y),$$

which can be small if the correlation coefficient ρ_{YX} is close in value to RS_Y/S_X. This is equivalent to saying that X and Y are roughly proportional to each other or that the regression of X on Y is linear and through the origin.

In large samples, we can utilize asymptotic results to find the following approximate forms:

$$E(r_2) = \frac{\overline{X}}{\overline{Y}} = \frac{X_T}{Y_T}$$

and

$$\mathrm{Var}(r_2) = \frac{1-f}{n\overline{Y}^2} \sum_{i=1}^{N} \frac{(X_i - RY_i)^2}{N-1}, \qquad (4.15)$$

where f is the sampling fraction n/N. Alternatively, we can write

$$\mathrm{Var}(r_2) = \frac{1-f}{n\overline{Y}^2} \{ S_X^2 - 2RS_{YX} + R^2 S_Y^2 \}, \qquad (4.16)$$

which explicitly includes the population covariance, S_{YX}. Further details on the adequacy of the approximate form for the variance of r_2 are given in Cochran (1977, Chapter 6).

Of course, we again encounter the problem that the variance of our estimator is expressed in terms of *population* characteristics, which will be unknown. So once more we will need to estimate Var (r_2) from our data, and it is usual to employ the direct sample analogue

$$S^2(r_2) = \frac{(1-f)}{n\overline{y}^2} \sum_{i=1}^{n} \frac{(x_i - ry_i)^2}{n-1}.$$

The exact distribution of r_2 is highly complex, but it approaches normality in large samples (when sampling from large populations). Thus, for large samples, we can construct confidence intervals for R. If $s(r_2)$ is the sample estimate of the standard error of r_2, that is, of $[\mathrm{Var}(r_2)]^{1/2}$, we have an approximate $100(1-\alpha)$ symmetric two-sided confidence interval for R of the form

$$r_2 - z_\alpha s(r_2) < R < r_2 + z_\alpha s(r_2).$$

4.3.2 Ratio estimator of a population total or mean

We can use our results on estimation of a population ratio in an interesting way to estimate a population total (or mean) under appropriate circumstances. Specifically, if our interest is in a variable X and we simultaneously sample an associated, concomitant, variable Y which is quite strongly correlated with X, we may be able to exploit this relationship to obtain improved estimators of the total (or mean) of X using what is known as the *ratio estimate* of the population total (or mean). It seems reasonable, in the local area expenditure example of Section 4.3.1, that expenditure on environmental issues will be linked to overall social reform expenditure – the two may even be roughly in proportion to each other.

Thus, suppose X_i denotes expenditure on environmental issues for area i, whilst Y_i is expenditure on all social reform matters in a given year, and we sample both measures simultaneously, at random from all the areas in our study region, to obtain a simple random sample of size n:$(x_1, y_1), \ldots, (x_n, y_n)$. Suppose that the total social reform expenditure for the whole population, Y_T, is known fairly accurately – for example, because of statutory returns. We will also know the number, N, of local areas in the population.

Although Y_T is known, we could have *estimated it* from the sample by means of the usual estimator $y_T = N\bar{y}$, where $\bar{y}$ is the simple random sample mean. Similarly, the total expenditure X_T (the characteristic of principal interest) can be estimated by $x_T = N\bar{x}$.

The estimate y_T is of no interest in its own right (since we know Y_T), but it can be used to reflect the representativity of the sample. If y_T is very much less than Y_T, then in view of the rough proportionality of X_i and Y_i we would expect that x_T underestimates X_T; if y_T is too large, so is x_T likely to be. If the proportionality relationship were *exact*, we would have

$$X_i = RY_i \quad (i = 1, \ldots, N),$$

where R is the population ratio, X_T/Y_T or $\bar{X}/\bar{Y}$, discussed above. Thus, $X_T = RY_T$. This suggests that we could estimate X_T by replacing R with the sample estimate, r_2, to obtain an estimate of the population total, X_T, in the form

$$x_{TR} = r_2 Y_T = \frac{Y_T}{y_T} x_T. \tag{4.17}$$

Note that we do not contemplate using the sample average ratio r_1 to estimate R in this context. We will only use the ratio of the sample averages, r_2, which will subsequently be denoted r, suppressing the subscript.

The estimator x_{TR} is called the simple random sample *ratio estimator of the population total*. We can see that it produces just the compensation effects we require for values of y_T, which happen to be larger (or smaller) than the known value, Y_T, where it reduces (or increases) our estimate of Y_T, accordingly.

Of course, we do not expect exact proportionality between X and Y, but we can anticipate the same compensatory effects whenever there is 'rough proportionality' between the variable of interest, X, and the ancillary (concomitant) variable, Y. In such cases we will use the ratio estimator, x_{TR}.

To estimate the population mean $\bar{X}$, rather than the total, X_T, similar arguments support the use of the *ratio estimator of the population mean*,

$$\bar{x}_R = r\bar{Y} = \frac{\bar{Y}}{\bar{y}} \bar{x}. \tag{4.18}$$

Such ratio estimators have a clear intuitive appeal, but we need to examine just when they will provide tangible efficiency gains over the simple random sample total, x_T, or mean, $\bar{x}$. This will depend on what is meant by 'rough proportionality'.

Note that the only sample statistic that is used is the *sample ratio, r,* whose properties have been discussed in some detail in the previous section (as r_2, the ratio of the sample averages). Consider the estimator $\bar{x}_R$. It is asymptotically unbiased; in certain circumstances it proves to be unbiased for all sample sizes, as we shall see.

From results for r in Section 4.3.1, the approximate variance of $\bar{x}_R$ (for a large sample) will be

$$\text{Var}(\bar{x}_R) = \frac{1-f}{n} \sum_{i=1}^{N} \frac{(X_i - RY_i)^2}{N-1}$$

$$= \frac{1-f}{n}(S_X^2 - 2RS_{YX} + R^2 S_Y^2)$$

$$= \frac{1-f}{n}(S_X^2 - 2R\rho_{YX}S_Y S_X + R^2 S_Y^2), \qquad (4.19)$$

where $\rho_{YX} = S_{YX}/S_Y S_X$ is defined to be the *population correlation coefficient.* For exact proportionality, $\text{Var}(\bar{x}_R)$ would of course be zero. In practice this will not be so, but $\text{Var}(\bar{x}_R)$ will be smaller, the larger the (positive) correlation between X and Y in the population.

For the estimation of X_T, we have analogous results for x_{TR}. Again we will need to estimate $\text{Var}(\bar{x}_R)$, or $\text{Var}(x_{TR})$, from the sample data, using the most accessible forms

$$\frac{1-f}{n(n-1)}\left(\sum_{i=1}^{n} x_i^2 - 2r\sum_{i=1}^{n} y_i x_i + r^2\sum_{i=1}^{n} y_i^2\right),$$

and

$$\frac{(1-f)N^2}{n(n-1)}\left(\sum_{i=1}^{n} x_i^2 - 2r\sum_{i=1}^{n} y_i x_i + r^2\sum_{i=1}^{n} y_i^2\right),$$

respectively.

Using the large-sample forms for the variances, and the asymptotic normality of the estimators, we can again obtain approximate confidence intervals for $\bar{X}$ or X_T.

A crucial question is whether $\bar{x}_R$ (or x_{TR}) is more or less efficient than $\bar{x}$ (or x_T). It turns out that either possibility can arise, depending on the values of ρ_{YX} and of the coefficients of variation C_Y and C_X.

Clearly $\text{Var}(\bar{x}_R)$ will be less than $\text{Var}(\bar{x})$ (and the ratio estimator will be the more efficient) if

$$R^2 S_Y^2 < 2R\rho_{YX}S_Y S_X,$$

that is, if

$$\rho_{YX} > \frac{1}{2}(C_Y/C_X). \qquad (4.20)$$

Thus a gain in efficiency is not in fact guaranteed; we need the population correlation coefficient to be sufficiently large. (In practice, we would need to assess the criterion for greater efficiency of the ratio estimator in terms of sample estimates of ρ_{YX}, C_Y, and C_X.)

Note, however, that however large ρ_{YX} turns out to be, we still need not necessarily obtain a more efficient estimator by using $\bar{x}_R$ (or x_{TR}). If $C_Y > 2C_X$, the ratio estimator $\bar{x}_R$ (or x_{TR}) cannot possibly be more efficient than $\bar{x}$ (or x_T), even with essentially perfect correlation between the Y and X values.

Thus we conclude that two matters control whether or not ratio estimators are more efficient than the corresponding simple random sample ones: the variability of the auxiliary variable Y must not be substantially greater than that of X and the correlation coefficient ρ_{YX} must be large and positive. Nonetheless, many practical situations are encountered where the appropriate conditions hold and ratio estimators offer substantial improvement over $\bar{x}$ or x_T.

In summary, we need:

- to be able to observe simultaneously two variables X and Y which appear to be roughly proportional to each other (i.e. which have high positive correlation);
- the auxiliary variable Y not to have a substantially greater coefficient of variation than the principal variable X;
- the population mean $\overline{Y}$, or total Y_T, to be known exactly.

'Rough proportionality' implies a more or less linear relationship through the origin. If Y and X were essentially linearly related, but the relationship did not pass through the origin, then we would need to consider alternative estimators (such as the *regression estimator*) which exploit the relationship between principal and concomitant variables but in a somewhat different way. See, for example, Barnett (2002a, pp. 82–93) for more details.

Example 4.5 In an environmental survey, questions were to be asked about the attitudes of householders in a particular town to the possibility that a new landfill site for domestic and industrial waste would be built the edge of the town. A list was drawn up of 12 560 householders in the town, and an interview-based survey was to be conducted of a simple random sample of 800 householders. Two topics were of major interest: how much, X, in extra local annual taxation would be acceptable to an individual to avoid the 'hazard' of the landfill facility, and how much, Y, per annum the individual contributes to local charities. It was felt that X and Y should be positively correlated.

A pilot study, and previous experience, combined to suggest approximate values for certain population characteristics as follows:

$$C_X = 0.5, \quad C_Y = 0.3, \quad \rho_{YX} = 0.60.$$

Suppose that Y_T is known from local authority records so that a ratio estimator could in principle be used. It is of interest to determine whether there is any advantage in using a ratio estimator, rather than the simple random sample total, to estimate X_T.

Firstly, consider estimating X_T by x_T (the simple random sample total). We have, for sample size $n = 800$,

$$\text{Var}(x_T) = (12\,560)^2(1 - 800/12\,560)S_Y^2/800.$$

Now, we have $0.6 = \rho_{YX} > 0.3 = \frac{1}{2}C_Y/C_X$ so that there must, from (4.20), be an efficiency gain from using the ratio estimator. We have

$$\text{Var}(x_{TR}) = 12\,560^2(1 - 800/12\,560)(S_X^2 - 2R\rho_{YX}S_YS_X + R^2S_Y^2)/800.$$

So $\text{Var}(x_{TR})/\text{Var}(x_T) = 1 - 2\rho_{YX}C_Y/C_X + C_Y^2/C_X^2 = 0.64$, and we conclude that use of the ratio estimator will yield an efficiency gain, $\text{Var}(x_T)/\text{Var}(x_{TR}) - 1$, of 56%.

4.4   STRATIFIED (SIMPLE) RANDOM SAMPLING

If the population falls into natural groups, or can be so divided, it is called a *stratified population*. For example, with a human population we might have two groups or *strata*, male and female, or we might have geographic or social-class stratification. An example with two *levels* of stratification might be a study of the differential effects, on men and women living in five different regions of a city, of road traffic exhaust fumes on respiratory illness. Here we have stratification with respect to both sex and locality.

There are, in fact, three possible advantages of such a stratified population structure:

- The stratification can improve the efficiency of estimation under appropriate conditions.
- The stratification may be administratively convenient and facilitate the drawing of a sample.
- We may want to estimate characteristics of the separate strata as well as of the overall population – for example, to distinguish the effects of traffic fumes on men and women separately.

It turns out that we may be able to estimate some population characteristics more efficiently by sampling from each stratum separately than by sampling from the population at large. This will be so, broadly speaking, if the variable of interest, X, shows less variation within each stratum than in the total population. We cannot guarantee to meet these conditions. Stratification chosen for administrative convenience (say, to ease access or sampling procedures) will not necessarily yield the required relative within-stratum homogeneity. It does often do so, however.

Let us consider this prospect in more detail in the context of estimating the population mean. We will seek to estimate the mean, $\overline{X}$, of the values $X_1, X_2, \ldots, X_N$ in a finite population. Suppose we have a stratified population, divided into k non-overlapping strata of sizes $N_1, N_2, \ldots, N_k$, where $\sum_{i=1}^{k} N_i = N$, with members $X_{ij}(i = 1, \ldots, k; j = 1, \ldots, N_i)$. If we think of each stratum as a subpopulation, we can use the earlier notation and denote the stratum means and variances by $\overline{X}_1, \overline{X}_2, \ldots, \overline{X}_k$, and $S_1^2, S_2^2, \ldots, S_k^2$, respectively.

The means and variances for the overall population, $\overline{X}$ and S_X^2, can be expressed in terms of characteristics of the strata (the subpopulations) as follows:

$$\overline{X} = \frac{1}{N} \sum_{i=1}^{k} N_i \overline{X}_i = \sum_{i=1}^{k} W_i \overline{X}_i \qquad (4.21)$$

(where $W_i = N_i/N$ is the *weight* of the ith stratum), and

$$S_X^2 = \frac{1}{N-1} \left\{ \sum_{i=1}^{k} \sum_{j=1}^{N_i} (X_{ij} - \overline{X}_i + \overline{X}_i - \overline{X})^2 \right\}$$

$$= \frac{1}{N-1} \left\{ \sum_{i=1}^{k} (N_i - 1)S_i^2 + \sum_{i=1}^{k} N_i(\overline{X}_i - \overline{X})^2 \right\}, \qquad (4.22)$$

respectively.

We will consider *stratified (simple) random sampling* where a sample of size n is chosen by taking a simple random sample of predetermined size from each stratum. The stratum sample sizes will be denoted $n_1, n_2, \ldots, n_k$ $\left(\sum_{i=1}^{k} n_i = n \right)$, and the sample members are $x_{i1}, x_{i2}, \ldots, x_{in_i}$, with sample means $\bar{x}_i = (1/n_i) \sum_{j=1}^{n_i} x_{ij}$, variances $s_i^2 = (1/(n-1)) \sum_{j=1}^{n_i} (x_{ij} - \bar{x}_i)^2$ and *sampling fractions* $f_i = n_i/N_i, i = 1, 2, \ldots, k$.

The usual estimator of $\overline{X}$ is the *stratified sample mean*, defined as $\bar{x}_{st} = \sum_{i=1}^{k} W_i \bar{x}_i$, where we assume that the stratum weights $W_i = N_i/N$ are known precisely. This is not the same as the *overall sample average* $\bar{x}' = (1/n) \sum_{i=1}^{k} n_i \bar{x}_i$ of the stratified random sample, except when the sampling fractions, $f_i = n_i/N_i$, are identical for all strata – a special and intuitively important form of stratified sampling called *proportional allocation* (where the sample sizes are proportional to the stratum sizes).

So how good is the stratified sample mean? We need to know its mean and variance, which are readily obtained. Clearly, $\mathrm{E}(\bar{x}_{st}) = \sum_{i=1}^{k} W_i \overline{X}_i = \overline{X}$, so that $\bar{x}_{st}$ is unbiased, reflecting the unbiasedness of the stratum sample means, $\bar{y}_i$. The *overall sample average $\bar{y}'$ will be unbiased only in the case of proportional allocation.*

The variance of $\bar{x}_{st}$ is

$$\mathrm{Var}(\bar{x}_{st}) = \sum_{i=1}^{k} W_i^2 (1 - f_i) S_i^2 / n_i \qquad (4.23)$$

provided that $\text{Cov}(\bar{x}_i, \bar{x}_j) = 0$ for $i \neq j$, that is, that the simple random sample means for the different strata are uncorrelated. With proportional allocation (4.23) takes the form

$$\text{Var}(\bar{x}_{st}) = \frac{1-f}{n} \sum_{i=1}^{k} W_i S_i^2. \tag{4.24}$$

Directly analogous results hold for estimating the population total, X_T.

We again have to face the problem that population and stratum variances will not be known, so that to obtain a standard error for the estimator $\bar{x}_{st}$, or to construct an approximate confidence interval for $\bar{X}$ (using the approximate normality of the $\bar{x}_{st}$ in large samples), we will need to estimate the S_i^2. The strata are just subpopulations. Thus the sampled values in each stratum constitute a simple random sample and

$$s_i^2 = \frac{1}{n_i - 1} \sum_{j=1}^{n_i} (\bar{x}_{ij} - \bar{x}_i)^2 \qquad (i = 1, 2, \ldots, k)$$

are (*unbiased*) estimators of the stratum variances $S_i^2 (i = 1, 2, \ldots, k)$. So we have as an *unbiased estimator of* $\text{Var}(\bar{x}_{st})$,

$$s^2(\bar{x}_{st}) = \sum_{i=1}^{k} W_i^2 (1 - f_i) s_i^2 / n_i$$

$$= \frac{1}{N^2} \sum_{i=1}^{k} N_i (N_i - n_i) s_i^2 / n_i.$$

Approximate $1 - \alpha$ confidence intervals for $\bar{X}$, or for X_T, take the form

$$\bar{x}_{st} - z_\alpha s(\bar{x}_{st}) < \bar{X} < \bar{x}_{st} + z_\alpha s(\bar{x}_{st})$$

or

$$N[\bar{x}_{st} - z_\alpha s(\bar{x}_{st})] < X_T < N[\bar{x}_{st} + z_\alpha s(\bar{x}_{st})],$$

respectively.

As before, such approximations will only be reasonable if the conditions are satisfied (in terms of sample size, and so on) for $\bar{x}_{st}$ to have a distribution which is essentially normal, and for $s^2(\bar{x}_{st})$ to be close in value to $\text{Var}(\bar{x}_{st})$.

If a proportion P of the overall stratified population possesses a particular attribute, there is no difficulty in estimating P from a stratified simple random sample. Each stratum sample member can now be thought of as having a value $x_{ij}(i = 1, \ldots, k; j = 1, \ldots, n_i)$ which is 1 or 0 depending on whether or not the sampled individual possesses the defined attribute.

Clearly, $P = \sum_{i=1}^{k} \sum_{j=1}^{N_i} X_{ij}/N = \bar{X}$ (the population mean X-value) and the stratified sample mean

$$\bar{x}_{st} = \sum_{i=1}^{k} W_i \bar{x}_i = \sum_{i=1}^{k} W_i p_i = p_{st}, \tag{4.25}$$

where p_i is the sampled proportion (just $\bar{x}_i$) in the ith stratum, will as reasoned above be an unbiased estimator of P. If the actual proportions in the population strata are $P_i (i = 1, \ldots, k)$, then the variance of p_{st} is seen to be

$$\mathrm{Var}(p_{st}) = \sum W_i^2 (N_i - n_i) P_i (1 - P_i) / [(N_i - 1) n_i]. \qquad (4.26)$$

If we ignore terms in $1/n_i$, this is just

$$\mathrm{Var}(p_{st}) = \sum W_i^2 (1 - f_i) P_i (1 - P_i) / n_i,$$

where f_i is the sampling fraction n_i / N_i in the ith stratum $(i = 1, \ldots, k)$. An unbiased estimator of $\mathrm{Var}(p_{st})$ is given by $\sum W_i^2 (1 - f_i) p_i (1 - p_i) / (n_i - 1)$.

4.4.1 Comparing the simple random sample mean and the stratified sample mean

One of the motives for considering stratification of the population (other than administrative convenience) was that we might increase the efficiency with which we can estimate population characteristics such as $\bar{X}$ or X_T. Let us examine if this can happen by comparing the properties of $\bar{x}$ (the simple random sample mean) and of $\bar{x}_{st}$ in the same situation. Both are unbiased estimators, and $\mathrm{Var}(\bar{x}) = (1 - f) S^2 / n$.

To ease the comparison, suppose initially that the stratified sample has been chosen with proportional allocation. Then we find, from (4.24), that

$$\mathrm{Var}(\bar{x}) - \mathrm{Var}(\bar{x}_{st}) = \frac{1 - f}{n} \left(S^2 - \frac{1}{N} \sum_{i=1}^{k} N_i S_i^2 \right).$$

However, for large enough stratum sizes (where the $1/N_i$ are negligible)

$$S^2 \approx \frac{1}{N} \left\{ \sum_{i=1}^{k} N_i S_i^2 + \sum_{i=1}^{k} N_i (\bar{Y}_i - \bar{Y})^2 \right\},$$

and we then find that

$$\mathrm{Var}(\bar{x}) - \mathrm{Var}(\bar{x}_{st}) = \frac{1 - f}{nN} \sum_{i=1}^{k} N_i (\bar{X}_i - \bar{X})^2$$

$$= \frac{1 - f}{n} \sum_{i=1}^{k} W_i (\bar{X}_i - \bar{X})^2,$$

which is strictly non-negative, except for the pathological case where the $\bar{X}_i$ are all equal. Hence we conclude that the stratified sample mean will always be more efficient than the simple random sample mean, the more so the larger the variation in the stratum means.

However, the assumption that the $1/N_i$ are negligible turns out to be critical. If we retain terms in $1/N_i$ we obtain a more accurate form:

$$\text{Var}(\bar{x}) - \text{Var}(\bar{x}_{st}) = \frac{1-f}{n(N-1)} \left\{ \sum_{i=1}^{k} N_i(\overline{X}_i - \overline{X})^2 - \frac{1}{N} \sum_{i=1}^{k} (N - N_i)S_i^2 \right\}. \quad (4.27)$$

This need no longer necessarily be positive. We now conclude that $\bar{x}_{st}$ *will be more efficient than* $\bar{x}$ *if*

$$\sum_{i=1}^{k} N_i(\overline{X}_i - \overline{X})^2 > \frac{1}{N} \sum_{i=1}^{k} (N - N_i)S_i^2. \quad (4.28)$$

We can interpret this more readily if we go to one further level of specialization. Suppose all the strata have the same variance, S_w^2. Then we require

$$\frac{1}{k-1} \sum_{i=1}^{k} N_i(\overline{X}_i - \overline{X})^2 > S_W^2.$$

Thus, *the stratified sample mean will be more efficient than the simple random sample mean if variation between the stratum means is sufficiently large compared with within-strata variation*; the greater this advantage, the greater the efficiency of $\bar{x}_{st}$ relative to $\bar{x}$.

Summarizing the results of this section, we can informally conclude that the more disparate are the stratum means, and the more internally homogeneous are the separate strata, the greater is the potential gain from using the stratified sample mean $\bar{x}_{st}$ (rather than $\bar{x}$) for estimating $\overline{X}$. The same will be true, of course, for the estimation of X_T.

4.4.2 Choice of sample sizes

Once more, with stratified sampling as with simple random sampling, we have the problem of deciding how to choose sample sizes to meet some specified requirement for precision or cost of estimation. This is now more complicated. Not only is the overall sample size relevant, but at a given total sample size the relative allocation of sampling effort to the different strata (each possibly with different internal variability) can also influence the accuracy of estimation of population characteristics. Furthermore, it may no longer suffice to put pre-scriptions on variances. Different sampling costs in different strata mean that we must try to take more direct account of cost factors in determining a reasonable allocation of stratum sample sizes.

Simple cost models have been proposed. The simplest assumes an overhead cost, c_0, of carrying out the survey, in addition to which individual observations from the ith stratum each cost an amount c_i. Thus the total cost is

$$C = c_0 + \sum_{i=1}^{k} c_i n_i. \quad (4.29)$$

Minimum variance for fixed cost

Suppose we adopt such a cost model (4.29) and try to determine what allocation of stratum sample sizes, $n_1, n_2, \ldots, n_k$, should be adopted to minimize
$\mathrm{Var}(\bar{x}_{st})$ for a given total cost C.

We need to choose $n_1, n_2, \ldots, n_k$ to minimize

$$\mathrm{Var}(\bar{x}_{st}) = \sum_{i=1}^{k} W_i^2 S_i^2 / n_i - \frac{1}{N} \sum_{i=1}^{k} W_i S_i^2$$

subject to the cost constraint.

$$\sum_{i=1}^{k} c_i n_i = C - c_0.$$

We can readily show that this requires $n_i \propto W_i S_i / \sqrt{c_i}$, and thus it turns out that
optimum allocation for fixed total cost is given by the allocation

$$n_i = \frac{(C - c_0) W_i S_i / \sqrt{c_i}}{\sum_{i=1}^{k} W_i S_i \sqrt{c_i}}. \tag{4.30}$$

Thus the total sample size will be

$$n = \frac{(C - c_0) \sum_{i=1}^{k} W_i S_i / \sqrt{c_i}}{\sum_{i=1}^{k} W_i S_i \sqrt{c_i}}. \tag{4.31}$$

This implies that the stratum sample sizes need to be proportional to the
population stratum sizes, proportional to the population stratum standard
deviations, and inversely proportional to the square root of the unit sampling
costs in each stratum.

One special case merits attention, where the unit sampling costs c_i are the same
in all strata, so that $C = c_0 + nc$, where c is the (constant) unit sampling cost.

Optimal allocation now needs

$$n_i = \frac{W_i S_i}{\sum_{i=1}^{k} W_i S_i} n,$$

where

$$n = (C - c_0)/c.$$

Clearly this allocation, which is known as *Neyman allocation*, can also be
regarded as optimum for fixed sample size, ignoring variation in costs of
sampling from one stratum to another, in the sense that, given n, it minimizes
$\mathrm{Var}(\bar{x}_{st})$. The resulting minimum variance is

$$\text{Var}_{\min}(\bar{x}_{st}) = \frac{1}{n}\left(\sum_{i=1}^{k} W_i S_i\right)^2 - \frac{1}{N}\sum_{i=1}^{k} W_i S_i^2, \tag{4.32}$$

where the second term arises from the finite-population correction.

Sample size needed to yield some specified $\text{Var}(\bar{x}_{st})$ with given sampling weights
Suppose we have specified the stratum sample weights w_i and we then need
to ensure that $\text{Var}(\bar{x}_{st}) = V$. For example, we might declare a margin of error, d,
and an acceptable probability of error, α, in the sense that we need
$P\{|\bar{x}_{st} - \bar{X}| > d\} \le \alpha$. On the normal assumption, this is equivalent to requiring that $V = (d/z_\alpha)^2$. Equating $\text{Var}(\bar{x}_{st})$ to V gives

$$n = \sum_{i=1}^{k} (W_i^2 S_i^2 / w_i) \Big/ \left(V + \frac{1}{N}\sum_{i=1}^{k} W_i S_i^2\right)$$

as the required sample size.

For the special case of proportional allocation, we have as the first approximation to the required sample size, and the more accurate form,

$$n_0 = \frac{1}{V}\sum_{i=1}^{k} W_i S_i^2 \quad \text{and} \quad n = n_0(1 + n_0/N)^{-1};$$

for Neyman allocation, we have

$$n_0 = \frac{1}{V}\left(\sum_{i=1}^{k} W_i S_i\right)^2 \quad \text{and} \quad n = n_0\left(1 + \frac{1}{NV}\sum_{i=1}^{k} W_i S_i^2\right)^{-1}.$$

4.4.3 Comparison of proportional allocation and optimum allocation

It must be (*ipso facto*) that optimum allocation is more efficient than proportional allocation. But is the improvement tangible and worthwhile? We will
consider just one case: comparing proportional allocation and Neyman allocation (which we have seen is the optimum allocation for constant unit sampling
costs in different strata).

Thus, denoting $\text{Var}(\bar{x}_{st})$ by V_P and V_N for proportional and Neyman allocation, respectively, it must be true that $V_P \ge V_N$. More specifically, we find
that

$$V_P - V_N = \frac{1}{n}\sum_{i=1}^{k} W_i(S_i - \bar{S})^2,$$

where $\bar{S} = \sum_{i=1}^{k} W_i S_i$. So we conclude that the extent of the potential gain from
optimum (Neyman allocation) compared with proportional allocation depends
on the variability of the stratum variances: the larger this is, the greater the
relative advantage of optimum allocation.

4.4.4 Optimum allocation for estimating proportions

Special forms of the results of Section 4.4.3 for choice of overall sample size n and of allocations $n_1, \ldots, n_k$ to achieve required accuracy and for optimum choice of allocation apply in the case of estimating proportions. In particular, with proportional allocation, we will have

$$
\mathrm{Var}(p_{st}) = \frac{N - n}{n} \sum_{i=1}^{k} \frac{W_i^2}{(N_i - 1)} P_i(1 - P_i),
$$

$$
= \frac{(1 - f)}{n} \sum_{i=1}^{k} W_i P_i(1 - P_i), \tag{4.33}
$$

if we replace the $N_i - 1$ in the denominators by N_i.

Adopting this latter approximation, the optimum allocations for fixed sample size ignoring costs (for Neyman allocation) and for fixed cost, $C = c_0 + \sum_{i=1}^{k} (c_i n_i)$, are

$$
n_i = \frac{n W_i \sqrt{P_i(1 - P_i)}}{\sum\limits_{i=1}^{k} W_i \sqrt{P_i(1 - P_i)}}
$$

and

$$
n_i = \frac{(C - c_0) W_i \sqrt{P_i(1 - P_i)} / c_i}{\sum\limits_{i=1}^{k} W_i \sqrt{P_i(1 - P_i) c_i}}
$$

respectively. The potential advantages of such optimum allocations over arbitrary or proportional allocations can be readily assessed and demonstrated.

In practice, we will again not know the P_i and must use appropriate sample estimates. For example, we can estimate the S_i^2 by the unbiased estimators $n_i p_i (1 - p_i) / (n_i - 1)$.

4.5 DEVELOPMENTS OF SURVEY SAMPLING

We have only scratched the surface of methods for opinion polls, sample surveys and finite-population sampling. We have briefly examined the principles of sampling according to a defined probability sampling scheme and examined some introductory schemes based on simple random sampling. Even at this level there is much more to investigate in terms of more complex structures. We can carry out stratified simple random sampling with multifactor stratification, as is the basis for the widely used method of *quota sampling* encountered in the wide range of opinion polls carried out by the major national organizations.

Then again, we might sample in *clusters* where we sample the *whole* of each of a randomly chosen *subset* of small population strata (e.g. households or farms) – rather than just parts of all of a smaller set of larger population strata. This distinquishes *cluster sampling* and stratified sampling. More complex surveys may involve several stages of clustering and stratification – for example, households in different wards of towns in different regions of a country.

Many sampling schemes go beyond simple random sampling. Population members may not be chosen with equal probabilities, but perhaps with probabilities related to their individual characteristics, as with *sampling with probability proportional to size*.

In principle, the problems of *outliers* can also arise in finite populations, although there are fundamental difficulties concerning just what is meant by *contamination* in this context. Little has been published on this matter, but see Barnett and Roberts (1993) for some discussion on outlier tests in sample surveys.

For further reading on survey principles and methods, see Barnett (2002a), Cochran (1977), Hedayat and Sinha (1991), Levy and Lemeshow (1991) and Thompson (2002). Lehtonen and Pahkinen (2004) and Skinner *et al.* (1989) consider the special problems of *complex surveys*.

Apart from the statistical properties of sampling schemes, there is a wide variety of problems in survey sampling which have to do with implementation of the survey – for example, how we ensure we obtain a random sample, whether we use personal contact, telephone, internet, or panel groups, and how we measure and allow for the effects of non-response. Various matters concerning implementation of surveys are considered by Dillman (2000), Groves (1989) and Moser and Kalton (1999); see also Barnett (2002a, Chapter 6).

CHAPTER 5

Inaccessible and Sensitive Data

Now you see it, now you don't.

There is an essential need to constantly monitor the environment for changes in levels of pollutants, industrial by-products, etc. This process can take the form of regular sampling of a fixed set of sites, often arranged roughly on a grid or network. *Network sampling* is relevant to this problem. There is much interest in the factors that need to be taken into consideration in setting up an *environmental change network* to provide a regular supply of data on the physical and, possibly, social environment. For example, we have to decide where best to locate the network sites: spatial regularity is intuitively appealing but may not yield the statistically most informative allocation. Furthermore, there need to be principles against which to judge which site to delete if economies are sought or where best to add new sites. With spatially distributed sites we must also take account of inevitable spatial correlations between readings at adjacent sites. Such matters are briefly reviewed by Cressie (1993, Section 4.6.2); Urquhart *et al.* (1993) compare sampling designs for ecological and environmental networks. Caselton and Zidek (1984) and Zidek *et al.* (2000) discuss the issues in designing and monitoring environmental networks. More will be said about network sampling in Section 11.6 when we examine spatial methods.

There is also a need to take samples for specific purposes: to examine some particular aspect of the current (rather than the continuing or developing) state of the environment. Here we might need sampling methods which could examine finite populations and might be based on simple random sampling, stratification, clustering etc. (as discussed in Chapter 4).

Often, however we have to take what is to hand (this is called *encounter sampling*) or we may have to ensure optimum use of scarce resources either by economizing in our number of observations or by exploiting any form of circumstantial information that is available. The aim will be to obtain efficient estimates at minimum, or acceptable, cost. Some of the methods of survey sampling (as described in Chapter 4) can on occasion help in the quest for sample economy or use of concomitant information (e.g. optimum stratification and ratio estimators, respectively).

However, there are quite different sampling approaches which offer special advantages in these regards. Composite sampling, adaptive sampling and

Environmental Statistics   V. Barnett
© 2004 John Wiley & Sons, Ltd   ISBN: 0-471-48971-9 (HB)

ranked-set sampling are particular examples which we will pursue in this chapter. Additionally, the special problems of accessibility and of estimating abundance can be studied by transect sampling or capture–recapture methods, especially for geographically distributed population members: these we will go on to discuss in Chapter 6.

5.1   ENCOUNTERED DATA

We need always to be concerned with how we are to obtain access to environmental data and on the look-out for possible bias. Let us start with such problems of access and of sources of bias.

Data are the essential raw material of any environmental study. To investigate an environmental issue we need objective observations of what is going on. We have seen (in Chapter 4) that the classical methods of survey sampling can be important. The long-established methods of designed experiments can also be valuable, as we will observe in Chapters 8 and 11 in different contexts. But such approaches need observations to be *taken at random and under prescribed circumstances*. This is often not possible with environmental problems – we have instead to make do with what forms and limited numbers of observations can be obtained, and on the occasions and at the places they happen to arise.

For example, climatological variables observed over time, and especially in the past, have to be limited to what was collected by the meteorological station; inundations and tornadoes occur when they occur! Measured pollution levels tend to be taken and published spasmodically and selectively, for example when site visits are made, perhaps because of a suspicion that levels have become rather high. Accessibility is an important factor here; measurements can only be taken when they are allowed to be taken, to the extent to which they can be afforded, when equipment is available, when they happen to have been taken, and so on.

Many such data satisfy the description of *encountered data:* they are what we happen to encounter – 'the investigator goes into the field, observes and records what he observes . . . what he encounters' (Patil, 1991). The term has been in sporadic use for about 15 years (see, for example, Patil and Taillie, 1989; Patil, 1991) as has the allied notion of *encounter sampling* as that procedure by which we obtain encountered data. Patil and Taillie (1989) provide one of the earliest discussions of encountered data, illustrated for different problem areas, with data that are inevitably non-random and limited by the circumstances in which they are obtained; see also Barnett (2002b). Note that encounter sampling is not a statistical sampling method *per se* – by the very definition of encountered data we cannot pursue a formal and statistically interpretable process to collect or analyse such data.

A more limited concept of 'encountered data' arises when we employ a defined statistical sampling procedure which by its very nature limits or distorts the scope or import of the data. For example, with *transect sampling* (examined in detail in Section 6.3) we traverse a 'transect line' observing the subject of interest (the golden oriole!) as we move along. Clearly we are more likely to see subjects close to the line but, on the other hand, we may have frightened them off before we see them.

Again, if we sample fish in a pond by catching them in a net, there will be another form of 'encounter bias' (more usually called *size bias*). This is because the mesh size will have the effect of lowering the incidence of the smaller fish in the catch – some will slip through the net. We might have to try to compensate by using *weighted distribution methods* (which we discuss in the next section).

Encounter sampling is also used to describe survey sampling where individuals are chosen by casual encounter as an interviewer traverses a random route through a survey area; see Otis *et al.* (1993).

The term 'encounter sampling' is sometimes inappropriately used to describe the collection of environmental data 'in bulk', for example from dredging a river bed or sampling a grain hopper. Such sampling is better described as *bulk sampling*; if properly and effectively carried out it can be thought of as a form of *composite sampling* and can be amenable to formal statistical processing.

5.2   LENGTH-BIASED OR SIZE-BIASED SAMPLING AND WEIGHTED DISTRIBUTIONS

We will return now to the sort of difficulties which have to be faced in the problem just described where we sample fish in a pond by catching them in a net. Related conceptual difficulties can arise if we were to sample harmful industrial fibres (in monitoring adverse health effects) by examining fibres on a plane sticky surface by *line–intercept* methods. In the two cases our data would consists of the sizes of fish that are caught in the net, and the lengths of fibres crossed by the intercept line (as shown in Figure 5.1), respectively.

One interest will be in the *distribution of sizes* (of the fish or of the fibres), but the sampling methods just described are clearly likely to produce seriously biased results, since almost all small fish will slip through the net – and the

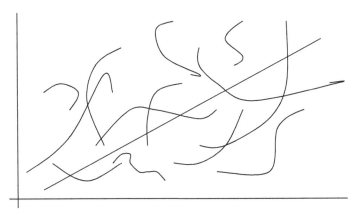

Figure 5.1   Line–intercept sampling of fibres.

longer the fibre the larger the probability of it crossing the intercept line, that is, of our sampling it. So we are bound to obtain what are known as *length-biased* or *size-biased samples*, and statistical inferences drawn from such samples will be seriously flawed because they relate to the distribution of *measured sizes* and not to the population at large, which is our real interest. Thus, we will typically overestimate the mean both in the fish and in the fibre examples, possibly to a serious extent. The typical relationship between populations of *measured values* and of *values at large* is as shown in Figure 5.2.

5.2.1   Weighted distribution methods

Interest in such concepts can be traced back 70 years (Fisher, 1934) although specific reference to length-biased sampling probably did not occur before the 1960s (Cox, 1962; see Patil *et al.* 1988). To provide a unified approach to such problems, Rao (1965) introduced *weighted distribution methods* which can help to remove such biases. Such methods also have wider application to forms of sample bias not just based on observation length or size.

Suppose X is a non-negative random variable with p.d.f. $f_\theta(x)$, but an observation x drawn from this distribution is only retained in the collected sample in accordance with a relative retention distribution having p.d.f. $g_\beta(x)$. Thus what we actually sample is a random variable X^* with a distribution having p.d.f.

$$f^*_{\theta,\beta}(x) = kf_\theta(x)g_\beta(x), \tag{5.1}$$

where k is the normalizing constant satisfying

$$k^{-1} = \int f_\theta(x)g_\beta(x)\mathrm{d}x.$$

X^* is said to be a *weighted version* of X with *weight function* $g_\beta(x)$. If we know the form of $g_\beta(x)$ in a given situation, we can try to adjust for its effect.

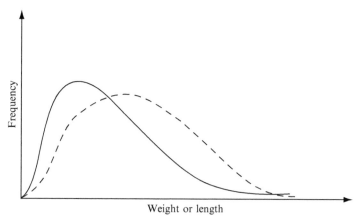

Figure 5.2   Size or weight biasing (observed population, dashed curve; actual population, solid curve).

In the special case where $g_\beta(x) \propto x$, X^* follows what is called the *sized-biased distribution* with p.d.f.

$$f_\theta^*(x) = x f_\theta(x)/\mu, \tag{5.2}$$

where $\mu = E(X)$. The variable actually sampled has expected value

$$E(X^*) = \int [x^2 f_\theta(x)/\mu] dx = \mu(1 + \sigma^2/\mu^2), \tag{5.3}$$

where $\sigma^2 = \text{Var}(X)$. So if we take a random sample of size n, $\bar{x}^*$ (the sample mean of the observed data) is biased upward by a factor $1 + \sigma^2/\mu^2$. This could be vast in its effect. For example, if $\sigma^2/\mu^2 = 9$ (a coefficient of variation of just 3) we find that $E(X^*) = 10\mu$ rather than $E(X^*) = \mu$ as we would hope!

But we have another problem: σ^2 and μ are typically unknown, so we have no direct way of correcting the bias. If we knew μ we would not be estimating it; if we knew σ we still could not make proper allowance for the bias of $\bar{x}^*$ in view of (5.3).

A different approach shows an interesting effect. Suppose we consider the reciprocal of the random variable X^*. It is interesting to note from (5.2) that

$$E(1/X^*) = \int f_\theta(x) dx/\mu = \frac{1}{\mu}, \tag{5.4}$$

so that in this size-biased case (but certainly not generally) its mean is the reciprocal of the mean of X.

This has implications for inference about μ. It is clear that the sample mean reciprocal value $\left(\sum_{i=1}^n 1/x_i^*\right)/n$ is unbiased for $1/\mu$. More importantly, however, since

$$E(X^*)E(1/X^*) = 1 + \frac{\sigma^2}{\mu^2}, \tag{5.5}$$

we note that $g = \bar{x}^* \left(\sum_{i=1}^n 1/x_i^*\right)/n$ provides an intuitively appealing estimate of the bias factor $1 + \sigma^2/\mu^2$ without the need to know μ or σ. Thus we can seek, at least informally, to compensate for the bias in (5.3) by using the estimator $\bar{x}^*/g = n\left(\sum_{i=1}^n 1/x_i^*\right)^{-1}$.

Let us consider a more specific situation. Suppose now that X has a Poisson distribution with mean μ, that is $X \sim P(\mu)$, and again we obtain observations subject to size-biased sampling. Then the distribution of X^* has probability function $q(x^*)$ proportional to $xp(x)$ where $p(x) = e^{-\mu}\mu^x/x!$ for $x = 1, 2, \ldots$. Thus

$$q(x^*) \propto e^{-\mu}\mu^{x^*}/[\mu(x^* - 1)!] = e^{-\mu}/\mu^{x^*-1}/(x^* - 1)!, \tag{5.6}$$

so that $X^* - 1$ has a Poisson distribution with mean μ.

So size-biased sampling here has a straightforward and interpretable effect. The observed outcomes X^* have the distributional form $X + 1$ where $X \sim P(\mu)$. (Clearly X^* cannot take the value 0 since a population member of value $x = 0$ will be suppressed totally by the size-biasing process.) Thus we have an unbiased estimator of μ in the form $\bar{x}^* - 1$.

We confirm this also from examination of (5.3). Since $\sigma^2/\mu = 1$ for the Poisson distribution $P(\mu)$, (5.3) shows that $\bar{x}^*$ has expectation $1 + \mu$ and hence that $\bar{x}^* - 1$ is unbiased for μ.

Let us now consider a more structured and substantial example of the use of weighted distribution methods which is particularly relevant to environmental concerns.

Example 5.1 Suppose an environmental population (of fish or animals, say) appears in groups (shoals or herds) of size $X = 1, 2, \ldots$ with probability function $p_\theta(x)$. Each member of any group has independently a probability p of possessing a distinct characteristic A, and only groups with at least one member with A can be observed. For example, the characteristic A may mean that the member has a radio-detectable tag from a previous capture–recapture exercise (see Section 6.2). Suppose we observe groups of size $Y = 1, 2, \ldots$.
Consider

$$p_\theta^*(y) = P(observe\ y) \propto p_\theta(y) \frac{P(group\ has\ at\ least\ one\ A)}{P(at\ least\ one\ observed)}.$$

Then

$$p_\theta^*(y) = \frac{\{1 - (1-p)^y\}p_\theta(y)}{\sum_{t=1}^{\infty}\{1 - (1-p)^t\}p_\theta(t)} = \frac{1 - (1-p)^y p_\theta(y)}{1 - E\{(1-p)^X\}}$$

Now suppose again that $X - 1 \sim P(\theta)$, so that

$$p_\theta(x) = \frac{e^{-\theta}\theta^{x-1}}{(x-1)!} \qquad (x = 1, 2, \ldots).$$

We would have

$$E\{1 - (1-p)^X\} = 1 - (1-p)e^{-\theta}\left(1 + (1-p)\theta + \frac{(1-p)^2\theta^2}{2!}\cdots\right)$$

$$= 1 - (1-p)e^{-\theta}e^{(1-p)\theta}$$

$$= 1 - (1-p)e^{-p\theta} = K,$$

say. So

$$p_\theta^*(y) = \frac{1 - (1-p)^y e^{-\theta}\theta^{y-1}}{(y-1)!K}.$$

Now consider using the sample mean $\bar{y}$ to estimate $E(X)$. We have

$$E(\bar{Y}) = E(Y) = \left\{\sum_{y=1}^{\infty}\frac{ye^{-\theta}\theta^{y-1}}{(y-1)!} - \sum_{y=1}^{\infty}\frac{ye^{-\theta}(1-p)^y\theta^{y-1}}{(y-1)!}\right\}/K$$

$$= \left\{\sum_{y=1}^{\infty}\frac{ye^{-\theta}\theta^{y-1}}{(y-1)!} - (1-p)e^{-p\theta}\sum_{y=1}^{\infty}\frac{ye^{-(1-p)\theta}[(1-p)\theta]^{y-1}}{(y-1)!}\right\}/K$$

$$= \{E(X) - (1-p)e^{-p\theta}E(Z)\}/K,$$

where $Z - 1 \sim P((1-p)\theta)$. So

$$E(\overline{Y}) = \{\theta + 1 - (1-p)e^{-p\theta}[(1-p)\theta + 1]\}/K$$

$$= (\theta + 1)\frac{1 - (1-p)e^{-p\theta}}{1 - (1-p)e^{-p\theta}} + \frac{p(1-p)\theta e^{-p\theta}}{1 - (1-p)e^{-p\theta}}.$$

$$= E(X) + \frac{p(1-p)e^{-p\theta}[E(X) - 1]}{1 - (1-p)e^{-p\theta}}$$

So $\overline{Y}$ is biased upwards. This is as we would expect, since we are increasing the probabilities of encountering larger groups (by the factor $1 - (1-p)^y \to 1$ as $y \to \infty$, but which is much less than 1 when y is small).

Many other weighted distribution methods have been studied and used. For instance, in the fish example with square mesh of size x_0, the weight function is more like a *truncation* and we would have

$$g(x) = \begin{cases} 0 & x \le x_0, \\ 1 & x > x_0. \end{cases}$$

So now

$$f_\theta^*(x) = f_\theta(x) \bigg/ \int_{x_0}^\infty f_\theta(x)\mathrm{d}x. \qquad (5.7)$$

However, since θ is unknown this will often not be easy to handle. In one case it is straightforward.

Example 5.2 Consider sampling from an exponential distribution with $f_\theta(x) = \theta e^{-\theta x}$. Then if $g(x)$ has the trunction form above, we find that

$$f_\theta^*(x) = \frac{\theta e^{-\theta x}}{e^{-\theta x_0}} = \theta e^{-\theta(x-x_0)} \qquad (x > x_0).$$

So $X^* - x_0$ is exponential with parameter θ, that is, $E(X^*) = x_0 + 1/\theta$. Thus $E(\overline{X}^*) = x_0 + 1/\theta$ and we can now readily correct for the bias in estimating the mean $1/\theta$. The bias is just x_0 (in the fish example this is the mesh size) so we use $\overline{X}^* - x_0$.

Patil and Rao (1977) illustrate the use of weighted distribution methods in a wide range of situations, beyond size biasing or truncation. For example, they show how such methods can be applied to problems with *missing data* or with *damaged data*. Rao (1965) dealt with an interesting application to the data on numbers of males (or females) in families of different sizes.

Patil and Rao (1977) also review *bivariate weighted distribution* methods. If (X, Y) is a pair of non-negative random variables with joint p.d.f. $f(x,y)$ and $g_\beta(x, y)$ is a non-negative weight function whose expectation exists, then we have a *weighted version* of $f(x,y)$ in the form $f(x, y) = g_\beta(x, y)f(x, y)/E[g_\beta(x, y)]$ for the weighted random variable $(X, Y)^*$.

Such an approach has been examined and applied, for example, by Arnold and Nagaraja (1991).

5.3 COMPOSITE SAMPLING

In Section 1.3.1 we considered an example where we needed to identify those individuals in a group of size n who possessed some rare characteristic but where the characteristic at issue was of a sensitive type or where it was difficult or expensive to test each individual separately. We approached this problem by mixing together sample material from each member and conducting just a single test on the combined material. If the characteristic is not found in the composite sample, all individuals are clearly free of it. This sampling method is known as *composite sampling*.

Whilst it can be used, as just indicated, to screen a group of individuals for possession of the characteristic of interest, it can also be employed for estimation or statistical testing of parameters characterizing the distribution of relevant values in the population of individuals at large. Thus, for example, we might want to estimate or test the proportion possessing the characteristic of interest.

Further, if there is some relevant quantitative random variable in terms of which the characteristic is defined, we can sometimes estimate or test parameters in the distribution of such a random variable. This would be possible if the random variable has the property that its observed value for a set of individuals is the sum of the values for each of them separately, for example for counts of numbers of pollutant spores on leaves of plant specimens of some species.

Applications of composite sampling cover a broad range, from testing for presence of disease to examining if materials fail to reach safety limits. Specific examples include: remedial clean-up of contaminated soil; geostatistical sampling; examining foliage and other biological materials; screening for dangerous chemicals; groundwater monitoring; and air quality (testing for allergens). Discussions of various practical aspects of the technique can be found in the early pioneering papers of Watson (1936) and Dorfman (1943). More recently, Carter and Lowe (1986) were concerned with examining forest floor properties, Green and Strawderman (1986) with estimating timber yields, Baldock *et al.* (1990) with investigating sheep infection, and Barnett and Bown (2002c) with setting environmental standards. Lovison *et al.* (1994) give further practical examples. Review discussions include Barnett (2002a, Section 7.2), Boswell and Patil (1987), Patil (2002), Rohde (1976), USEPA (1995) and van Belle *et al.* (2002).

5.3.1 Attribute sampling

For attribute data, the composite sampling method in its simplest form works as follows. Suppose a population has a proportion p with characteristic A and we take a random sample of size n, but, instead of observing the individual values separately, we test the overall sample (in composite form) once only for the presence of the characteristic A in at least one of the sample members. Then

$$P(A \text{ encountered}) = 1 - (1 - p)^n. \tag{5.8}$$

If we do not find A, we conclude that no members of the sample of size n have the characteristic A. If we find A, and we need to identify precisely which sample members have the characteristic, we must examine the sample in more detail. The most obvious approach is to retest each separate sampled member. Note that this requires some care. If we have used all the sample material for the composite test, we could not subsequently examine individuals separately without resampling. It would be more prudent, and this is common practice, to use only some of the material in the composite test and to retain some from each individual (so-called *audit samples*) for later use if necessary.

In the *full retesting* approach just described (see also Section 1.3.1), it is clear that we will need either one test (if negative) or $n + 1$ tests (if positive) to identify precisely which sample members are affected (i.e. have the characteristic A). When p is small, this can be a highly economical approach compared with full testing at the outset. Simple probability arguments show that, in general, we need on average $(n + 1) - n(1 - p)^n$ tests. So, as we observed in Section 1.3.1, if $p = 0.0005$ and $n = 20$, just 1.2 tests are required on average.

In Section 1.3.1 we noted various other strategies for identifying all affected members after the initial test of the composite sample shows that A is present. Full retesting of all n members is just one possibility; other prospects include *group retesting* and *cascading*. Thus, instead of testing each individual sample member in the case when A is detected, we might retest in *composited subsamples,* smaller than the whole sample (see Sobel and Groll, 1966; Elder *et al.,* 1980). Alternatively (Sterrett, 1957), individuals might be tested separately but individual retesting ceases once A is encountered, and the remainder of the sample is retested in composite form. A version of this might be termed *sudden death*. Here, following a positive first test, we would test the individuals one at a time until we find the first affected individual and then conduct a composite test on the remainder. If the composite test is positive we repeat the process, and so on (see Figure 5.3).

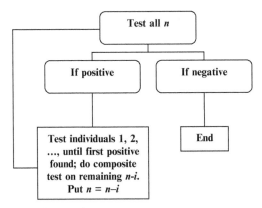

Figure 5.3 'Sudden death' retesting.

More complex multi-cycle methods have also been proposed; a sample of size n_1 is divided into g_1 groups for composite testing, the groups revealing A are pooled to form a sample of size n_2, which is divided into g_2 groups, and so on. Kannan and Raychaudhuri (1994) present a two-stage approach in which a more accurate form of test is used at the second stage than at the first, whilst Hwang (1972) considers the case where an upper bound is known for the total number of population members possessing the characteristic A.

Example 5.3 On an urban site previously used for chemical processing, 128 soil samples were chosen from different locations to test for the presence of a particularly noxious substance which may be present independently in each soil sample with (unknown but small) probability, p. A composite sample is formed, retaining separate samples in case they are needed later. The composite sample tests positive for the substance of interest and we wish to identify which of the 128 locations are contaminated.

We consider two alternatives for testing: full retesting, as outlined above, and cascading, where we divide the 128 samples into subgroups of 64 and do composite tests on each, then (as necessary) into subgroups of 32, and so on.

If we opt for full retesting, we know that on average we must carry out $(n+1) - n(1-p)^n = 129 - 128(1-p)^{128}$ tests. Under cascading, it is more complicated to work out the average number of tests required. We might (in the extreme case where all soil samples are contaminated) need to carry out $2^8 - 1 = 255$ tests. At the other extreme (where only one soil sample is contaminated) we would need to carry out just 15 tests.

Which alternative is more economical will depend on the value of p. The larger p is the more likely we are in cascading to go towards the extreme case of 255 tests. So for small enough p there may be economies to be achieved by using the cascading method rather than the full-retesting method.

Often the compositing process is essentially inevitable due to the non-discreteness of the medium being sampled. For example, if we are sampling bags of fertilizer, or water containing organisms, or pesticide residues in soil, the test can be expensive and it is sensible *to bulk* the samples together *(bulk sampling)* to reduce the number of tests needed. (*Grab sampling* is a related approach.) See, for example, Rohde (1976).

Composite sampling is concerned with estimating p as well as with identifying sample members with the characteristic A. Interesting applications have included screening servicemen for syphilis (Dorfman, 1943) and estimating timber volume (Green and Strawderman, 1986).

If estimation is of interest we can proceed as follows. We could test m composite samples. Suppose r of them exhibit characteristic A, then r is binomial B$[m, 1 - (1-p)^n]$ so that, since r/m is the maximum likelihood estimator of an estimator of $1 - (1-p)^n$, an estimator of p is provided by

$$p^* = 1 - (1 - r/m)^{1/n}. \tag{5.9}$$

This estimator is easily shown to be biased: it tends to overestimate p. Boswell and Patil (1987) – who review earlier and more general work by Garner et al. (1988), Rohde (1976), Elder et al. (1980) and others – claim that the bias is likely to be less than $0.07p$.

5.3.2 Continuous variables

A modified composite sampling scheme can be employed if X is a continuous variable. For example, X may measure pollution levels in a river, and we want to know if any observed x_i in a sample of size n are illegally high values above some control value, or standard, x_H. Thus characteristic A is now defined by $x_i > x_H$, say. We suppose that if we measure X for a composite sample, its value will be the sum of the constituent x_i (e.g. X measures counts of some organism). Suppose the value of X for the composite sample is x and put $\bar{x} = x/n$, which is, of course, the equivalent of the sample mean of the distinct values making up the composite sample.

Note that if any $x_i > x_H$, then for the whole sample we would be bound to have $\bar{x} > x/n$. This reflects the fact that even for the minimal case of violation where all but one of the x_i is zero and just one $x_i > x_H$ we would still have $\bar{x} > x_H/n$.

The condition $\bar{x} > x_H/n$ has been proposed as a (highly conservative) basis for declaring that the composite sample indicates violation: it is known as the 'rule of n'. Its highly conservative nature needs re-emphasizing and is easily demonstrated. If we had a situation where all but one $x_i = x_H/n$ and only one is just a little larger – a totally innocuous set of values – we would still have $\bar{x} > x_H/n$ and would wrongly declare the sample to be in violation.

On this 'rule of n' composite sampling proceeds as follows. If $\bar{x} < x_H/n$, we declare all observations to be satisfactory. If $\bar{x} > x_H/n$, we would need to retest all sample members (or smaller composite subsamples), with analogous results to the attribute sampling case above.

Specific proposals based on this approach are considered by Boswell and Patil (1987), Rohde (1976) and Brumelle et al. (1984). Barnett and Bown (2002c) present a less conservative approach to composite sampling, in the context of setting and testing standards for water quality – see Chapter 9 below.

5.3.3 Estimating mean and variance

Let us now consider how we might use composite sampling to estimate summary measures of a quantitative variable X. Suppose the variable X has mean μ and variance σ^2 and we decide to use composite sampling to estimate μ by taking n observations in k composite samples of size $m(n = mk)$. We assume that X is an aggregate variable in the sense that its value $X_{m,i}$ on the ith composite sample is

$$X_{m,i} = \sum_{j=1}^{m} X_{j,i}$$

(the *sum* of the values that would have been obtained on the individual items in the composite sample).

Thus, to estimate μ we could take *just one* composite sample of size n and use

$$\tilde{\mu}_n = X_{n,1}/n$$

as the estimator. But this approach provides no estimate of σ^2, hence no estimate of the standard error of the estimator.

Consider two composite samples each of $n/2$ items. We would then estimate μ by

$$\tilde{\mu}_{n/2} = \left(X_{n/2,1} + X_{n/2,2}\right)/n$$

and can now estimate σ^2 by

$$\tilde{\sigma}_{n/2}^2 = \left(X_{n/2,1} - X_{n/2,2}\right)^2/2n$$

For $k \geq 3$, we can also construct corresponding unbiased estimates of σ^2 and can seek to choose k to balance cost and efficiency needs.

Consider the general case where we have k composite samples of size m – compared with a single sample of size k (assuming measurement of composite sample values is essentially the same in cost and effort as single samples). For composite sample $i(i = 1, 2, \ldots, k)$ we observe $x_{m,i} \equiv \sum_{j=1}^{m} x_{j,i}$ and then $\bar{x}_{m,i} = x_{m,i}/m$ will have expected value μ and sample variance σ^2/m. So if $\tilde{\mu}_k = 1/k \sum_{i=1}^{k} \bar{x}_{m,i}$ is used as the overall estimate of μ, it is unbiased with sampling variance $\sigma^2/km = \sigma^2/n$.

Thus, on the above assumption concerning measurement costs and if all measurements cost the same amount (and, in practice, costs of measurement are often expensive compared with costs of sample collection), the composite sample approach needs to be compared with taking just k single samples $x_i(i = 1, 2, \ldots, k)$, yielding an estimate $\bar{x}$ with variance σ^2/k. So we have relative efficiency

$$e(\tilde{\mu}_k; \bar{x}) = \left(\sigma^2/k\right)/\left(\sigma^2/km\right) = m. \qquad (5.10)$$

Thus for any $m > 1$, we achieve major (order of magnitude) efficiency gains for estimating μ by the composite sample mean $\tilde{\mu}_k$. But what about estimating σ^2?

From the single-sample case we have

$$s_1^2 = \frac{1}{k-1} \sum (x_i - \bar{x})^2,$$

which is known to be unbiased with sampling variance

$$V\left(s_1^2\right) \sim \frac{1}{k}\left(\mu_4 - \frac{k-3}{k-1}\sigma^4\right).$$

From the composite sample we use

$$s_2^2 = \frac{m}{k-1} \sum_{i=1}^{k} \left(\bar{x}_{m,i} - \tilde{\mu}_k\right)^2,$$

which is again unbiased and can be shown to have sampling variance

$$V(s_2^2) \sim \frac{1}{k} \left\{ \frac{\mu_4}{m} + \frac{3(m-1)}{m} \sigma^4 - \frac{k-3}{k-1} \sigma^4 \right\}. \tag{5.11}$$

So

$$V(s_1^2) - V(s_2^2) = \frac{(m-1)\sigma^4}{mk} \left(\frac{\mu_4}{\sigma^4} - 3 \right) = \frac{(m-1)\sigma^4}{km} \gamma_2, \tag{5.12}$$

where γ_2 is the coefficient of kurtosis (flatness). Note that for the normal distribution we have $\gamma_2 = 0$. So the composite sample estimator of σ^2 (i.e. s_2^2) is also more efficient than s_1^2 if $\gamma_2 > 0$ (i.e. for leptokurtic distributions, flatter than the normal distribution).

Brumelle *et al.* (1984) discuss these matters against the background of monitoring industrial effluent, discussed earlier by Nemetz and Dreschler (1978). A comprehensive bibliography on composite sampling is given by Patil *et al.* (1996).

Example 5.4 Suppose we test blood specimens for presence of a disease by pooling them and testing the pooled (composite) sample. The separate blood specimens are set out in an $r \times c$ array of r rows and c columns, and it is possible to automatically carry out tests on the composite samples made up of rows and of columns: $r + c$ in all.
Possible identification schemes might be:

(i) Test all rows and all columns and then test individual samples at the intersections of positive rows and columns (although there is clearly some redundancy in this process).
(ii) Test all rows and proceed if any row is positive. If one row is positive test all members of that row; if more than one row is positive proceed as in (i).

Phatarfod and Sudbury (1994) examine for some different values of r, c and p the expected number of required tests, which is shown to be smaller than for the basic scheme which tests the overall composite sample of size rc and then tests individual samples if the composite is positive.

5.4 RANKED-SET SAMPLING

Although one of the major activities in environmental statistics is that of obtaining relevant data for statistical investigation, we have to face the problem that the circumstances for obtaining data by means of censuses, sample surveys or designed experiments as described in Chapter 4 (and elsewhere, e.g. in Chapters 7, 8 and 11) are likely to be rather rare. Quite often we have to take what limited data are to hand ('encountered data'; see Section 5.1 above) which may be difficult to analyse using formal methods. If we are to collect even limited data for our purposes we may need to abandon such hallowed principles as strict randomization, not only in view of access constraints but also to contain costs and to improve efficiency.

In the many areas of environmental risk such as radiation (soil contamination, disease clusters, air-borne hazard) or pollution (water contamination, nitrate leaching, root disease of crops) we commonly find that the taking of measurements can involve substantial scientific processing of materials and correspondingly high attendant costs. It becomes particularly appropriate that we should seek to draw statistical inferences as expeditiously as possible with regard to containing the sample costs. That is to say, we need to look for highly efficient procedures. One way of doing this is to use what is known as *ranked-set sampling*.

A simple example of the problem arises even when we wish to estimate as basic a quantity as a population mean. Suppose we are interested in the mean pollution level of the bathing water around an inland lake used for recreational purposes. We might decide to take a random sample of modest size at regular intervals of time and to use the sample mean to estimate the mean pollution level of the lake on each occasion. But with the attendant costs even of such a simple monitoring process it is desirable to keep the sample as small (and cheap) as possible to achieve the desired level of assurance.

Ranked-set sampling is becoming widely used for sampling in such contexts, where measuring environmental risk factors is expensive. It could operate here in the following way. If we want a sample of size 5 we would choose five sites at random, but rather than measuring pollution at each of them we would ask a local expert which would be likely to give *the largest value*. (We might, alternatively, choose the candidate for highest value by a cheaply observed concomitant variable such as the opacity of the water.) We then repeat the process by selecting a second random set of five sites (and a second expert to guide us) and seeking to measure *the second largest pollution level* amongst these, and so on, until we seek the *lowest pollution level* in the final random set of five sites. The resulting *ranked-set sample* of size 5 is then used for the estimation of the mean.

Such an approach can also be used to estimate a measure of dispersion, a quantile or even to carry out a test of significance or to fit a regression model. The gains can be dramatic: efficiencies relative to simple random sampling may reach 300%.

The aim is to employ concomitant (and cheaply and readily available, sometimes subjective) information to seek to 'spread out' the sample values over their possible range. It is this 'spreading out' that can produce such a dramatic increase in efficiency over simple random sampling. The method has been around for nearly 50 years since it was first mooted in an agricultural/environmental context by McIntyre (1952) for estimating mean pasture yields.

Takahasi and Wakimoto (1968) provided mathematical support for the informally developed ideas of McIntyre, showing in particular that the ranked-set sample mean is unbiased for the population mean and more efficient than the random sample mean (for the same sample size). Dell and Clutter (1972) relaxed an earlier assumption (that observations in ranked-set samples were *correctly ordered*) to examine the implications of imprecise or imperfect ordering. Again ranked-set sampling yielded an unbiased estimator and provided useful efficiency gains as long as there was non-zero correlation between sample values and attributed order. Stokes (1977) formalized this latter situation in the case where an easily ranked concomitant variable was present; see also Stokes (1980). Ridout and Cobby (1987) further developed the modelling basis for the imperfect-ranking situation. Muttlak and McDonald (1990) examined a different prospect: applying ranked-set sampling methods as a second-phase sampling procedure following probability-proportional-to-size sampling, in the finite-population context.

Further modifications continue apace to improve efficiency and applicability for different distributional forms of the underlying random variable and different types of inference, and we will review some of these later.

5.4.1   The ranked-set sample mean

The ranked-set sampling approach may be described in the following way. We first consider a set of n *conceptual random samples*, of observations of a random variable X. These would yield observations, say, of the form

$$x_{11}, x_{21}, \ldots, x_{1n}$$
$$x_{21}, x_{22}, \ldots, x_{2n}$$
$$\vdots$$
$$x_{n1}, x_{n2}, \ldots, x_{nn}$$

if we were to actually measure the outcomes. Instead each subsample yields only *one* measured observation $x_{i(i)}$: the ith ordered value in the ith sample $(i = 1, 2, \ldots, n)$. The *ranked-set sample* is then defined as

$$x_{1(1)}, x_{2(2)}, \ldots, x_{n(n)}.$$

Early studies of ranked-set sampling proposed that the mean μ of the underlying distribution should be estimated by

$$\bar{\bar{x}} = \frac{1}{n} \sum_{i=1}^{n} x_{i(i)} \qquad (5.13)$$

which is, for obvious reasons, known as the *ranked-set sample mean*.

The estimator $\bar{\bar{X}}$ needs to be compared in terms of efficiency and other properties with $\bar{X}$ (the mean of a random sample – but of size n, *not of size* n^2), *since measurement* is again (as in estimating the mean from a composite sample which we have just discussed) assumed to be of overriding effort compared with selection and ordering.

It is readily shown that $\bar{\bar{X}}$ is unbiased and that

$$\text{Var}\left(\bar{\bar{X}}\right) \leq \text{Var}\left(\bar{X}\right);$$

indeed $\bar{\bar{X}}$ can have variance *markedly less* than that of $\bar{X}$ for a range of different sample sizes and distributions.

In forming the ranked-set sample we assume, for the moment, that we have correctly ordered the potential observations in each conceptual subsample. This is termed *perfect ordering*. If ordering is not entirely correct, then, as remarked above, benefits still arise (unbiasedness and increased efficiency compared with $\bar{X}$) provided there is non-zero correlation between the observed sample values and their claimed order. The advantage disappears once the ranking becomes equivalent to mere randomization (but we still do as well as random sampling). We examine such prospects in more detail later.

Suppose we adopt a more structured approach and assume that observations come from a distribution of random variable X with distribution function in the form $F[(x - \mu)/\sigma]$, having mean μ_x and variance σ_x^2. We know that $\bar{X}$ is unbiased, with expected value μ_x and variance σ_x^2/n. If is symmetric we have the special case where $\mu_x = \mu$ and $\sigma_x^2 = \sigma^2$.

Consider $\bar{\bar{X}}$. The ranked set sample values $x_{1(1)}, x_{2(2)}, \ldots, x_{n(n)}$ are the values of the *order statistics* $X_{(1)}, X_{(2)}, \ldots, X_{(n)}$ in a sample of size n *except that*, because each comes from a separate sample, they are *independent* (rather than correlated as is usually the case). Consider the *reduced order statistics*

$$U_{(i)} = (X_{(i)} - \mu)/\sigma$$

with means and variances denoted α_i and v_i respectively ($i = 1, 2, \ldots, n$), which are independent of μ and σ. Then

$$E(X_{(i)}) = \mu + \sigma\alpha_i \tag{5.14}$$

and

$$\text{Var}(X_{(i)}) = \sigma^2 v_i \tag{5.15}$$

Thus

$$E(\bar{\bar{X}}) = \mu + \sigma\bar{\alpha} \tag{5.16}$$

and

$$\text{Var}(\bar{\bar{X}}) = \left(\sum v_i\right)\sigma^2/n^2, \tag{5.17}$$

where $\bar{\alpha} = (\sum \alpha_i)/n$. Clearly, we have the simple relationship,

$$\sum X_i = \sum X_{(i)},$$

so that by taking expectations of each side we have

$$n\mu_x = E\left(\sum X_i\right) = E\left(\sum X_{(i)}\right) = n\mu + \sigma\sum \alpha_i. \tag{5.18}$$

Thus

$$\mu_x = \mu + \sigma\bar{\alpha} \tag{5.19}$$

which shows that $\bar{\bar{X}}$ is unbiased for μ_x. (If X is symmetric, $\mu_x = \mu$, so that we must have $\bar{\alpha} = 0$.)

Now consider the identity

$$\left[\sum(X_i - \mu)\right]^2 = \left[\sum(X_{(i)} - \mu)\right]^2.$$

Taking expectations, we now obtain

$$n\sigma_x^2 = E\left\{\left[\sum(X_{(i)} - \mu - \sigma\alpha_i + \sigma\alpha_i)\right]^2\right\}$$
$$= E\left[\sum(X_{(i)} - \mu - \sigma\alpha_i)^2\right] + \sigma^2\sum\alpha_i^2,$$

so that

$$\frac{\sigma_x^2}{n} = \frac{\sigma^2}{n^2}\sum v_i + \frac{\sigma^2}{n^2}\sum\alpha_i^2. \tag{5.20}$$

That is,

$$\text{Var}(\bar{X}) = \text{Var}(\bar{\bar{X}}) + \frac{\sigma^2}{n^2}\sum\alpha_i^2, \tag{5.21}$$

and the essentially positive nature of the term $(\sigma^2/n^2)\sum\alpha_i^2$ shows, as required, that

$$\text{Var}(\bar{X}) \geq \text{Var}(\bar{\bar{X}}).$$

The relative efficiency of $\bar{\bar{X}}$ and $\bar{X}$ is thus

$$e = \frac{\text{Var}(\bar{X})}{\text{Var}(\bar{\bar{X}})} = 1 + \frac{\sum\alpha_i^2}{\sum v_i}. \tag{5.22}$$

If X is symmetric $\sigma_x^2 = \sigma^2$, so that from (5.20) we have $n = \sum v_i + \sum\alpha_i^2$ and the relative efficiency can be written in terms of the α_i as

$$e = \left(1 - \sum\alpha_i^2/n\right)^{-1}. \tag{5.23}$$

It is possible to show that $1 < e \leq (n+1)/2$, so that the potential efficiency gains from using the ranked-set sample mean can be very high.

Example 5.5 Consider random and ranked-set samples of sizes 5 and 10 from normal, uniform and exponential distributions. The high efficiencies, e, of the ranked-set sample mean relative to the random sample mean are as follows:

	Sample size	
Distribution	5	10
Normal	2.77	4.79
Uniform	3.00	5.00
Exponential	2.19	3.41

We should note that when using ranked-set samples we are unlikely to take single samples of substantial size, since the ordering process of prospective samples is likely to be difficult when n is as large as about 10. Thus, in practice, ranked-set sampling often uses replication of smaller samples. A sample of size 50, say, may be made up of 10 independent ranked set samples each of size 5 (or, less likely, of 5 samples of size 10).

Thus replication is of some interest. Suppose, that instead of taking just *one* value $X_{i(i)}$ at each i, we were to take m_i values ($i = 1, 2, \ldots, n$). So the total sample size is now $\sum_{i=1}^{n} m_i$; for example, if we have equal replication with $m_i = m$ (all i), then the total sample size is nm and we have m smallest values in a conceptual sample of size n, m second smallest and so on. In the general case the ranked-set sample mean is now defined to be

$$\bar{\bar{X}} = \frac{1}{n}\sum \bar{X}_{(i)},$$

where $\bar{X}_{(i)}$ is the average of the m_i ith largest values. Thus

$$\mathrm{Var}(\bar{\bar{X}}) = \sigma^2 \sum (v_i/m_i)/n^2 \tag{5.24}$$

and

$$e = \frac{1}{N}/\left\{\sum (v_i/m_i)/n^2\right\}, \tag{5.25}$$

where $N = \sum_{i=1}^{n} m_i$.

There can be advantages in taking the m_i *unequal in value* with regard to aspects of sampling design. See Kaur *et al.* (1997) and Barnett (1999b) for discussion of the gains that can be achieved under some circumstances when we do so. See also Barnett (2002c).

5.4.2　Optimal estimation

Let us now return to the unreplicated case where $m_i = 1 (i = 1, 2, \ldots, n)$ and consider a more general situation.

The ranked-set sample mean $\bar{\bar{X}} = n^{-1}\sum x_{i(i)}$ is just a specific case of the *general linear estimator* (L-estimator) based on ordered values of the form

$$\tilde{\mu} = \Sigma \gamma_i X_{i(i)} \tag{5.26}$$

If $\sum \gamma_i = 1$, then $\tilde{\mu}$ is unbiased for μ_x so why should we not choose the γ_i for optimal effect – for example, to minimize $\mathrm{Var}(\tilde{\mu})$ – rather than just putting $\gamma_i = 1/n$ as for $\bar{\bar{X}}$? This extended prospect has been studied with some interesting results. Let us examine the properties of the optimal unbiased estimator, $\tilde{\mu}$.

We start with a more general problem distinct from ranked-set sampling. For a random variable X with d.f. $F[(x - \mu)/\sigma]$, it is well known (see, for example, Lloyd, 1952) that the best linear unbiased order statistics estimator of the parameter vector $\boldsymbol{\theta} = (\mu, \sigma)$, based on observed order statistics $\mathbf{x}' = (x_{(1)}, x_{(2)}, \ldots, x_n)$, takes the form

$$\boldsymbol{\theta}^* = \left(\mathbf{A}'\mathbf{V}^{-1}\mathbf{A}\right)^{-1}\mathbf{A}'\mathbf{V}^{-1}\mathbf{x}, \tag{5.27}$$

with variance/covariance matrix

$$\mathrm{Var}(\boldsymbol{\theta}^*) = \sigma^2\left(\mathbf{A}'\mathbf{V}^{-1}\mathbf{A}\right)^{-1}, \tag{5.28}$$

where $\mathbf{A} = (\mathbf{1}, \boldsymbol{\alpha})$ with $\boldsymbol{\alpha}' = \{\alpha_i\} = \{E(U_{(i)})\} = \{E[(X_{(i)} - \mu)/\sigma]\}$, and $\mathbf{V} = \{v_{ij}\}$ is the variance–covariance matrix of the reduced order statistics $U_{(i)} = (X_{(i)} - \mu)/\sigma$. These earlier results have been applied to the ranked-set sample problem by Sinha *et al.* (1996), Stokes (1995) and, in more generality and detail, by Barnett and Moore (1997).

When carried over to ranked-set sampling, we can exploit an important simplifying feature of the ranked-set sample $(X_{1(1)}, X_{2(2)}, \ldots, X_{n(n)})$: namely, that the sample members are uncorrelated, because they arise from independent samples. Thus, the optimal L-estimator of $\boldsymbol{\theta} = (\mu, \sigma)$ which we will call the *ranked-set BLUE*, will take the form of $\boldsymbol{\theta}^*$ as in (5.27), with variance–covariance matrix $\mathrm{Var}(\boldsymbol{\theta}^*)$ as in (5.28), provided that we replace $\mathbf{V}$ with the diagonal matrix

$$\mathbf{W} = \mathrm{diag}(v_i) \tag{5.29}$$

where v_i is what is denoted as v_{ii} in the more general case described above.

Specifically, if we write the optimal estimators as

$$\mu^* = \sum_{j=1}^{n} \gamma_i X_{i(i)}, \qquad \sigma^* = \sum_{j=1}^{n} \eta_i X_{i(i)}, \tag{5.30}$$

we will obtain

$$\gamma_i = \frac{(1/v_i)\left[\sum_{j=1}^{n}\left(\alpha_j^2/v_j\right) - \alpha_i \sum_{j=1}^{n}\left(\alpha_j/v_j\right)\right]}{\Delta} \tag{5.31}$$

and

$$\eta_i = \frac{(1/v_i)\left[\alpha_i \sum_{j=1}^{n}(1/v_j) - \sum_{j=1}^{n}\left(\alpha_j/v_j\right)\right]}{\Delta} \tag{5.32}$$

where

$$\Delta = \sum\left(\frac{\alpha_j^2}{v_j}\right)\sum\left(\frac{1}{v_j}\right) - \left[\sum\left(\frac{\alpha_j}{v_j}\right)\right]^2. \tag{5.33}$$

We also find that

$$\mathrm{Var}(\mu^*) = \sigma^2 \frac{\sum_{i=1}^{n}\left(\alpha_i^2/v_i\right)}{\Delta}$$

$$\mathrm{Var}(\sigma^*) = \sigma^2 \frac{\sum_{i=1}^{n}(1/v_i)}{\Delta}. \tag{5.34}$$

For the special case where X is *symmetric* these results are much simpler in form. We have

$$\mu^* = \frac{\sum (X_{i(i)}/v_i)}{\sum (1/v_i)},$$

$$\sigma^* = \frac{\sum (\alpha_i X_{i(i)}/v_i)}{\sum (\alpha_i^2/v_i)}, \tag{5.35}$$

with

$$\text{Var}(\mu^*) = \frac{\sigma^2}{\sum (1/v_i)} \tag{5.36}$$

and

$$\text{Var}(\sigma^*) = \frac{\sigma^2}{\sum (\alpha_i^2/v_i)}. \tag{5.37}$$

Noting that we have found, in (5.17), that $\text{Var}(\overline{\overline{X}}) = \sigma^2 \sum v_i/n^2$, we find, in the symmetric case, that the efficiency of μ^* relative to $\overline{\overline{X}}$ is

$$e(\overline{\overline{X}}, \mu^*) = \text{Var}(\overline{\overline{X}})/\text{Var}(\mu^*)$$

$$= \frac{(\sum v_i) \sum (1/v_i)}{n^2}. \tag{5.38}$$

5.4.3 Ranked-set sampling for normal and exponential distributions

We will now apply these optimal forms to some special cases, namely when X is normally distributed or is exponentially distributed with unknown origin (a model of some practical importance in, for example, environmental issues involving life-testing and reliability considerations).

Normal distribution
If X is normally distributed with mean μ and variance σ^2, we can obtain μ^*. The α_i and v_i terms are tabulated up to $n = 20$ in Pearson and Hartley (1976) and can be used to determine the coefficients γ_i and η_i. In particular, we can determine the variances $\text{Var}(\overline{\overline{X}})/\sigma^2$ and $\text{Var}(\mu^*)/\sigma^2$ as well as the relative efficiency $e(\overline{\overline{X}}, \mu^*) = \text{Var}(\overline{\overline{X}})/\text{Var}(\mu^*)$ to explore the potential gain in efficiency of μ^* compared with $\overline{\overline{X}}$. For larger values of n, corresponding quantities were obtained in the discussion by Barnett and Moore (1997) from results of Ross (1995).

Table 5.1 gives the modified values of the two variances $n^2\text{Var}(\overline{\overline{X}})/\sigma^2$ and $n^2\text{Var}(\mu^*)/\sigma^2$, and the corresponding relative efficiencies, for $n = 2, 3, 5, 8, 10, 15$ and 20. The results show the extent of the improvement of μ^* over $\overline{\overline{X}}$. While modest at small values of n, it is still valuable and becomes impressive for $n > 10$ (ranging from about 9% when $n = 10$, as shown in Table 5.1, to 50% when $n = 1000$).

Table 5.1 $n^2\text{Var}(\bar{\bar{X}})/\sigma^2$, $\text{Var}(\mu^*)/\sigma^2$ and $e(\bar{\bar{X}}, \mu^*)$ for the normal distribution.

n	$n^2\text{Var}(\bar{\bar{X}})/\sigma^2$	$n^2\text{Var}(\mu^*)/\sigma^2$	$e(\bar{\bar{X}}, \mu^*)$
2	1.363	1.363	1.000
3	1.568	1.551	1.011
5	1.805	1.739	1.038
8	2.001	1.862	1.074
10	2.086	1.906	1.094
15	2.229	1.966	1.134
20	2.322	1.996	1.163

We might ask to what extent the coefficients γ_i in μ^* differ from the values $1/n$ in $\bar{\bar{X}}$. Some examples are shown in Table 5.2, which shows the γ_i terms for $n = 2, 4, 6, 10$ and 15 (only the values for $i = 1, \ldots, [n/2] + 1$ need to be shown, because of the symmetry of the γ_i values). We see that the optimal weights are far from constant (as they are for $\bar{\bar{X}}$). The extreme observations are markedly downweighted. For example, when $n = 15$, the median has about three times the weight of the extreme order statistics.

Table 5.2 γ_i coefficients in μ^* for normal distribution $(\gamma_i = \gamma_{n-i})$.

i	$n = 2$	$n = 4$	$n = 6$	$n = 10$	$n = 15$
1	0.500	0.211	0.120	0.055	0.029
2		0.289	0.178	0.089	0.049
3			0.202	0.109	0.062
4				0.121	0.071
5				0.126	0.078
6					0.083
7					0.085
8					0.086

Exponential distribution

The efficiency gains are even more impressive when X has the (skew) exponential distribution with

$$F\left(\frac{x - \mu}{\sigma}\right) = 1 - \exp\left(-\frac{x - \mu}{\sigma}\right), \qquad x > \mu. \tag{5.39}$$

In fact, the mean is not μ of course, but

$$E(X) = \xi = \mu + \sigma, \tag{5.40}$$

that is to say, the mean is a combination of the two parameters, $\mu + \sigma$.

We can obtain the optimal estimator as $\xi^* = \mu^* + \sigma^*$ and its variance as $\text{Var}(\mu^*) + \text{Var}(\sigma^*) + 2\text{Cov}(\mu^*, \sigma^*)$ from the results in Section 5.4.2 above.

Alternatively, we can seek the ranked-set BLUE of ξ directly. (Note that if $\mu = 0$, we then revert to the usual one-parameter exponential distribution, in which the mean is just σ.) Suppose we estimate ξ by

$$\xi = \sum \delta_i x_{i(i)}. \tag{5.41}$$

This yields the ranked-set BLUE ξ^* of ξ in the form of (5.41), with

$$\delta_i = \frac{(1/v_i)\left[\sum\left(\alpha_j^2/v_j\right) - \sum(\alpha_j/v_j)\right] + (\alpha_i/v_i)\left[\sum(1/v_j) - (1/\alpha_i)\sum(\alpha_j/v_j)\right]}{\Delta}, \tag{5.42}$$

which is just $\mu^* + \sigma^*$, as would be obtained from the results of Section 5.2.1. The variance of ξ^* can be readily obtained, and it reduces to the form

$$\text{Var}(\xi^*) = \sigma^2 \frac{\sum(\alpha_i^2/v_i) - 2\sum(\alpha_i/v_i) + \sum(1/v_i)}{\Delta}. \tag{5.43}$$

Of course, $\overline{\overline{X}}$ is unbiased for $E(X)$ with variance $\sigma^2 \sum v_i/n$, so that the relative efficiency $e(\overline{\overline{x}}, \xi^*)$ is now given by

$$e(\overline{\overline{x}}, \xi^*) = \Delta \frac{\sum v_i}{n^2}\left[\sum\left(\frac{\alpha_i^2}{v_i}\right) - 2\sum\left(\frac{\alpha_i}{v_i}\right) + \sum\left(\frac{1}{v_i}\right)\right]. \tag{5.44}$$

Corresponding to Tables 5.1 and 5.2 for the normal distribution, we have Tables 5.3 and 5.4 which give respectively the relative efficiency values for ξ^* and $\overline{\overline{X}}$, and the coefficients δ_i in the optimal form ξ^* for a choice of values for n. The improvement in efficiency of ξ^* is much more marked than for the normal distribution (with 25% improvement even at $n = 10$). Note the markedly asymmetric form of the coefficients η_i and the negative value of η_1 for $n = 10$ (and in fact for $n > 10$).

5.4.4 Imperfect ordering

In many practical situations it may not be possible precisely and correctly to order the sample values. So we cannot be sure that $x_{i(i)}$ is truly the ith ordered

Table 5.3 $n^2\text{Var}(\overline{\overline{X}})/\sigma^2, n^2\text{Var}(\xi^*)/\sigma^2$ and $e(\overline{\overline{x}}, \xi^*)$ for the exponential distribution.

n	$n^2\text{Var}(\overline{\overline{x}})/\sigma^2$	$n^2\text{Var}(\xi^*)/\sigma^2$	$e(\overline{\overline{x}}, \xi^*)$
2	1.500	1.500	1.000
3	1.833	1.784	1.028
5	2.283	2.082	1.097
8	2.718	2.279	1.192
10	2.929	2.348	1.248
15	3.318	2.438	1.361
20	3.598	2.480	1.451

Table 5.4 η_i coefficients in ξ^* for exponential distribution.

i	$n = 2$	$n = 4$	$n = 6$	$n = 10$
1	0.500	0.137	0.003	−0.100
2	0.500	0.341	0.214	0.091
3		0.319	0.250	0.143
4		0.202	0.235	0.160
5			0.189	0.161
6			0.111	0.153
7				0.138
8				0.117
9				0.089
10				0.049

value in the ith conceptual sample. However, it is known (see, for example, Dell and Clutter, 1972) that, in this case, $\overline{\overline{X}}$ will still be unbiased and will continue to yield an efficiency gain over $\overline{X}$, provided only that there is non-zero correlation between the sample values and the attributed order.

The X_{ij} terms are often ranked in terms of a concomitant variable Y (rather than from subjective 'expert opinion') in order to identify the ith 'ordered' value in the ith conceptual sample (Stokes, 1977, 1980). Provided X and Y are correlated (X might be concentration of algae in a pond, Y the level of sunlight radiation) we will preserve some correlation ρ between the sample values and their attributed order. Hence there will still be an advantage in using the ranked-set sample. It is possible to be more specific and to determine the optimal ranked-set BLUE in the modified situation of imperfect ordering (see Barnett and Moore, 1997).

Let us examine some of the effects of imperfect ordering, using the notion of *concomitants* (see, for example, Barnett et al., 1976) to derive the results. In a random sample $(X_1, Y_1), (X_2, Y_2), \ldots, (X_n, Y_n)$ the X-value, $X_{i[i]}$, corresponding with the ith order statistic of the Y-values is known as the ith *contaminant*. It is these $X_{i[i]}(i = 1, 2, \ldots, n)$ for n separate conceptual samples of size n that constitute the ranked-set sample based on the concomitant variable Y.

If the correlation coefficient between X and Y is ρ, we can readily confirm that

$$\mathrm{E}(X_{i[i]}) = \mu + \rho\sigma\alpha_i \tag{5.45}$$

and that

$$\mathrm{Var}(X_{i[i]}) = \sigma^2(1 - \rho^2) + \sigma^2\rho^2 v_i = \sigma^2\omega_i. \tag{5.46}$$

Again, due to the $X_{i[i]}$ coming from separate samples, we will have $\mathrm{Cov}(X_{i[i]}, X_{j[j]}) = 0$ for $i \neq j$.

The results (5.45) and (5.46) are basically similar to (5.14) and (5.15), respectively, so that the form and properties of the optimum estimators based on such a concomitant variable approach with correlation $\rho < 1$ are readily established following parallel arguments to those used in Section 5.4.2 above; see Barnett and Moore (1997) for details.

To illustrate the effects of imperfect ordering , Table 5.5 shows, for selected sample sizes and values of ρ, in the normal and exponential cases, the relative efficiency of the optimal estimator of the mean, compared with $\overline{\overline{X}}$. We see that the optimal estimator of the mean remains more efficient than $\overline{\overline{X}}$.

It is possible to go much further with ranked-set sampling. For example, we can estimate a scale parameter (Barnett and Moore, 1997) or a variance (Stokes, 1980), carry out tests of means or fit regression lines (Barreto and Barnett, 1999). Baretto and Barnett (2001) consider estimation of the mean of a Poisson distribution, whilst Barnett (1999b) extends earlier work (e.g. by Kaur *et al.*, 1997) on rank-set sampling methods which use unequal numbers of the different $x_{i[i]}$ (e.g. using more minima and maxima than intermediate values).

There are many developments of ranked-set sampling that merit further study. For example, the ultimately extreme case of unequal representation of each $X_{i[i]}$ is where we use *only one such ordered value*. Suppose we take from each potential sample of size n just the median $m_i (i = 1, 2, \ldots, n)$ in the case of a symmetric distribution. We might estimate the mean μ by what could be termed the *memedian* (the mean median value),

$$\overline{m} = \left(\sum m_i\right)/n$$

(see Barnett, 2002c). This can be even more efficient than the ranked-set BLUE, as we see in Barnett (2002c) who shows that, for a sample of size 5, the memedian exhibits major efficiency gains over the ranked-set BLUE for normal and double exponential distributions, has similar efficiency for the triangular distribution and is noticeably less efficient for the uniform distribution.

Example 5.6   Motivated by interest in characteristics of root growth (particularly root weight) for the experimental plant *Arabidopsis thaliana*, ranked-set sampling methods have been used to estimate means, since the measurement of root weight without soil attached is so laborious and maximum efficiency of estimation is therefore to be desired.

Table 5.5   Relative efficiency of the optimal estimator of the mean, compared with $\overline{\overline{X}}$, for normal and exponential distributions.

	Normal				Exponential			
ρ	$n = 5$	10	15	20	$n = 5$	10	15	20
1.0	1.038	1.094	1.134	1.163	1.097	1.248	1.361	1.451
0.9	1.014	1.022	1.023	1.022	1.034	1.062	1.072	1.075
0.8	1.006	1.008	1.007	1.006	1.015	1.025	1.028	1.029
0.6	1.001	1.001	1.001	1.001	1.003	1.005	1.005	1.005
0.0	1.000	1.000	1.000	1.000	1.000	1.000	1.000	1.000

The ranked-set sample mean and the best linear unbiased estimator, for perfect and for imperfect ordering, were considered for collected root data. The distribution of root weights is unknown.

A. thaliana is used widely as a model plant in plant biology, especially for molecular and genetic studies where it is useful to be able to detect phenotypic differences between plants that could be linked to genetic changes. These differences are usually small. Hence large sample sizes are needed to increase precision of the estimates. However, using a large sample size is not feasible when the characteristics of root systems are of interest, since they are difficult and time-consuming to measure.

During a pilot study, ranked-set sampling was examined as an alternative method to random sampling to increase precision of estimation of mean root weight of *A. thaliana*.

The pilot study consisted of 25 replicate plants randomly allocated into five sets each of five plants grown in three typical treatment environments (75 plants in all). The environments were large pots in low or high temperatures and small pots at the low temperature. Both root weight and shoot weight were measured for each plant (these data are given in Barnett and Moore, 1997) and a set of simulation experiments was conducted on these data to investigate the precision of various estimators of root weight (since distributional forms were unknown).

Ten thousand samples of size $n = 5$ were taken from each of the three treatment populations of root weight by ranked-set sampling with perfect ordering (i.e. the root weights themselves were used to order the sets) and by ranked-set sampling with imperfect ordering (i.e. the shoot weights were used as a concomitant variable to order the sets). The ranked-set sample mean was calculated for each of the two sampling methods as well as the BLUE estimate for each of the ranked-set samples on the assumption of a normal underlying distribution (but see below); thus, there were four estimates for each simulated sample. Sampling variances for these four estimates (based on the 10 000 observations) were calculated and are presented in Table 5.6. This table also gives the sampling variances of the sample mean for simple random sampling without replacement, as a standard for comparison.

We would expect to find that ranked-set sampling (which selects a sample that is spread across the range of population values) will substantially increase the precision of the estimated mean when compared with choosing a simple random sample. The results of Table 5.6 confirm this. Improvement occurs from use of $\overline{\overline{X}}$ or of the ranked-set BLUE estimator even when the ordering is imperfect, although the degree of this improvement is highly dependent upon the correlation between shoot weight and root weight. In this latter regard, for example, consider from Table 5.6 the variance of the estimate of mean root weight for the treatment consisting of plants grown in large pots at a low temperature when the samples are selected by ranked-set sampling with imperfect ordering (where $\rho = 0.438$). This is fairly close to the value obtained for simple random sampling, which demonstrates the need to choose appropriate concomitant variables with care.

Table 5.6 Estimates of the variance of the various estimators of the mean.

	Large pots		Small pots
	High temp.	Low temp.	Low temp.
Simple random sampling without replacement			
	1.640	15.732	16.160
Ranked-set sampling			
$\overline{\overline{X}}$			
Imperfect	0.9137	13.82	7.265
Perfect	0.2624	2.910	3.480
BLUE			
Imperfect	0.9844	15.02	6.847
Perfect	0.2990	2.845	2.928

CHAPTER 6

Sampling in the Wild

Catch me if you can.

In this chapter we will consider a number of sampling methods which are particularly suitable for examining living things. We thus regard these methods as suitable for 'sampling in the wild', although we will see that they also have applications beyond flora and fauna. We will consider such approaches as quadrat sampling, capture–recapture or mark–recapture, transect sampling and adaptive sampling.

The aim is usually to assess abundance: how dispersed or concentrated are the members of a species in a defined area. Thus we might seek, through statistical sampling, to estimate the density of the population of pheasants in Philadelphia, of *Digitalis* on Dartmoor, of springbok in the Serengeti or of whales off the coast of Wales. An interesting introductory report by Manly and McDonald (1996) on 'sampling wildlife populations' describes the historical setting and offers some illuminating examples of the application of many of the methods developed in this chapter to such problems as assessing the survival prospects of the endangered spotted owl of the Pacific Northwest, and of polar bears in Alaskan waters. They also study population characteristics of salmon in the Snake and Columbia rivers of Washington state in the USA, and of yellow-eyed penguins in New Zealand. Stehman and Overton (1994) provide a general review of approaches to environmental sampling which briefly covers various topics discussed in this chapter and in Chapters 4 and 5. See also Pollock (1991) for a wide-ranging review of methods and models, particularly on the capture–recapture theme.

6.1 QUADRAT SAMPLING

We start with a general technique for environmental sampling particularly for ecological studies, rather than with a specific sampling method. If we wish to count the numbers of one, or of several, species of plant in a meadow (to estimate population size or assess biodiversity) we might throw a *quadrat* at random and do our count, or counts, within its boundary. Correspondingly, for aquatic wildlife we might cast a net of given size into a pond, river or sea and count what it trawls.

Environmental Statistics   V. Barnett
© 2004 John Wiley & Sons, Ltd   ISBN: 0-471-48971-9 (HB)

A quadrat is usually a square (or round) light wood or metal frame of (often) a metre or several metres side (or diameter). Where it lands defines the search area in which we take appropriate measures of numbers of individual plants, biomass or extent of ground cover. Quadrat sampling can be used with a variety of sampling designs, especially of the more classical type such as simple random sampling or stratified simple random sampling (see Chapter 4). Such a procedure can of course form part of many more-structured sampling methods, such as capture–recapture, mark–recapture or adaptive sampling, as described below – see also Seber (1986). It can also be particularly useful in *spatial sampling* to give indications of non-randomness, clumpiness, or anisotropy – see Chapter 11 and Pielou (1974).

6.2 RECAPTURE SAMPLING

A wide range of sampling methods are based on the principle of initially 'capturing' and 'marking' a sample of the members of a closed (finite) population and subsequently observing, in a later (or separate) independent random sample drawn from the population, how many marked individuals are obtained. The term *capture–recapture* is usually used for animals or insects, while *mark–recapture* is often reserved for when studying plants. The sample information is then used to infer characteristics of the overall population, principally its total size. Many forms and variations are possible on this theme, and particular assumptions have to be made to justify statistical methods for analysing the resulting data:

- The marking process may be multi-stage, with separate capture and recapture episodes.
- Individuals in the samples may be chosen *with* or *without replacement.*
- The marking process may sometimes contaminate the population, with the effect that marked individuals may be selectively removed from the population (e.g. rejected by the unmarked members, as happens for example with the Chillingham white cattle in Northumberland, UK, when members have been in contact with humans).
- Individuals may be 'trap shy' (they avoid contact with outside agencies) or even 'trap happy' (they are curious or eager for contact, perhaps seeking food); animals or plants may even die due to the intervention process (this has been observed for the Chillingham white cattle). Either prospect can distort the reflection we obtain of the population. In these cases the sample chosen for marking, or subsequently recaptured, may be far from random.
- The population may change from capture to recapture due to births, deaths or inward or outward transfer.

In the simplest approaches to capture–recapture we assume *randomness of the samples* with *constant capture probabilities* in a *fixed population* (no births, deaths, etc.) and *no capture-related effects* (of being 'trap shy' or 'trap happy' or marks being lost).

It is by no means easy to ensure that such conditions hold. Pollock (1991) provides an incisive review of the different qualitative aspects of capture–recapture methods, with particular concern for the effects of departures from the assumptions of the simplest approaches. His discussion progresses from *closed population models* to *open population models* where the population may change between 'captures' or be affected by the catching, marking and recapture processes. See also Seber (1986, 1992) for methodological detail and extensive bibliographic coverage. A well-illustrated lay review of sampling wildlife populations is, as remarked above, given by Manly and McDonald (1996).

What do we mean by marking? This might be achieved by rings on the legs of birds, small radio transmitters inserted under the skin of animals, radioisotope marking, nicks or cuts on the fins of fish, colour marking of skin or fur, cutting patterns in the bark of trees, tying a coloured plastic or material marker to an individual. Seber (1986) refers to efficient and small radio transmitters, remote sensing thermography (heat detection) and underwater accoustics as being crucial technological advances in capture–recapture methods for studying wildlife populations.

Fields of application of capture–recapture methods have proliferated over the years – from animal and plant studies to further aspects of biological (including human and societal) concern. Other issues include proof-reading of texts, attribution of authorship, computer software errors, industrial quality control and various sociological concerns such as homelessness, criminality, illness and so on.

6.2.1 The Petersen and Chapman estimators

We start with the basic closed population assumptions of the simplest approaches. Typically, an initial random sample is selected ('captured') and then 'marked' in such a way that, if any of its members arise in a subsequently chosen independent random sample, they are identifiable as having been chosen at the earlier stage. Capture probabilities are taken to be constant, and the population is assumed to remain the same throughout (with no entry or exit from it). This assumption of a homogeneous population is crucial and failures of the closed population methods arise predominantly because of population inhomogeneity in one or more of its many manifestations.

Let us examine this approach. We will suppose that an initial random sample of size n is chosen from a population of size N. Each of the members of the sample is 'marked' and the sample is then returned to the population. A second random sample of size m is then taken and turns out to contain m' of the originally marked individuals. How can we use these data to estimate N?

It is assumed that the samples have been chosen at random *without replacement* (no individual can appear twice in the sample) and with no distorting influences (e.g. trap shyness) to render the sample unrepresentative. Then the probability $p(m')$ of finding m' of the originally marked individuals in the second sample is obtained from a *hypergeometric distribution* and will have the form

$$p(m') = \frac{\binom{N-n}{m-m'}\binom{n}{m'}}{\binom{N}{m}}. \tag{6.1}$$

If N happens to be large in relation to n, and in most practical applications this will be so, then the effects of sampling *without replacement* are negligible and we can instead employ the binomial approximation

$$p(m') \sim \binom{m}{m'}\left(\frac{n}{N}\right)^{m'}\left(1 - \frac{n}{N}\right)^{m-m'}, \tag{6.2}$$

or the Poisson approximation

$$p(m') \sim \frac{e^{\frac{-mn}{N}}}{(m')!}\left(\frac{mn}{N}\right)^{m'}. \tag{6.3}$$

Thus, for known n and m, estimation of N corresponds with estimation of the reciprocal of the mean of the underlying hypergeometric (or binomial, or Poisson) distribution. The binomial and Poisson means are each mn/N.

The estimation of N in the simplest case arises from observing that if the second random sample is independent of the first we should expect the sample proportion m'/m to be (on the binomial approximation) a reasonable estimate of the corresponding population proportion n/N of marked members. Thus

$$\frac{m'}{m} \sim \frac{n}{N},$$

yielding the so-called *Petersen estimator* of population size N:

$$\tilde{N} = mn/m'. \tag{6.4}$$

It was during the 1890s that Carl Petersen, a Danish fisheries biologist, proposed this estimation method in his study of plaice. In another context (waterfowl populations) Frederick Lincoln suggested the same technique in the 1930s, and the approach is sometimes known as the *Lincoln–Petersen model*, although the method is also attributed to Laplace (for estimating the population of France in 1783). See Seber (1982), White *et al.* (1982), Laplace (1786), Lincoln (1930) and Manly and McDonald (1996).

Let us examine the Petersen estimator further. For the binomial approximation $B(m, n/N)$ for m', the binomial probability $p = n/N$ is usually estimated by the maximum likelihood estimator $\hat{p} = m'/m$, and hence we are estimating $N = n/p$ by $\tilde{N} = n/\hat{p}$ which is also the maximum likelihood estimator, using the invariance property of maximum likelihood estimators.

Standard techniques for approximating the sampling properties of $1/\hat{p}$ provide a first-order approximate form for an estimate of the sampling variance of $\tilde{N}$, in terms of observed quantities, as

$$\text{Var}(\tilde{N}) \sim \frac{nm(n - m')(m - m')}{(m')^3}. \tag{6.5}$$

The sampling distribution of $\tilde{N}$ can be shown to be markedly skew, and $\tilde{N}$ is typically biased for N, so that the approximate form for the variance is of limited use. This is compounded by the fact that we might in practice obtain *no marked members* in the second sample ($m' = 0$), so that $\tilde{N}$ could have infinite estimated variance. Specifically, the bias in $\tilde{N}$ takes the approximate form

$$E(\tilde{N}) = N\left(1 + \frac{m - m'}{(m')^2}\right). \tag{6.6}$$

Various attempts have been made to produce estimators of N with lower bias and approximately unbiased estimators of their variances (see, for example, Seber, 1982).

In particular, the so-called *Chapman estimator* has been proposed (see Chapman, 1951, 1955) in which each observation is increased by 1, producing an estimator of the form

$$\hat{N} = \frac{(n + 1)(m + 1)}{(m' + 1)} - 1. \tag{6.7}$$

This turns out to be unbiased if $n + m > N$ (and approximately so otherwise), and an approximately unbiased estimator of its variance is given by

$$\text{Var}(\hat{N}) \approx \frac{(n + 1)(m + 1)(n - m')(m - m')}{(m' + 1)^2(m' + 2)}, \tag{6.8}$$

which remains finite even in the limiting case $m' = 0$.

Example 6.1 Suppose we are interested in estimating the number N of fish in a pond by capturing by trawl net a random sample of them, tagging them, releasing them and recapturing a second random sample. The initial sample was of size 300. The second sample, of size 200, contained 50 previously tagged specimens. Here we have $n = 300, m = 200$ and $m' = 50$.
The Petersen estimator gives

$$\tilde{N} = mn/m' = 4 \times 300 = 1200,$$

with estimated variance

$$\text{Var}(\tilde{N}) \sim \frac{nm(n - m')(m - m')}{(m')^3} = 18\,000.$$

In contrast, the Chapman estimator gives

$$\hat{N} = \frac{(n + 1)(m + 1)}{(m' + 1)} - 1 = 1185,$$

with estimated variance

$$\text{Var}(\hat{N}) = \frac{(n + 1)(m + 1)(n - m')(m - m')}{(m' + 1)^2(m' + 2)} = 16\,775.$$

Approximate 95% confidence intervals for N are (937, 1463) and (931, 1439) respectively, little different in this case.

Consider a more general situation where we have initially marked a number of individuals in a region and we then make *several* passes through the region, on each of which we take random recapture samples. Skalski and Robson (1982) and Seber (1986) discuss this situation. For example, we might take k transect lines through a forest and on each pass catch and observe a sample from the transect line (transect sampling will be discussed in more detail in Section 6.3). We thus have k capture–recapture samples, with m_i observed animals, of which m_i' turn out to be marked, in sample $i (i = 1, 2, \ldots, k)$. Suppose n were initially marked, out of a population of N. We have $m = \sum m_i$ animals, of which $m' = \sum m_i'$ are marked.

The corresponding Petersen estimate (6.4) of N can now be re-expressed as

$$n(k\bar{m}/k\bar{m}') = n\bar{m}/\bar{m}',$$

where $\bar{m} = \sum m_i/k$ and $\bar{m}' = \sum m_i'/k$, which is just the classical *finite-population ratio estimate* (see Section 4.3.2). So from the results for ratio estimates, $\tilde{N}$ is now *approximately unbiased*. Correspondingly, we can estimate $\mathrm{Var}(\tilde{N})$ by

$$\hat{\mathrm{Var}}(\tilde{N}) = \left[\sum (m_i - rm'x_i)^2/k(k-1) \right] (n/m')^2,$$

where $r = m/m' = \bar{y}/\bar{x}$ (see the results in Section 4.3.2). If the m_i and m_i' are roughly proportional (as might arise with an even spread of marked animals throughout the region of study), then $\tilde{N}$ will be a highly efficient estimator (i.e. it will have low variance), as we know from our earlier examination of ratio estimators.

Another extension of capture–recapture for the closed population model is known as the *Schnabel census* (Schnabel, 1938) where a series of samples is taken and in each case all unmarked specimens are given individually indentifiable marks and the sample returned to the population. Estimation of population size is now possible even if capture probabilities vary with trap response, with time since previous capture (or capture attempt) or due to inherent heterogeneity (where each specimen has the same capture probability in each sample but the capture probability may vary from specimen to specimen). This extended set of prospects has been widely examined – see, in particular Darroch (1958), Otis *et al.* (1978) and Seber (1986).

6.2.2 Capture–recapture methods in open populations

We need now to consider the effects of our study population possibly changing from the initial capture to any subsequent recapture. This might happen in a natural way due to growth, decay, birth, death, immigration, and so on. On the other hand, the process of capture, marking and recapture may itself critically affect the population through such mechanisms as internal rejection of marked members or such members becoming 'trap shy' or 'trap happy'. So we now have what is called an *open population* to deal with, and we will need to consider how to take account of changes in the population from capture to recapture, or over time.

Seber (1986) and Pollock (1991) provide informative reviews of capture–recapture methods for *open populations* of wildlife. Detailed study of the wide range of models and methods that are available can be pursued from the extensive lists of references in these reviews, ranging widely over such practical applications as 'trap shyness' of yellow wagtails, the longevity of the little brown bat, and model-based developments such as 'a versatile growth model with statistically stable parameters'.

The basic approach to capture–recapture in open populations extends the notion of multiple and repeated recapture sampling which we discussed at the end of Section 6.2.1. When population size can change due to birth, immigration or death, the *Jolly–Seber* (JS) *model* has been widely and successfully applied (see Seber, 1982; Seber and Manly, 1985) Effects of differential trap response, loss of tag, etc., have also been incorporated (sometimes resorting to non-intrusive 'sightings' of animals or birds rather than 'captures' or 'recaptures' can reduce the contamination of the population). For some details of these extended prospects, see Pollock (1991) and Seber (1992).

In its simplest form the JS model operates as follows. A random sample is chosen from the population at the first of a set of sampling times, marked and returned. At a set of later sampling times, a random sample is chosen and the numbers of marked members are recorded; previously unmarked members might also be marked before the sample is returned to the population. It is assumed that at each of the sampling times:

- each member of the population is equally likely to be chosen;
- each marked member present after a sampling time is equally likely to survive to the next sampling time;
- no markings are lost or become impossible to identify;
- each sample is obtained and released instantaneously.

Using this approach we can now allow for possible additions and removals from the population by means of births, deaths and immigrations. As noted above, marking takes place at the original capture but can also (or might need to) take place on subsequent recaptures.

The JS model allows us to estimate the population size at each sampling time (effectively using Petersen-type estimators) and to estimate the rates of survival (in terms of net additions and removals) between sampling times. In its basic form the JS model does not enable us to distinguish immigration from birth and death. Early formulations of the approach are described by Jolly (1965) and Seber (1965), while that of Darroch (1959) only allows birth and immigration.

More sophisticated ways of handling open populations are manifold and allow, *inter alia*, for short-term effects of marking (Brownie and Robson, 1983) and for survival and capture probabilities to depend on past captures or on age (Robson, 1969; Pollock, 1975, 1981). These are just some of the prospects for extending the approach to accommodate more complex assumptions. To find out about other forms of extended models and to examine computational and design implications, see the reviews by Pollock (1991) and Seber (1992), as well as the work to which they in turn refer.

6.3   TRANSECT SAMPLING

We will now consider a distinct class of sampling methods also developed principally for biological applications, with the aim of estimating the density, or the number, of a species of animal, fish or plant (say) distributed over a geographic region.

In its simplest form, known as *line-transect sampling*, a line is drawn at random across a search region and the objects (be they tigers or tiger lilies) are sought by moving along the line and noting how many of the target objects are observed as one goes from one end of the line to the other. Several lines may be drawn and the sample data accumulated from traversing all the lines.

Estimation of the abundance of animals or plants is known to be difficult and time-demanding, and line-transect sampling is as efficient and effective an approach as is likely to be found, for most types of problem. Often the search is restricted to a *strip* of equal width either side of the line: hence *strip-census methods* (Hayne, 1949; Robinette et al., 1974). The statistical sampling methods are inevitably based on assumptions which are major simplifications of the real-life circumstances (Eberhardt, 1968). For example, the search objects are assumed (in the simplest applications):

- to be stationary (i.e. not moving);
- to be similarly and independently able to be observed;
- to be seen at right angles to the transect line;
- to be seen on one occasion only,
- to be unaffected (e.g. neither repelled nor attracted) by the observer.

Of course, these may not all hold in practice, but methodology developed on such assumptions should at least provide a reasonable initial basis for sampling and estimation.

Further reviews of and references to work on line-transect methods include Thompson's (2002) detailed treatment which includes distance sampling for animal abundance), Seber (1982, 1986, 1992) on surveys of animal abundance estimation, Ramsey et al. (1988) which includes reference to marine applications, Burnham et al. (1980) on density estimation for biological applications and Burnham et al. (1981) which gives special attention to birds.

Other possible modifications include the prospects of:

- observing objects other than at right angles to the transect line; we may see them ahead of us (see Section 6.3.4 for further details);
- taking into account the fact that larger objects may be more visible than smaller ones, at any specific distance (we will shortly consider such a prospect and examine its implications further also in Section 6.3.4);

- deciding to take observations in all directions *from a single point* rather than by traversing a transect line.

The latter approach is known as *point-transect sampling*. Several points may be chosen and a period of time is then spent at each point recording all observed objects of the type sought. This can be a particularly useful approach for birds and for elusive animals. Some relevant reviews include Ramsey and Scott (1979), Burnham *et al.* (1980), Buckland (1987) and Quang (1993). See Section 6.3.5 below for more details.

The use of line-and point-transect methods can operate on various scales – from an aerial survey of an African National Park at the one extreme, to minute inspection of plant species at the other. The plant species survey might be approached by means of quadrat sampling (see Section 6.1), with quadrats thrown from the line or point.

The review of *distance sampling methods* by Buckland *et al.* (2002) describes many variations on the transect sampling theme, including covariate models for aiding detectability, recording evidence of presence rather than the subject itself, and *double-platform methods* where line-transect methods may be combined with mark–recapture or capture–recapture (as discussed in Section 6.2).

Many of the published accounts of line-and point-transect sampling present real-life or experimental data to illustrate their results. Otto and Pollock (1990) describe an interesting set of artificial data where two transect lines of lengths $l_1 = 200$ m and $l_2 = 160$ m were used. For each line, a number r of different observers looked for objects of interest (in this case, beer-can clusters) of increasing levels of visibility (1, 2, 4 or 8 cans) placed within 20 m of the line. In fact, $r = 9$. There were about 30 objects of each of the four cluster sizes around each transect line. The data shown in Table 6.1 and Figure 6.1 for the 200 m transect are a useful resource to illustrate estimation methods and detectability characteristics in the discussion of the following subsections. For example, it is clear from Figure 6.1 (and intuitively expected) that more of the larger clusters are observed than of the smaller, and out to further distances. The implicit notion of *detectability* underlying this effect will prove to be important.

6.3.1 The simplest case: strip transects

Suppose a line transect of length l is chosen at random, extending over an observation region which contains individual specimens of some object of interest distributed at random at *rate* λ (or with *density* δ). We assume for the moment that all objects will be sighted if they lie within a strip of width w either side of the transect line, and that no objects outside the strip will be sighted. Thus if we observe n objects, our estimate of the density δ of the objects over the region is

$$\tilde{\delta} = \frac{n}{2wl}. \tag{6.9}$$

Table 6.1   The Otto and Pollock (1990) beer-can data for transect 1. For each observed cluster of cans, the perpendicular distance from the transect line is presented, together with how many of the nine observers sighted it.

			Group size				
1		2		4		8	
Dist.	Seen by	Dist.	Seen by	Dist.	Seen by	Dist.	Seen by
0.11	8	0.93	9	0.88	9	0.14	9
2.18	9	1.11	9	1.77	9	0.26	9
2.19	9	1.33	9	3.35	9	0.30	9
3.51	1	1.69	9	4.35	9	0.74	9
3.56	0	2.31	9	5.98	0	1.12	9
3.75	9	4.15	8	7.24	0	2.68	9
4.05	8	4.26	1	8.56	0	2.91	9
4.09	8	4.35	9	8.84	0	4.13	8
4.10	2	5.24	1	8.97	8	5.58	7
5.16	1	5.28	8	9.58	7	6.03	6
7.26	0	5.48	2	9.61	8	7.27	4
7.34	0	5.87	0	10.34	0	7.63	9
7.80	0	7.20	6	10.88	0	7.73	0
8.00	0	7.31	7	10.98	0	7.90	9
8.42	1	9.52	0	11.40	0	8.35	1
8.70	0	9.53	0	12.19	0	8.51	9
8.82	0	11.59	0	12.57	0	9.20	0
9.28	0	11.69	0	12.91	0	9.69	2
9.75	0	11.78	0	13.69	0	11.79	1
10.46	0	12.80	0	13.95	0	11.88	0
10.54	0	13.03	0	15.99	0	13.04	0
12.17	0	13.63	0	16.20	0	13.49	0
12.74	0	14.54	0	16.54	0	14.60	0
13.85	0	15.59	0	16.63	0	14.80	0
14.57	0	15.86	0	16.72	0	14.86	0
14.75	0	16.20	0	16.94	0	17.37	0
15.27	0	16.44	0	17.13	0	17.68	0
15.28	0	17.01	4	17.15	0	17.82	0
15.75	0	17.44	0	17.90	0	18.84	0
17.94	0	18.57	0	18.00	0	18.94	0
18.89	0	18.94	0	18.04	0	18.96	0
18.91	0	19.59	0	18.25	0	19.26	0
19.75	0	19.63	0	18.71	0	19.64	0

If the observation region has area A, we can estimate the total number of objects in the region as

$$\tilde{T} = \frac{An}{2wl}. \tag{6.10}$$

This approach is somewhat artificial (in assuming perfect sighting within the strip) and likely to be inefficient (in assuming that no objects are sighted outside

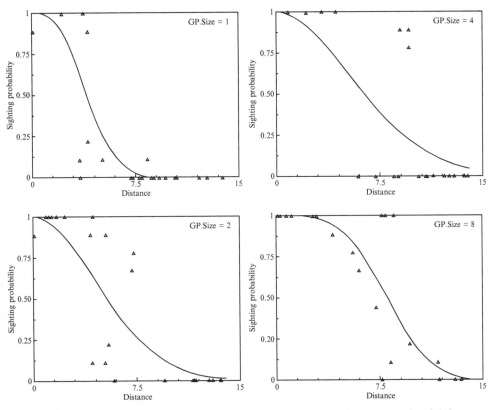

Figure 6.1 Plots of beer-can data showing empirical exponential power series sighting function versus perpendicular distance for each group size on transect 1. The triangles give the proportion of observers who sighted each object (Otto and Pollock, 1990).

the strip or in ignoring any that are sighted outside the strip). Also, how in practice are we to determine the strip width w?

We could try dealing with these matters conceptually by considering *all sightings* irrespective of distance and drawing a histogram of the numbers of sightings at the different distances.

If, as shown in Figure 6.2, there is a distance w_0 at which sightings effectively tail off, we might choose to act as if we had an 'effective strip' or 'equivalent strip' of width $2w_0$ in which we observed all objects and then use the count of the sightings within the distance w_0 as if they arose from the 'strip' in the method above. We would effectively assume that no other sightings had been made – estimating δ or T as shown above in (6.9) or (6.10), respectively

Example 6.2   Consider the beer-can data of Table 6.1 for group size 1 (see also Figure 6.1). Essentially all objects out to a distance 4.1 m are sighted, and very few are seen beyond this distance. The 'detectability' seems to be more or less constant (at unit value) out to $y = 4.1$ and to decay rapidly to zero beyond this

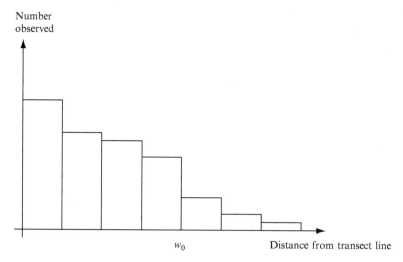

Figure 6.2   Histogram of all sightings.

point. Thus, although we do not have a strip-transect sample, it is as if we did and with $w_0 = 4.1$. So we have an 'equivalent strip' of width $2w_0 = 8.2$ and we can estimate δ as

$$\tilde{\delta} = \frac{n}{2w_0 l_1 r} = \frac{54}{1640 \times 9} = 0.00366/\text{m}^2,$$

where r is the number of independent replicated attempts to find the objects (here $r = 9$).

It will clearly be important for a more rigorous and general approach to recognize also that objects which are present may or may not be observed and that this prospect may depend on their distance y from the line, governed perhaps by a *detectability function* $g(y)$ which takes the value $g(0) = 1$ (perfect sighting on the line) and where $g(y) \to 0$ as $y \to \infty$.

This leads to a broader concept of 'equivalent strip' which will be introduced in the following Section 6.3.2.

6.3.2   Using a detectability function

To develop a more formal approach (as discussed, for example, by Burnham and Anderson, 1976), we must start by declaring a set of assumptions under which we will operate. We will again assume that

- objects are distributed at random rate λ over a region of area A;
- we take observations perpendicularly on either side of a randomly placed transect line of length l.

Whether or not objects which are present are actually observed is governed by the detectability function

$$g(y) = P(\text{object observed}| \text{ object is at distance } y)$$

with $g(0) = 1$ (so that all objects on the line will be observed).

We again concentrate on a strip of width w on either side of the line. Any object in the strip will have a perpendicular distance y from the line. We will assume that y is an observation of a random variable Y which is uniformly distributed on $(0, w)$. There is an important implicit assumption here: namely, that there is some distance w from the line beyond which objects will not be encountered, even though they are assumed to be uniformly distributed within this distance. Such a value w is purely notional and is introduced to aid the theoretical development. In practice, either we find that w does not enter explicitly into the development (as in the estimation of δ below) or we will be concerned only with what happens as w increases indefinitely in size.

If there are n_w objects in the strip, n_w is an observed value of a random variable N_w with $E(N_w) = 2wl\delta$. But we will be unlikely to *observe* all n_w objects which are in the strip. In fact, we have

$$P(\text{observe object which is in strip}) = P_w = \int_0^w \frac{g(y)}{w}\,dy = E[g(Y)] \quad (6.11)$$

So if

$$\mu_w = \int_0^w g(y)dy, \quad (6.12)$$

then

$$P_w = \frac{\mu_w}{w}. \quad (6.13)$$

Suppose we actually observe n objects at distances $z_1, z_2, \ldots, z_n$. These are the Y-values of the *observed objects*. (Other objects, with their Y-values, may be in the strip but not observed.) For the random variable N (of which we have the observed value $N = n$) we have

$$E(N) = E_{N_w}\{E(N|N_w)\} = E_{N_w}\{P_w N_w\} = 2wl\delta P_w = 2l\delta\mu_w. \quad (6.14)$$

So, given n, we can estimate δ as

$$\tilde{\delta} = \frac{n\tilde{m}_w}{2l}, \quad (6.15)$$

where $\tilde{m}_w$ is an estimate of μ_w^{-1}.

We can use the actual distances $z_1, z_2, \ldots, z_n$ of the observed objects to estimate μ_w^{-1}. Let them be observations of a random variable Z with d.f.

$$F(z) = P(Z \le z) = P(Y \le z | \text{object observed})$$
$$= \frac{P(Y \le z \text{ for an observed object})}{P(\text{object observed})}$$
$$= \int_0^z \frac{g(y)}{w} \, dy \bigg/ \int_0^w \frac{g(y)}{w} \, dy$$
$$= \int_0^z \frac{g(y)}{w} \, dy / \mu_w$$

So the p.d.f. of Z is just $g(z)/\mu_w$ (for $0 \le z \le w$) with $g(0) = 1$, and hence

$$f(0) = \frac{1}{\mu_w}. \tag{6.16}$$

So from (6.15) for $\tilde{\delta}$ we can estimate δ as

$$\tilde{\delta} = \frac{n\tilde{f}(0)}{2l}, \tag{6.17}$$

and we have only to estimate $f(z)$ (the p.d.f. of distances of observed points from the transect line), and in particular $f(0)$, to estimate δ. Equivalently, from the above form for $\tilde{\delta}$, it is as if we had a strip of width $1/f(0)$ in which all objects were observed. This is sometimes known as the *equivalent strip*.

We need now to consider possible ways of estimating $f(z)$ and specifically $f(0)$ – and to examine possible forms for the detectability function $g(y)$.

6.3.3   Estimating $f(y)$

Several approaches to estimating $f(y)$ have been discussed in the literature on line-transect sampling. Central to these is the need to attempt to choose some appropriate parametric family of distributions $f_\theta(z)$ for the observations $z_1, z_2, \ldots, z_n$ and then to apply a relevant detectability function $g(y)$.

The most basic model is one which declares that *all* objects which are present will in fact be observed up to a distance w, and *none* beyond that distance, so that

$$g(y) = 1 \quad (0 \le y \le w)$$

and, from (6.11), $P_w = 1$ so that $\mu_w = w$. Here $f(z) = 1/w$ ($0 \le z \le w$) also, and we have $f(0) = w^{-1}$. Hence, we obtain $\tilde{\delta} = n/(2wl)$ as we originally obtained for the 'simplest case' discussed in Section 6.3.1 above; see (6.9).

But we would be better advised to seek to use the observed distances $z_1, z_2, \ldots, z_n$ to estimate the distribution of Z, and hence to estimate δ by means of (6.17) involving $\tilde{f}(0)$. We could form a grouped relative frequency distribution (cf. Figure 6.2) and fit a distribution to it, testing goodness of fit in some appropriate way (e.g. by a χ^2 goodness-of-fit test) if the sample size is large enough.

More complex methods of density estimation are also available for estimation of $f(0)$ and have been widely explored and reviewed. The following are works which deal with various aspects of this issue: Gates *et al.* (1968), Eberhardt (1968), Sen *et al.* (1974), Pollock (1978), Burnham *et al.* (1980), Seber (1982), Buckland and Turnock (1992), Buckland *et al.* (1993, 2002) and Chen (1996), all specific to transect sampling; and Silverman (1986) for general coverage of density estimation.

One particular alternative approach uses isotonic regression analysis (Burnham and Anderson, 1976), in which $f(0)$ is estimated by

$$\tilde{f}(0) = \max_i \{i/nZ_{(i)}\}, \tag{6.18}$$

where the $Z_{(i)}$ are the order statistics of the observed distances.

Example 6.3 In a line-transect sample chosen from a transect line of length 100 m through a woodland, a total of 28 birds are observed at the following distances.

0	0	0	0	1	1	3	4	7	9	9	11	12	15
16	18	19	20	21	23	23	27	29	31	38	41	49	54

Let us consider three approaches to estimating the total number of birds.

If we group the sighting distances in intervals of width 10 m we obtain the following frequency distribution:

Distance (m)	0–9	10–19	20–29	30–39	40–49	50–59
Frequency (n_i)	11	6	6	2	2	0

It is clear that there is no obvious marked drop-off in the sightings to justify the choice of an effective finite strip width of $2w_0$ within which detectability is more or less constant and outside which few birds are observed.

If we were nonetheless to choose $w_0 = 30$, then we would conclude that 23 birds were observed in a strip of width 60 m, and (6.9) would give an initial estimate of δ as $\delta_1 = 0.0038/m^2$. How does this compare with other approaches? We might try fitting $f(y)$ by an exponential distribution with mean estimated by $\bar{x} = 480/28 = 17.14$. This yields expected frequencies of 12.4, 6.9, 3.9, 2.2, 1.2 and 1.5 which are not inconsistent with the exponential model. Thus we obtain $\tilde{f}(0) = 0.0583$ and we now estimate δ by $\tilde{\delta}_2 = 28 \times 0.0583/200 = 0.0082/m^2$, a much higher rate than our primitive estimate above. We could also use the monotone regression approach just described, in which case we would obtain an estimate $\tilde{f}(0) = 0.0689$, giving $\tilde{\delta}_3 = 0.0096/m^2$.

With such a small number of sightings we cannot expect very precise estimates of δ. We might not expect $\tilde{\delta}_1$ to be appropriate under the circumstances, but $\tilde{\delta}_2$ and $\tilde{\delta}_3$ are reasonably similar, implying a density of about 1 bird per 100 m^2.

The expression $\tilde{\delta} = n\tilde{f}(0)/(2l)$ is the basic form for estimating δ and attempts have been made to study the properties of $\tilde{\delta}$. Burnham and Anderson (1976) derive an approximation to the large-sample variance, $V(\tilde{\delta})$, which they show is of order l^{-1}. For the situation where Var $[\tilde{f}(0)] \sim \sigma^2/n$, they obtain an explicit approximation for $V(\tilde{\delta})$ in the form

$$V(\tilde{\delta}) = \frac{\delta(1 + \mu^2\sigma^2)}{2\mu l}, \tag{6.19}$$

where $\mu = \lim_{w \to \infty} \mu_w$.

6.3.4   Modifications of approach

There are many possible departures from the basic assumptions outlined in the introduction to this section. In particular, objects may be sighted 'ahead' rather than perpendicular to the transect line. In this case we would have observed values (r, θ) for distance and angle of sighting for an object whose perpendicular distance from the line is y. Thus we observe (r, θ) rather than y and will need to use a detectability function $g(r, \theta)$, which reflects this added dimensionality (see Figure 6.3).

Then again, objects at a given distance may have different probabilities of being observed due to their physical characteristics; for example some may be larger or more brightly coloured than others. We would then observe a sighting distance y (or (r, θ)) and a 'physical' or size measure, ϕ. This situation was illustrated in Example 6.1.

Thus line-transect data may be multivariate. Observations may consist of pairs (r, θ) for sighting ahead – or (y, ϕ) for perpendicular sighting and size effects – or even triples (r, θ, ϕ) – sighting ahead and size effects. Burnham and Anderson (1976) present some results for the (r, θ) sighting-ahead situation, and this has been extensively followed up by others; see for example, Buckland *et al.* (1993).

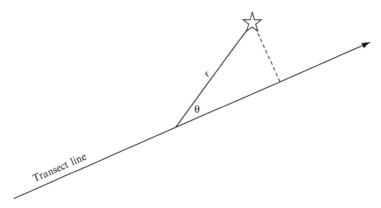

Figure 6.3 'Sighting ahead'; object at distance r and at angle θ.

The problem of differential sighting probabilities at any given distance is referred to by Chen (1996) as 'school size effects', presumably by analogy with sampling schools of fish. Indeed, Chen refers to a satellite-operated Global Positioning System (GPS) which is used to detect schools of tuna in South Australia where in the summer the schools bask close to the surface and GPS can measure perpendicular distance from a transect line whilst spotters in a light aircraft also estimate school size s (in tonnes). Larger schools are clearly easier to spot than smaller ones and this needs to be taken into account in the estimation process, by using a *bivariate* detection function $g(y,s)$. Chen (1996) examines two kernel smoothing methods (with fixed and varying bandwidths) for fitting the bivariate line-transect data (y,s) for the joint purposes of estimating school abundance and mean school size. The results are illustrated for an interesting data set on the aerial survey of Australian southern bluefin tuna, containing $n = 119$ schools at distances up to 10 miles from a transect line of length in excess of 2750 miles. The mean school size was 115.5 tonnes.

Drummer and McDonald (1987) consider different parametric models for $g(y,s)$, whilst Buckland *et al* (1993) and Quang (1991) also explore this size-effect problem by regressing a distance-based detectability function on the size measure, and by using nonparametric Fourier methods, respectively.

6.3.5 Point transects or variable circular plots

Instead of observing objects whilst traversing a transect line, we might choose to take observations in all directions whilst stationary at a point (possibly repeating this for many points). This process is known as a *point-transect* or *variable circular plot* survey. Similar general principles apply to those examined for line transects, but the details will be somewhat different.

The point-transect approach may be more convenient in certain applications. This could apply, for instance, in bird surveys where observing in all directions from a fixed 'hide' may provide less disruption, especially in rough terrain where traversing a line transect might frighten away all prospective sightings.

Again we will need to postulate a detectability function which in its simplest form, $g(r)$, will depend only on radial distance r of an object from the observer in the region and not on its orientation from the observation point. (A more complex formulation involving orientation with detectability function $g(r, \theta)$ could be developed, but it is difficult to contemplate how specific forms might be modelled in the real-life situation.)

Thus suppose we have detectability function $g(r)$ and observe n objects at distances $z_1, z_2, \ldots, z_n$.. Again, others may be present, but are not observed. So we have for any observed object (see the discussion following (6.15) in Section 6.3.2),

$$F(z) = P(Z < z) = \int_0^z \frac{g(r)\mathrm{d}r}{w} \bigg/ \int_0^\rho \frac{g(r)\mathrm{d}r}{w} = \int_0^z g(r)\mathrm{d}r/\mu_\rho, \qquad (6.20)$$

where ρ is the disc radius within which, by analogy, an object at distance R is assumed to be uniformly distributed and, again by analogy with the earlier results,

$$\mu_\rho = \int_0^\rho g(r)\mathrm{d}r.$$

Relevant methodological and applications-oriented results on point transects can be found in Ramsey and Scott (1979), Buckland (1987), Ramsey et al. (1987) and Quang (1993).

Returning to general issues in transect sampling, it is important to draw attention to an operational problem: namely, how we should in practice choose the transect line or transect points over a study region. This is an aspect of sampling design. Some interesting preliminary comments and procedures on this theme are described by Thompson (2002, pp. 210–216) who also extends the discussion into more detailed study of the effects of imprecise information on detectability.

6.4 ADAPTIVE SAMPLING

A common element in almost all approaches to sampling is the need to ensure that successive observations are taken independently of each other and (usually) by the same basic selection principle. Thus, for example, in simple random sampling (Section 4.1) a first sample member is chosen at random from the population, a second at random from those remaining, and so on. This ensures 'fair play': it makes the resulting sample interpretable in probabilistic terms, and we can readily assess the statistical properties of the resulting inferences.

And yet there are occasions when it seems tempting to depart from such an even-handed approach and to follow our instinct, especially when we are trying to find somewhat elusive population members. In a transect sampling approach to observing woodland mammals, the randomly chosen transect line through a wood may not seem the obvious best choice, especially if it runs across the plain where ramblers commune with nature! Might it not be better to go through the hilly area at the edge of the wood and if we find one of our subjects of interest continue to look in the same area (if the mammal is gregarious) or move away to another wooded area (if the mammal is solitary in nature). Either way, we could let the current outcome of our sampling scheme colour its subsequent progress and this might be advantageous. This is sometimes done, and the obvious example of such a principle of sampling is to be found in what is termed *adaptive sampling*.

An informative coverage of adaptive sampling methods, with emphasis on sampling wildlife, is provided by Thompson and Seber (1996). Buckland et al. (2002) refer to *adaptive distance sampling* in which line- or point-transect surveys are carried out adaptively.

6.4.1　Simple models for adaptive sampling

If we are to depart from strict randomness in sampling, we will still need to be able to assess the statistical and probabilistic properties of any estimators. This requires the formulation of some model to represent both the essentially non-random deployment of the individuals in the population and the sampling approach to them. Figure 6.4 shows one possible scheme for adaptive sampling: suppose a woodland has a superimposed square grid, in the cells of which we will search for small mammals. A simple random sample of n cells is chosen without replacement. Suppose the chosen cells are those shown as A, B, C, F, ... Some of these cells (e.g. A, B and C) do not contain one of the mammals but others do. Suppose cell F is a 'hit' (and contains 2 of our specimens). We now depart from randomness and search the *neighbouring cells* F_1, F_2, F_3 and F_4, finding 0, 2, 1 and 3, respectively. We then continue to search all cells neighbouring those *where we have found specimens*, but only F_{31} now provides a 'hit' (just one of our specimens) and the process terminates. We now move on, and again conduct random searches of the initially chosen cells until we find another cluster of specimens.

　　This is a basic form of adaptive sampling which is commonly modelled under the label of *adaptive cluster sampling*. Essentially, an initial random sample of n units is chosen (with or without replacement) from the population of N units. Our interest is in estimating some population characteristic of a variable Y: perhaps its mean $\overline{Y}$ or total Y_T. The sampling scheme operates as follows. It is

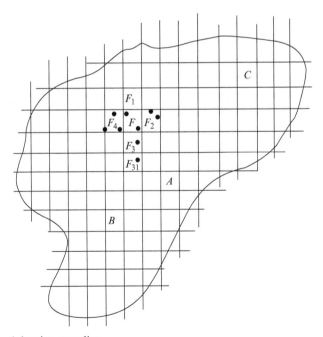

Figure 6.4　Adaptive sampling.

assumed that any unit *i* lies in a defined neighbourhood of units. If the observation y_i satisfies some predefined condition, then the sample is augmented by all other units in its neighbourhood. If any of the selected neighbourhood units satisfies the condition, the sample is further augmented by the units in its neighbourhood, and so on.

How does this work out in our woodlands mammals example above? Here the units can be defined as the geographic regions defined by the grid squares. The initial sample is taken without replacement. The variable Y will be the number of relevant mammals in a grid region. The unit neighbourhood is the set of four grid squares sharing an edge with the chosen unit. The condition will be that $Y \geq 1$, that is, the square contains at least one of the mammals. This is clearly in the mould of the more general form of adaptive cluster sampling just described.

Consider the position of any unit *i*. This will be included in the sample if it is chosen in the initial random sample of size *n*; it will also be in the sample if it is in any neighbourhood ('network') of a selected unit (including unit *i* itself) which satisfies the occupancy condition. Thus the probability of inclusion π_i of any unit *i* in the chosen sample will vary with the relationship between the clustering process and the conformance with the occupancy condition.

If these inclusion probabilities were known for all the units included in the sample, we could fruitfully employ the Horwitz–Thompson estimator (see Barnett, 2002a, pp. 53–54) – but they will typically not be known. Nonetheless some of the advantages of the Horwitz–Thompson approach can be realized if we adapt it to apply to a reduced sample which only contains observations not satisfying the condition when they happen to have been contained in the initial sample of *n* observations. Thompson (2002, Chapter 24) gives details of this approach and demonstrates that such adaptive cluster sampling is more efficient than simple random sampling provided that the within-neighbourhood variance of Y is large enough. Thompson (2002, Chapter 25) and Thompson and Seber (1996) develop more sophisticated models and methods for adaptive sampling, including multistage and transect sampling, and clearly show the advantages it can confer for clustered populations.

PART III

Examining environmental effects: Stimulus–response relationships

CHAPTER 7

Relationship: regression-type models and methods

Like father, like son.

In environmental studies we frequently need to examine the links between variables or outcomes. For example, how does incidence of cancer relate to levels of radioactive contamination around a contaminated site? What levels of building damage can be anticipated from extreme weather conditions? How are fish stocks in the North Sea likely to decrease over the coming 10 years? In all cases we have a *stimulus variable*, sometimes called a *predictor variable* or *regressor variable* (contamination level, values of weather variables, year of study) and a *response variable* or *outcome variable* (incidence rate of cancer, level of building damage, level of fish stocks). The response relates to the stimulus and we need to examine different possible forms of model (sometimes called *stimulus–response models*), and corresponding inference methods, to express and explore this relationship.

Any relationship of interest in environmental studies will, typically, not be a *deterministic* one except in the most trivial cases; it must involve uncertainty and indeterminism. It will tend to be *associative* rather than *causal* (in the sense that the stimulus goes along with, rather than causes, the response). There is an apochryphal tale of how the number of births in villages in a small European country is almost proportional to the number of storks nesting in the village. No one suggests the storks cause the births.

But causality is a crucial issue. If lung cancer incidence in cigarette smokers can be seen to relate to the numbers of cigarettes smoked each day, does this mean that smoking *causes* cancer or just that both increases reflect some *jointly concomitant* variable? Is there separate medical evidence of a causal link? Thus whilst some observed relationships are causal, and some merely associative, the question of demonstrating causality beyond mere association is difficult and can be vitally important. More detailed discussion of these matters can be found in Cox (1992), Stone (1993) and Barnett (1999a, Section 8.7), which also gives further references.

Thus, we suppose that we have a stimulus variable X with typical value x and a response variable Y with typical value y. We might have scalar X and Y, or they could be multivariate. A common situation is one where X is multivariate and Y is scalar. In special cases X may measure time or spatial characteristics,

Environmental Statistics   V. Barnett
© 2004 John Wiley & Sons, Ltd   ISBN: 0-471-48971-9 (HB)

in which cases the models and methods are more appropriately described under the headings of *time series* or *spatial models* (see Chapters 10 and 11, respectively).

A number of texts are concerned with model construction and validation, often with emphasis on *linear* models. These include Box and Draper (1987), on empirical model-building and response surfaces, Linhart and Zucchini (1986) on model selection, and Stapleton (1995) on linear statistical models. We cannot seek to emulate the extensive treatments of such full texts; our aim here is just to reveal the 'bones' of the problem of model and relationship in its linear and non-linear forms and to indicate the related basic methodological, exploratory and transformation procedures.

Our model will typically declare that

$$H(Y) = G(X) \tag{7.1}$$

where $H(Y)$ and $G(X)$ are some appropriate functions (we will shortly examine what is 'appropriate') and where X and Y may enter as random variables with their full variational characteristics or may be constrained in some way. A typical example covers the wide-ranging *regression models* which are conditional on a particular observed value x and where

$$E(Y \mid x) = G(x). \tag{7.2}$$

Models of this kind are often completed by stating that

$$Y = G(x) + \varepsilon,$$

where ε is a random variable (the *error* or *residual*) with a prescribed, or modelled, form possibly involving unknown parameters. The nature of the model can spring from some general formulation; for example, for bivariate $(X, Y) \sim N(\mu, \Sigma)$ it is easily confirmed that the regression model for Y is *linear with normal error structure*. Thus we have

$$Y = \alpha + \beta x + \varepsilon, \tag{7.3}$$

where ε is normally distributed with zero mean and constant variance.

Alternatively, we may choose to employ the relationship (7.3) as an *empirical model* even without the support of a bivariate normal structure and possibly without making any assumptions about the distributional form of ε. This is so for the class of linear models familiarly examined by *least-squares* methods.

7.1 LINEAR MODELS

We start by considering a range of *linear regression models* (and more general linear models) applied to a continuous response variable Y and a possibly multivariate stimulus variable X. We will assume that

$$y_i = \alpha_1 x_{1i} + \alpha_2 x_{2i} + \ldots + \alpha_p x_{pi} + \varepsilon_i \qquad (i = 1, \ldots, n)\ (n > p) \tag{7.4}$$

where the x_{ji} are the *ith* observed levels ($i = 1, \ldots, n$) of p stimulus variables and ε_i is a random error (or residual) variable. The ε_i might be assumed independent with constant variance (or to have more structure with differing variances, and covariances, and perhaps some specified distributional form, e.g. normal).

Note that if we take $x_{1i} \equiv 1$, we have a model with a constant intercept α_1. Our interest will be in estimating the *parameters* $\alpha_1, \alpha_2, \ldots, \alpha_p$, and their variational or distributional characteristics.

We can express (7.4) in matrix form as

$$\mathbf{y} = \mathbf{X}\boldsymbol{\alpha} + \boldsymbol{\varepsilon}, \tag{7.5}$$

where $\boldsymbol{\alpha}' = (\alpha_1, \alpha_2, \ldots, \alpha_p)$ is the vector of parameters, $\boldsymbol{\varepsilon}$ is a vector of residual errors and

$$\mathbf{X} = \begin{pmatrix} x_{11} & x_{12} & \cdots & x_{1p} \\ x_{21} & x_{22} & \cdots & x_{2p} \\ \cdots & \cdots & \cdots & \cdots \\ x_{n1} & x_{n2} & \cdots & x_{np} \end{pmatrix} \tag{7.6}$$

is of full rank p and is sometimes known as the *design matrix* (by analogy with the use of linear models in *experimental design*).

As expressed, the model (7.5) is not fully specified distributionally; we would need to make some assumptions about $\boldsymbol{\varepsilon}$, at least to declare that it consists of independent components of zero mean and constant variance. In this latter case, the variance–covariance matrix for $\boldsymbol{\varepsilon}$ would take the form

$$V(\boldsymbol{\varepsilon}) = \mathbf{I}_n \sigma^2, \tag{7.7}$$

where $\mathbf{I}_n$ is the $n \times n$ identity matrix. This would be the most basic *linear model*. Traditionally analysed by the so-called *method of least squares*, it has long been widely applied to practical problems, with antecedents in the work of Laplace and Gauss. We will examine this in more detail shortly.

The next stage of complexity would involve relaxing the assumption of independent components of $\boldsymbol{\varepsilon}$ and assuming instead that the components may be correlated and have differing variances. In this case and where, as is often feasible, we can specify the variance–covariance matrix of $\boldsymbol{\varepsilon}$ in the more general form

$$V(\boldsymbol{\varepsilon}) = \mathbf{V}\sigma^2, \tag{7.8}$$

for a prescribed $\mathbf{V} \neq \mathbf{I}_n$, we have what is called the *extended linear model* and can employ the *extended least-squares method* to estimate $\boldsymbol{\alpha}$ and σ^2.

Finally, in terms of complexity, we might specify a distributional form for $\boldsymbol{\varepsilon}$, such as multivariate normal, and we would then have a *normal linear model* to which we can also apply more formal estimation methods such as the principle of *maximum likelihood*.

We will now examine the linear model defined by (7.5) and (7.7), the extended linear model defined by (7.5) and (7.8), and the normal linear model where we add the assumption of normality of error structure, in more detail.

7.1.1 The linear model

For the linear model (7.5) with error structure (7.7), the method of least squares proposes that we estimate $\boldsymbol{\alpha}$ and σ^2 by minimizing the sum of squares of 'residuals' (deviations, errors)

$$R = \sum \varepsilon_i^2 = \boldsymbol{\varepsilon}'\boldsymbol{\varepsilon} = (\mathbf{y} - \mathbf{X}\boldsymbol{\alpha})'(\mathbf{y} - \mathbf{X}\boldsymbol{\alpha}).$$

This can be re-expressed as

$$\begin{aligned} R = &[\mathbf{y} - \mathbf{X}(\mathbf{X}'\mathbf{X})^{-1}\mathbf{X}'\mathbf{y}]'[\mathbf{y} - \mathbf{X}(\mathbf{X}'\mathbf{X})^{-1}\mathbf{X}'\mathbf{y}] \\ &+[\mathbf{X}'(\mathbf{X}'\mathbf{X})^{-1}\mathbf{X}'\mathbf{y} - \mathbf{X}\boldsymbol{\alpha}]'[\mathbf{X}'(\mathbf{X}'\mathbf{X})^{-1}\mathbf{X}'\mathbf{y} - \mathbf{X}\boldsymbol{\alpha}], \end{aligned} \tag{7.9}$$

in which we note that the first term does not involve $\boldsymbol{\alpha}$, whilst the second term is non-negative and is minimized at the value zero (thus minimizing R overall) if we set

$$\boldsymbol{\alpha} = \tilde{\boldsymbol{\alpha}} = (\mathbf{X}'\mathbf{X})^{-1}\mathbf{X}'\mathbf{y}. \tag{7.10}$$

Thus $\tilde{\boldsymbol{\alpha}}$ is the *least-squares estimator* of $\boldsymbol{\alpha}$.

We can readily establish that $\tilde{\boldsymbol{\alpha}}$ is unbiased and determine its variance–covariance matrix. We have

$$\mathrm{E}(\tilde{\boldsymbol{\alpha}}) = (\mathbf{X}'\mathbf{X})^{-1}\mathbf{X}'\mathrm{E}(\mathbf{y}) = (\mathbf{X}'\mathbf{X})^{-1}\mathbf{X}'\mathbf{X}\boldsymbol{\alpha} = \boldsymbol{\alpha},$$

thus confirming that $\tilde{\alpha}$ is *unbiased*. Furthermore, using standard properties of matrices, the variance–covariance matrix of $\tilde{\boldsymbol{\alpha}}$ can be shown to be

$$V(\tilde{\boldsymbol{\alpha}}) = (\mathbf{X}'\mathbf{X})^{-1}\mathbf{X}'V(\mathbf{Y})\mathbf{X}(\mathbf{X}'\mathbf{X})^{-1} = (\mathbf{X}'\mathbf{X})^{-1}\sigma^2. \tag{7.11}$$

But how good an estimator is $\tilde{\boldsymbol{\alpha}}$? We should note that it is entirely specific to the set of sample observations $\mathbf{y}$ and takes no account of any distributional characteristics of the underlying random variable $\mathbf{Y}$.

Nonetheless, we can show that $\tilde{\boldsymbol{\alpha}}$ has minimum possible variance in the class of linear unbiased estimators; it is what is called the *minimum-variance linear unbiased estimator* (MVLUE).

Suppose $\boldsymbol{\alpha}^* = \mathbf{B}\mathbf{y}$ is some other unbiased estimator of $\boldsymbol{\alpha}$. Then

$$\boldsymbol{\alpha} = \mathrm{E}(\mathbf{B}\mathbf{y}) = \mathbf{B}\mathrm{E}(\mathbf{y}) = \mathbf{B}\mathbf{X}\boldsymbol{\alpha},$$

so that $\mathbf{B}\mathbf{X} = \mathbf{I}$ and

$$V(\boldsymbol{\alpha}^*) = \mathbf{B}\mathbf{B}'\sigma^2 = (\mathbf{X}'\mathbf{X})^{-1}\sigma^2 + (\mathbf{B} - (\mathbf{X}'\mathbf{X})^{-1}\mathbf{X}')(\mathbf{B}' - \mathbf{X}(\mathbf{X}'\mathbf{X})^{-1})\sigma^2. \tag{7.12}$$

But the right hand term of (7.12) has diagonal elements which are squares and hence is non-negative. Thus the MVLUE must be that estimator which has variance–covariance matrix $(\mathbf{X}'\mathbf{X})^{-1}\sigma^2$, namely the least-squares estimator, $\tilde{\boldsymbol{\alpha}}$.

This is a limited form of optimality and is to be welcomed, but it leaves open the question of whether non-linear estimators might be more efficient.

Correspondingly, we can show that if we want to estimate some linear transformed version $\gamma = \mathbf{L}\alpha$ of α, then the least-squares estimator will be just $\mathbf{L}\tilde{\alpha}$ and this will also be the MVLUE of γ.

Example 7.1 Simple linear regression Earlier we considered (7.3) as the simple linear regression model which would enable us to examine the prospect of a straight-line fit to a set of n values of y and x. We can apply least-squares methods directly with

$$\alpha' = (\alpha, \beta) \quad and \quad \mathbf{X} = \begin{pmatrix} 1 & x_1 \\ 1 & x_2 \\ \vdots & \vdots \\ 1 & x_n \end{pmatrix},$$

so that

$$\mathbf{X}'\mathbf{X} = \begin{pmatrix} n & \sum x_i \\ \sum x_i & \sum x_i^2 \end{pmatrix}.$$

Thus

$$(\mathbf{X}'\mathbf{X})^{-1} = \begin{pmatrix} \sum x_i^2 & -\sum x_i \\ -\sum x_i & n \end{pmatrix} / \sum (x_i - \bar{x})^2 \quad and \quad \mathbf{X}'\mathbf{y} = \begin{pmatrix} \sum y_i \\ \sum x_i y_i \end{pmatrix}.$$

So

$$\tilde{\alpha} = (\mathbf{X}'\mathbf{X})^{-1}\mathbf{X}'\mathbf{y} = \begin{pmatrix} \sum x_i^2 \sum y_i - \sum x_i \sum x_i y_i \\ n \sum x_i y_i - \sum x_i \sum y_i \end{pmatrix} / \sum (x_i - \bar{x})^2$$

Thus we have

$$\tilde{\alpha} = \bar{y} - \bar{x} \sum (x_i - \bar{x})(y_i - \bar{y}) / \sum (x_i - \bar{x})^2,$$
$$\tilde{\beta} = \sum (x_i - \bar{x})(y_i - \bar{y}) / \sum (x_i - \bar{x})^2.$$

with variances

$$\sigma^2 \sum x_i^2 / \sum (x_i - \bar{x})^2 \quad and \quad n\sigma^2 / \sum (x_i - \bar{x})^2$$

respectively. These are the familiar results for classical linear regression analysis as will be found in any elementary statistics text; see, for example, Mood *et al.* (1974).

Before moving on to the extended linear model, let us consider the estimation of σ^2. The minimized residual sum of squares is seen from (7.9) to take the form $R_{min} = [\mathbf{y} - \mathbf{X}(\mathbf{X}'\mathbf{X})^{-1}\mathbf{X}'\mathbf{y}]'[\mathbf{y} - \mathbf{X}(\mathbf{X}'\mathbf{X})^{-1}\mathbf{X}'\mathbf{y}]$, which can be confirmed to have expected value $(n - p)\sigma^2$. Thus we can obtain an unbiased estimate of σ^2 as

$$S^2 = \frac{R_{min}}{n - p} = \frac{[\mathbf{y} - \mathbf{X}(\mathbf{X}'\mathbf{X})^{-1}\mathbf{X}'\mathbf{y}]'[\mathbf{y} - \mathbf{X}(\mathbf{X}'\mathbf{X})^{-1}\mathbf{X}'\mathbf{y}]}{n - p}. \tag{7.13}$$

7.1.2 The extended linear model

Now let us consider the case where the linear model (7.5) still holds but the error structure is more complex, reflecting dependencies (correlation and differing variances) between the component variables. Thus we now have $V(\boldsymbol{\varepsilon}) = \mathbf{V}\sigma^2$ (as in (7.8)), where $\mathbf{V} \neq \mathbf{I}_n$ takes a prescribed form. This is the *extended least-squares model*, and corresponding results can be obtained from those already derived using the method of least squares for the basic linear model. For this generalization to the case of dependent Y_i we use what is known as the *extended principle of least squares*, sometimes called, rather ambiguously, 'generalized least squares'.

We suppose that $y_1, y_2, \ldots, y_n$ are observations of $Y_1, Y_2, \ldots, Y_n$, where, as before, $E(\mathbf{Y}) = \mathbf{X}\boldsymbol{\alpha}$, $\mathbf{X}$ has full rank and we now assume that the variance–covariance matrix of $\mathbf{Y}$ takes the extended form $V(\mathbf{Y}) = \sigma^2 \mathbf{V}$, where $\mathbf{V}$ is a *known* positive definite symmetric $n \times n$ matrix.

Using standard results in matrix algebra (see, for example, Rao, 1973) we know that there exists a lower triangular matrix $\mathbf{T}$ such that $\mathbf{V} = \mathbf{T}\mathbf{T}'$. Suppose $\mathbf{T}$ has inverse $\mathbf{T}^{-1}$ and let $\mathbf{z} = \mathbf{T}^{-1}\mathbf{y}$ be an observation of the linearly transformed variable $\mathbf{Z} = \mathbf{T}^{-1}\mathbf{Y}$. Then $E(\mathbf{Z}) = \mathbf{T}^{-1}\mathbf{X}\boldsymbol{\alpha}$ and

$$V(\mathbf{Y}) = \mathbf{T}^{-1}\mathbf{V}(\mathbf{T}^{-1})'\sigma^2 = \mathbf{T}^{-1}\mathbf{T}\mathbf{T}'(\mathbf{T}^{-1})'\sigma^2 = \mathbf{I}\sigma^2.$$

Thus $\mathbf{Z}$ satisfies a linear model $E(\mathbf{Z}) = \mathbf{B}\boldsymbol{\alpha}$, where $\mathbf{B} = \mathbf{T}^{-1}\mathbf{X}$ and has uncorrelated components of constant variance. Thus, from the results above, the least-squares estimator of $\boldsymbol{\alpha}$ is

$$\tilde{\boldsymbol{\alpha}} = [\mathbf{X}'(\mathbf{T}^{-1})'\mathbf{T}^{-1}\mathbf{X}]^{-1}\mathbf{X}'(\mathbf{T}^{-1})'\mathbf{Z} = [\mathbf{X}'\mathbf{V}^{-1}\mathbf{X}]^{-1}\mathbf{X}'\mathbf{V}^{-1}\mathbf{Y}, \qquad (7.14)$$

with

$$V(\tilde{\boldsymbol{\alpha}}) = [\mathbf{X}'\mathbf{V}^{-1}\mathbf{X}]^{-1}\sigma^2. \qquad (7.15)$$

It is easily shown that this is the MVLUE. Equivalently, we can show that the estimator (7.13) arises from minimization of the *weighted residual sum of squares* $(\mathbf{y} - \mathbf{X}\boldsymbol{\alpha})'\mathbf{V}^{-1}(\mathbf{y} - \mathbf{X}\boldsymbol{\alpha})$.

Example 7.2 Suppose that $Y_1, Y_2, \ldots, Y_p$ is an independent random sample from a uniform distribution on $(0, \alpha)$ and we order them to obtain observed values of the order statistics $Y_{(1)}, Y_{(2)}, \ldots Y_{(p)}$. We want to estimate α from the ordered sample. This might be appropriate to the study of rounding errors that could arise in recording values of an environmental variable.

The p.d.f. of $Y_{(i)}$ is

$$f_i(y) = \frac{n!(x/\alpha)^i(1 - x/\alpha)^{n-i}}{(i-1)!(n-1)!},$$

and we can confirm that $E(Y_{(i)}) = i\alpha/(n+1)$.

Thus $E(\mathbf{Y}) = \mathbf{X}\alpha$, where $\mathbf{X}' = [1/(n+1), 2/(n+2), \ldots, n/(n+1)]$, and we can show that $V(\mathbf{Y}) = \mathbf{V}\alpha^2$, where

$$\mathbf{V} = (n+1)^{-1}(n+2)^{-2} \begin{bmatrix} n & (n-1) & (n-2) & \dots & 1 \\ (n-1) & 2(n-1) & 2(n-2) & \dots & 2 \\ (n-2) & 2(n-2) & 3(n-2) & \dots & 3 \\ \dots & \dots & \dots & & \dots \\ 1 & 2 & 3 & \dots & n \end{bmatrix}.$$

So the least-squares estimator of α is $(\mathbf{X}'\mathbf{V}^{-1}\mathbf{X})^{-1}\mathbf{X}'\mathbf{V}^{-1}\mathbf{Y}$ which will have variance $(\mathbf{X}'\mathbf{V}^{-1}\mathbf{X})^{-1}\alpha^2$. In fact,

$$\mathbf{V}^{-1} = (n+1)(n+2) \begin{bmatrix} 2 & -1 & 0 & \dots & 0 & 0 \\ -1 & 2 & -1 & \dots & 0 & 0 \\ 0 & -1 & 2 & \dots & 0 & 0 \\ \dots & \dots & \dots & & \dots & \dots \\ 0 & \dots & \dots & \dots & -1 & 2 \end{bmatrix}$$

So

$$\mathbf{X}'\mathbf{V}^{-1} = (0, 0, \dots, 0, (n+1)(n+2)) \quad and \quad \mathbf{X}'\mathbf{V}^{-1}\mathbf{X} = n(n+2).$$

Thus $\tilde{\alpha} = (n+1)Y_{(n)}/n$, with variance $\alpha^2/n(n+2)$; that is to say, we would estimate α from a somewhat enhanced largest sample value.

It is surprising how often we can express practical problems, and in the present context particularly environmental problems, in a straightforward linear model mould. A class of such problems relates to spatial interests (and we will pursue these more fully later; in Chapter 11) of which the simplest might involve concerns with surveying and with assessing shape and area. A simple illustration is given in the following example.

Example 7.3 In an environmental forestry survey we need to estimate the interior angles of a quadrilateral-shaped woodland copse (as shown below). Suppose the angles are $\alpha_1, \alpha_2, \alpha_3$ and $\alpha_4 = \pi - \alpha_1 - \alpha_2 - \alpha_3$, and we measure each of these independently but rather imprecisely on n occasions yielding observed mean values a_1, a_2, a_3 and a_4. We might assume that the individual measurements are independent with constant measurement error variances, σ^2. How would we estimate the true angles, $\alpha_1, \alpha_2, \alpha_3$ and α_4?

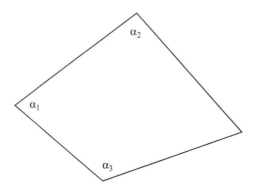

In fact, we again have a linear model. If $\mathbf{a}$ is the vector of transformed observations a_1, a_2, a_3 and $a_4' = \pi - a_1 - a_2 - a_3$ and $\boldsymbol{\alpha}$ is the vector of relevant parameters (i.e. α_1, α_2 and α_3), then

$$\mathbf{a} = \mathbf{X}\boldsymbol{\alpha} + \boldsymbol{\varepsilon},$$

where

$$\mathbf{X} = \begin{pmatrix} 1 & 0 & 0 \\ 0 & 1 & 0 \\ 0 & 0 & 1 \\ 1 & 1 & 1 \end{pmatrix} \quad \text{and} \quad V(\boldsymbol{\varepsilon}) = \mathbf{I}\sigma^2/n.$$

So

$$\mathbf{X'X} = \begin{pmatrix} 2 & 1 & 1 \\ 1 & 2 & 1 \\ 1 & 1 & 2 \end{pmatrix}, \quad (\mathbf{X'X})^{-1} = 1/4 \begin{pmatrix} 3 & -1 & -1 \\ -1 & 3 & -1 \\ -1 & -1 & 3 \end{pmatrix},$$

$$\mathbf{X'a} = \begin{pmatrix} a_1 + a_4' \\ a_2 + a_4' \\ a_3 + a_4' \end{pmatrix}.$$

Thus the least-squares estimator $\tilde{\boldsymbol{\alpha}}$ of $\boldsymbol{\alpha}$ is seen to be

$$\tilde{\alpha}_1 = 3a_1 - a_2 - a_3 + a_4'$$
$$\tilde{\alpha}_2 = -a_1 + 3a_2 - a_3 + a_4'$$
$$\tilde{\alpha}_3 = -a_1 - a_2 + 3a_3 + a_4'$$

with variance–covariance matrix

$$\begin{pmatrix} 3 & -1 & -1 \\ -1 & 3 & -1 \\ -1 & -1 & 3 \end{pmatrix} \sigma^2/n.$$

Finally, from (7.13), the least-squares estimate of σ^2/n is readily obtained as

$$(\mathbf{a} - \mathbf{X}\tilde{\boldsymbol{\alpha}})'(\mathbf{a} - \mathbf{X}\tilde{\boldsymbol{\alpha}})/(n - 3).$$

Suppose now that we are informed that the copse was originally set out in a diamond-shaped form, i.e. $\alpha_1 = \alpha_3$ and $\alpha_2 = \alpha_4$. We now have only one parameter, α_1, since $\alpha_3 = \alpha_1$ and $\alpha_2 = \alpha_4 = \pi/2 - \alpha_1$. Thus the estimation problem is greatly simplified.

If we redefine the observations as $b_1 = a_1, b_2 = \pi/2 - a_2, b_3 = a_3$ and $b_4 = \pi/2 - a_4$ then we have the simple estimation problem of four independent (mean) observations each with expected value α_1 and with common variance.

7.1.3 The normal linear model

As a further level of complexity, suppose that the linear model (7.5) continues to hold but we are now prepared to assume that the errors ε_i are *normally distributed* with zero mean and are either independent (as for the ordinary linear model) or have dependencies described by the variance–covariance matrix $V(\boldsymbol{\varepsilon}) = \mathbf{V}\sigma^2$ (see (7.8) for the extended linear model). Clearly, we can now use more structured methods of estimation such as the *method of maximum likelihood*.

The likelihood of the sample will be proportional to

$$\sigma^{-n}|\mathbf{V}|^{-1/2}\exp\left[-\frac{1}{2}(\mathbf{y} - \mathbf{X}\boldsymbol{\alpha})'\mathbf{V}^{-1}(\mathbf{y} - \mathbf{X}\boldsymbol{\alpha})/\sigma^2\right],$$

and we can immediately see that the maximum likelihood estimators will be the same as the least-squares and extended least-squares estimators derived above. Further, since these are linear estimators they will be normally distributed and the results of normal sampling theory will apply to the drawing of inferences about $\boldsymbol{\alpha}$ (and about σ^2).

Some special cases of linear models have traditionally been considered in their own rights in applications of statistical methods over the years. An obvious example is *simple linear regression* which we examined in Example 7.1 above. *Weighted linear regression* is an obvious extension of this when the errors do not have constant variance and may even be correlated. We saw an example of this earlier (in Section 2.2 on best linear order statistics estimators based on order statistics, and in Section 5.3 on ranked set sampling).

Other examples (usually with errors which have constant variance and are uncorrelated) include **multiple regression**, which is just another term for the general p-component linear model (7.4), and **polynomial regression**, which is a special case of this where the $x_{i,j}$ in the design matrix $\mathbf{X}$ take the special form $(x_i)^{j-1}(j = 1, 2, \ldots, p)$ *so that we are seeking to represent* E(Y) *as a power series of order $p - 1$ in a single regressor variable x*. Then again, in **designed experiments**, where we seek to explain variations in the outcome values of E(Y) at different discrete levels of influencing factors, we again have a linear model or extended linear model formulation but where the design matrix $\mathbf{X}$ is now typically composed only of entries 0 or 1 (see Hinkleman and Kempthorne, 1994).

7.2 TRANSFORMATIONS

The linear models of Section 7.1 can find useful application in many environmental problems, but there is of course a limit to their applicability. More complex relationships may need to be accommodated. To extend the range of problems where linear models may be applied, it is useful to consider whether transformations of the data may restore the linear structure which does not seem to apply to the raw data – for example, log responses may be linearly related to a stimulus or dose variable and even have normally distributed residuals or errors, whereas the actual (raw) responses may be far from linear in dose, and residuals may be patently non-normal.

As a first stage to considering in any particular problem whether data transformations might be necessary and useful, we could carry out a number of graphical examinations of the data, or *numerical data-screening* activities, to investigate whether the key assumptions of

- linearity of relationship,
- independence,
- constancy of residual variance and
- normally distributed errors or residuals

are separately or jointly reasonable for the study in hand, and whether there might be anomalous observations in the form of outliers.

If any of these are counter-indicated, we might then consider different possible forms of transformation which might restore the linear model form to an extent to which we can again apply the methods of Section 7.1.1.

Finally, transformations may *not* prove to be adequate for this purpose and we may have to reject all thought of a *linear model* in the sense discussed above, and proceed instead to consider a range of alternative non-linear models which do not fit the form of (7.5). An initial prospect is to use what are known as generalized linear models, which greatly extend the applicability and flexibility of the basic linear model. These are discussed in detail in, for example, Nelder and Wedderburn (1972) and McCullagh and Nelder (1989). We will return to this theme later but need to start with empirical data screening for examining basic assumptions.

7.2.1 Looking at the data

As a first stage in checking out the appropriateness of any model, say a linear model, we need to examine the empirical form of any indicated relationships. For the simple case of univariate Y and X, the starting point is to look at a *scatter plot* of the data. This is just the graphical representation of the sample points $(y_i, x_i), i = 1, 2, \ldots, n$, and it can immediately reveal gross departures from linearity.

Example 7.4 Chatterjee and Price (1991, Section 2.8) examine some data on the mortality of bacteria exposed to X-ray radiation. The numbers y_i (in hundreds) of marine bacteria which are measured on plate counts as surviving exposure to 200 kilovolt X-rays after different exposure times $x_i = i$ (in multiples of six hours) $(i = 1, 2, \ldots, 15)$ are plotted in Figure 7.1. It is clear that Y and X are not linearly related, and that any simple linear model such as (7.3) will not suffice.

In a more complex example with several stimulus variables $X_i(i = 1, 2, \ldots, p)$ we might correspondingly examine scatter plots of the y_i against one, many or indeed all of the sets of observed values of the separate stimulus variables, but care is needed in reaching even informal assessments of multilinearity

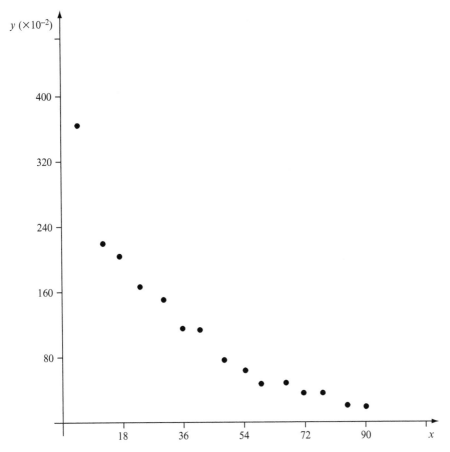

Figure 7.1   Plot of numbers of surviving bacteria (y) against time (x) from chatterjee and Price (1991, Section 2.8)

in such cases; see, for example, Cook (1998) on wide-ranging 'regression graphics', including the interpretation of scatter plots in 2 and 3 (and more) dimensions.

Beyond an initial look at response/stimulus scatter plots, we can obtain more information by fitting some simple (even contraindicated) model and then examining the behaviour of the *estimated residuals* in relation to that model. That is to say, we examine the differences, or departures, $\tilde{\varepsilon}_i = y_i - Y(x_i)$ of the observed values from those fitted by the model $Y(x)$ as an indication of what might constitute a more appropriate model.

Such *response plots* and *residual plots* may usefully takes various forms:

- the y_i against the observed stimulus values x_i (particularly useful for confirming non-linearity of fit);
- the estimated residuals $\tilde{\varepsilon}_i$ against the fitted response values $Y(x_i)$;
- the estimated residuals $\tilde{\varepsilon}_i$ against of the observed stimulus values x_i;

and so on. Sometimes we might consider plotting the estimated residuals $\tilde{\varepsilon}_i$ against the actual reponse observations y_i but there can be induced correlation between them which can readily confuse the message. Also it is common to 'standardize' the estimated residuals ε_i by dividing their values by their estimated standard errors under the fitted model.

The different forms of response and residual plots can informally reveal – as a guide to ultimate model choice – such aberrations as *heteroscedasticity* (non-constancy of residual variance), *autocorrelation, inadequacy of the linear model, non-normality* of residuals and the presence of *outliers*.

These methods are developed in many texts. Good introductory treatments are given by Chatterjee and Price (1991) and by Montgomery and Peck (1992); a much more detailed and extended coverage is found in Cook (1998), which has interesting examples drawn from the ecological and environmental fields (e.g. there is an extensive study in Section 14.6 of the effects of an environmental contaminant on an ecosystem) and thus chimes well with our interests in this book. An earlier ecological example in this same reference examines how the edible muscle mass of New Zealand horse mussels relates to various measures of shell size and site indicators. One plot is informative for our present purposes. Figure 7.2 shows the estimated standardized residuals plotted against a transformation of shell height in the regression of mussel mass on transformed shell height. Note how the plot 'fans out' as the fitted values increase in size – a clear indication of heteroscedasticity.

The range of diagnostic prospects is illustrated archetypically in Figure 7.3, where we note indications (again) of heteroscedasticity, of autocorrelation (non-independence), of linear model inadequacy, of non-normality and of outlying behaviour (or contamination). Let us consider these manifestations in more detail.

In Figure 7.3(a) we have the estimated residuals $\tilde{\varepsilon}_i$ plotted against the observations y_i of the response variable (we might have used the stimulus variable) and we note that the residuals seem to become more variable as the y-values increase – thus indicating heteroscedasticity (non-constant variance). But we must be careful of such plots. If the simple linear model (7.3) is valid, we will find nonetheless that the $\tilde{\varepsilon}_i$ do *not* have constant variance. They are more variable the closer the corresponding x_i is to the mean of the x-values; see Barnett and Lewis (1994, Section 8.1). But broad departures from homoscedasticity will indeed be observable in plots such as that in Figure 7.3(a).

Now consider the plot in Figure 7.3(b); here the $\tilde{\varepsilon}_i$ are plotted against the corresponding x_i and at successive x_i values they are positive and negative. This could in principle imply some very high-order polynomial fit but much more plausibly suggests serial correlation: overshoot of y_i at one x_i-value will be followed by undershoot at the next and vice versa, due to some compensating mechanism.

In the plot of estimated residuals $\tilde{\varepsilon}_i$ against the observations y_i of the response variable shown in Figure 7.3(c) we see large negative residuals at low y-values and large positive residuals at high y-values; clearly there is some second-order (non-linear) component in the relationship.

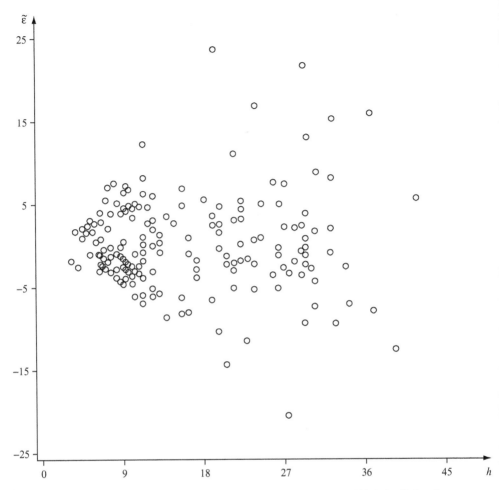

Figure 7.2 Plot of standardized residuals ($\tilde{\varepsilon}$) against transformed shell height (h) for the horse mussel data from Cook (1998, Section 2.2)

For the plot of estimated residuals $\tilde{\varepsilon}_i$ against the observations y_i of the response variable shown in Figure 7.3(d) we see a different effect; no indications of heteroscedasticity, dependence, or non-linearity, *but* clear non-normality of the estimated residuals (here it is more like a uniform distribution). Graphical study of non-normality is best pursued by means of *normal probability plots*, where the ordered estimated residuals are plotted against a transformation of their rank order on what is termed 'normal probability paper' with non-linearity of plot indicative of non-normality. See Montgomery and Peck (1992, Section 3.2.2) for more details and Barnett (1976b) for a more direct method.

Finally, consider point A in Figure 7.3(d). This seems a real 'loner'. The estimated residual is unreasonably large. It is an outlier and may be the effect of

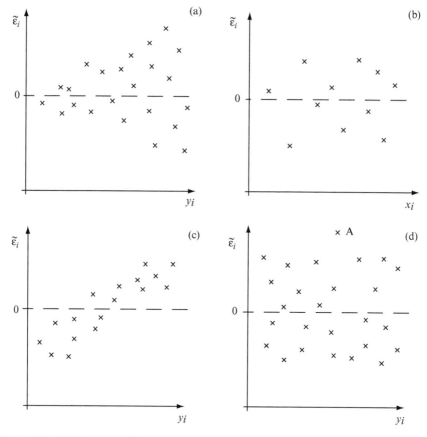

Figure 7.3   Different manifestations of residual plots.

contamination. We have considered such prospects in some detail in Chapter 3 (specifically Section 3.11 on outliers in linear models).

The practice of high-dimensional data plotting for revealing model structure has been around for some time; see, for example Barnett (1981) and, in particular, Tukey and Tukey (1981).

7.2.2   Simple transformations

In Example 7.4 we considered some data on marine bacteria exposed to X-rays from Chatterjee and Price (1991). Clearly, from Figure 7.1, the number of surviving bacteria is not linear in the X-ray dose level (in terms of time of exposure). A plot of the standardized residuals from a linear fit against dose level (see Figure 7.4) shows clear departure from a linear model in a highly specific manner and even hints at an outlier at minimum dose.

In this example, however, we have good scientific reason (the 'single-hit hypothesis') to expect a specific form of non-linear relationship:

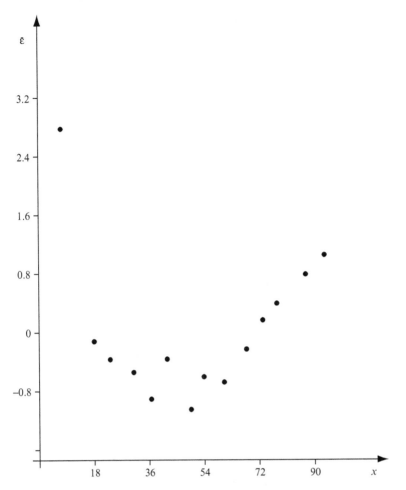

Figure 7.4 Plot of estimated standardized residuals ($\tilde{\varepsilon}$) against (x) for a linear regression of the Example 7.1 data, from Chatterjee and Price (1991, Section 2.8).

$$Y_i = \alpha \exp(\beta X_i), \tag{7.16}$$

or equivalently,

$$\ln Y_i = \ln \alpha + \beta X_i, \tag{7.17}$$

which reverts to a simple linear model, but relating the transformed variable $Y' = \ln Y$ (rather than Y) to X. Sure enough, if we plot $\ln y_i$ against x_i, we confirm the linear fit and a plot of the estimated residuals against the x_i suggests homoscedasticity; see Chatterjee and Price (1991, Section 2.8).

Thus we have an example where a simple logarithmic transformation has restored linearity and has yielded an apparently constant residual variance, so that we are able to proceed with a simple least-squares analysis. Sometimes such simple transformations (a logarithm or a power or a root) are variously supported by general scientific theory (as here) or by some plausible distributional assumptions, or may be merely indicated empirically by plotting the data.

The aim of the transformation of response and/or stimulus variables might be (again as in the previous example) to restore linearity of relationship, but it may be intended also to stabilize the variance or to yield normally distributed residual errors. As implied above, on occasions we may need to transform the stimulus variable(s), rather than the response variable. Let us pursue these various matters further.

In the case of a single stimulus variable X, forms of departure from the basic assumptions of simple linear regression (such as non-linearity, heteroscedasticity, autocorrelation, non-normality of residuals and contamination) manifest themselves typically as demonstrated in the architypal scatter plots shown in Figure 7.3.

Similar prospects can be considered for a multivariate stimulus variable, X (e.g. as in the multilinear regression model (7.4)) by plotting the response variable against the different univariate components of X, but care is needed in reaching an overall interpretation from such a set of plots.

A major problem is that of *multicollinearity*, where two or more components of X are so highly correlated as to essentially require the dimensionality of the model to be reduced if sense is to be made of a least-squares or maximum likelihood estimation approach to the problem. Montgomery and Peck (1992, Chapter 8 and elsewhere) examine the issue of multicollinearity in more detail. See also Chatterjee and Price (1991, Chapter 7).

7.2.3   General transformations

We will now consider three further matters of importance in seeking to make use of linear models.

To this end, we need firstly to examine what are the effects of an arbitrary transformation of a random variable on its distributional properties and the implications for transforming such a variable in a linear model analysis. We will then relate such ideas to the problem of seeking to transform a random variable specifically to *stabilize its variance* and also to an important general class of transformations known as Box–Cox transformations.

Consider a random variable Y with mean μ_Y and variance σ_Y^2. Suppose we define a new random variable, by means of a one-to-one transformation, $Z = f(Y)$. What can we say of the mean and variance of Z?

Using a Taylor series expansion, we can write

$$Z = f(Y) = f(\mu_Y) + (Y - \mu_Y)f'(\mu_Y) + \frac{(Y - \mu_Y)^2}{2}f''(\mu_Y) + \ldots. \qquad (7.18)$$

Taking expectations, we can see that

$$\mu_Z = \mathrm{E}[f(Y)] = f(\mu_Y) + f''(\mu_Y)\frac{\sigma_Y^2}{2} + \ldots. \qquad (7.19)$$

Thus

$$Z - \mu_Z = (Y - \mu_Y)f'(\mu_Y) + \frac{[(Y - \mu_Y)^2 - \sigma_Y^2]}{2}f''(\mu_Y) + \dots.$$

Squaring this and taking expectations, we have

$$\sigma_Z^2 = \text{Var}[f(X)] = [f'(\mu_Y)]^2\sigma_Y^2 \dots \tag{7.20}$$

When Y is a univariate single random variable, the residual terms, beyond $f(\mu_Y)$ and $[f'(\mu_Y)]^2\sigma_Y^2$ respectively, in (7.19) and (7.20) *do not necessarily become negligible*, although *sometimes* we can get fairly accurate results from the leading terms.

Example 7.5 Suppose Y is Poisson, $P(\mu_Y)$, and $Z = f(Y) = \sqrt{Y}$. Then $\sigma_Y^2 = \mu_Y$ and from (7.19) and (7.20) we get

$$\mu_Z = \sqrt{\mu_Y} - \frac{1}{8\sqrt{\mu_Y}} - \dots \approx \sqrt{\mu_Y}$$

and

$$\sigma_Y^2 = \frac{1}{4\mu_Y}\mu_Y + \dots = \frac{1}{4},$$

respectively, if μ_Y is large – taking account of the successively lower order of successive terms in the expansions.

Special cases of (7.19) and (7.20) arise when Y is a *statistic* based on a sample of size n, for example a mean $\bar{Y}$. Then the residual terms are of lower order in n than the leading terms of (7.19) and (7.20) which, for reasonable size n, will ensure that those leading terms alone will be quite accurate.

Example 7.6 Consider the sample variance $S^2 = \sum(y_i - \bar{y})^2/(n-1)$ for a random sample from a normal distribution, $N(\mu, \sigma^2)$. It is easy to show that

$$\text{Var}(S^2) = 2\sigma^4/n.$$

But, we might ask, what is $\text{Var}(S)$? Here we have $Y = \sqrt{S^2}$ as the transformation, so that

$$\text{Var}(S) = (2\sigma^4/n)/(4\sigma^2) + O(n^{-3/2}) = \sigma^2/(2n) + O(n^{-3/2}),$$

which is indeed known to be the case.

Stabilization of variance
We have seen that linear model analysis is particularly straightforward if the residuals have constant variance – that is, if the model is homoscedastic. Since the model asserts a relationship between the mean of the response variable and values of the stimulus variable(s), homoscedasticity essentially declares that the variance of the response variable is not related to its

mean. Thus we might be especially interested in examining whether particular transformations can ensure this property of independence of mean and variance. This leads us to examine what are called *variance-stabilizing transformations*.

Suppose we start with a random variable Y which has variance which is a function of its mean. An example would be a Poisson random variable Y with distribution $P(\mu)$ for which $\text{Var}(Y) = \sigma^2 = \mu$.

In general, suppose that $\text{Var}(Y) = V(\mu)$. Consider transforming Y to $Z = f(Y)$ with the purpose of seeking to eliminate the dependence of the variance on the mean. Then, from (7.20), we have

$$\text{Var}[f(Y)] \propto [f'(\mu)]^2 V(\mu).$$

If this is to be independent of μ we will need $f'(\mu) \propto [V(\mu)]^{-1/2}$, which requires us in turn to use the transformation

$$f(Y) = \int [V(Y)]^{-1/2} dY. \tag{7.21}$$

Thus, for example, in the Poisson case we are lead to the transformation $f(Y) = \int (1/\sqrt{Y}) dY \propto \sqrt{Y}$, which is precisely the transformation which we recently discovered as yielding a constant value for the variance independent of the value of the mean (see Example 7.5).

Box–Cox transformations

We recall that the linear model analysis is most conveniently carried out when the response variable Y is linearly related to the stimulus variable X and when the residuals are independent and normally distributed with constant variance. If this is not the case, transformations may help to achieve these aims. Indeed, there is a class of transformations due to Box and Cox (1964) – following Tukey (1957) – which seeks to deliver all these requirements – homoscedasticity, independence and normality of residuals – simultaneously.

Sometimes known as the *Box–Cox transformation* (or the *Box–Cox method*) it proposes a family of monotonic power transformations of the response variable in the form

$$Y^{(\lambda)} = \begin{cases} (Y^\lambda - 1)/\lambda & \lambda \neq 1, \\ \ln Y & \lambda = 1, \end{cases} \tag{7.22}$$

applicable when all observations y_i are positive (see Cook, 1998, Section 2.3).

A modified 'shifted' form uses

$$Y^{(\lambda)} = \begin{cases} [(Y + \eta)^\lambda - 1]/\lambda & \lambda \neq 1, \\ \ln (Y + \eta) & \lambda = 1, \end{cases} \tag{7.23}$$

where η is chosen to ensure that all observations yield positive observed values $y_i + \eta$, that is, all y_i exceed $-\eta$. There have been many subsequent modifications of the form of these transformations, often discussed in an environmental context (Manly, 1976); see also John and Draper (1980) and Bickel and Doksum (1981).

In practice, the challenge is to determine what value of λ is appropriate to use in transforming the observed values of the response variable. Box and Cox proposed using the *maximum likelihood method* (and a Bayesian approach) for estimation. Cressie (1978) suggested a graphical procedure for obtaining an informal estimate.

The maximum likelihood method proceeds as follows. We assume that the observations $y_i^{(\lambda)}$ satisfy a linear model (7.5) with homoscedastic, independent, normally distributed residuals and we estimate λ by maximizing the log-likelihood

$$l(\lambda) = -n\ln\{\mathbf{y}'^{(\lambda)}[\mathbf{I} - \mathbf{X}(\mathbf{X}'\mathbf{X})^{-1}\mathbf{X}']\mathbf{y}^{(\lambda)}\}/[2 + (\lambda - 1)\sum \ln y_i]. \qquad (7.24)$$

This is a fairly lengthy process usually conducted numerically by examining $l(\lambda)$ for a plausible range of possible values of λ and picking that which yields the maximum. Draper and Smith (1998, pp. 279ff.) further explain and illustrate the method and discuss modifications to keep the calculations tractable. They advocate, in view of its 'volume-preserving' properties, use of a modified form of the transformation (7.22),

$$Y^{(\lambda)} = \begin{cases} (Y^\lambda - 1)/(\lambda \dot{Y}^{\lambda-1}) & \lambda \neq 1 \,, \\ \dot{Y}\ln Y & \lambda = 1 \,, \end{cases}$$

where $\dot{Y}$ is the geometric mean of the n sample values Y_i.

See Dey (2001) for a review of the Box–Cox transformation approach.

7.3 THE GENERALIZED LINEAR MODEL

A further important generalization of the linear model arises arises when we modify (7.4) to produce

$$y_i = h(\alpha_1 x_{1i} + \alpha_2 x_{2i} + \ldots \alpha_p x_{p_i}) + \varepsilon_i \quad (i = 1, \ldots, n) \qquad (7.25)$$

for some prescribed non-identity function h. So

$$E(Y_i) = \mu_i = h(\alpha_1 x_{1i} + \alpha_2 x_{2i} + \ldots \alpha_p x_{p_i}). \qquad (7.26)$$

Usually, we assume that the Y_i are independent. As with the linear model, we might make different levels of assumption about the error structure as reflected in the distributional properties of the residual variable, ε.

At an *initial* level we assume that

$$\text{Var}(Y_i) = \sigma^2 V(\mu_i) \qquad (7.27)$$

where the variance function $V(\mu)$ is assumed to be known and where $\sigma^2 > 0$ is known as the *dispersion parameter* (or *scale parameter*).

At a *more detailed* level we may assume that ε has a specified *error distribution G* in the *exponential family* (as defined, for example, by Barnett, 1999a,

p. 149). Such a model is known as a *generalized linear model* (GLM). It is said to have *link function g*, where if

$$E(Y) = \mu_Y = h(\boldsymbol{\alpha}'\mathbf{x}) \tag{7.28}$$

then

$$g(\mu_Y) = g[h(\boldsymbol{\alpha}'\mathbf{x})] = \boldsymbol{\alpha}'\mathbf{x}, \tag{7.29}$$

and to have error structure specified in terms either of a variance function $V(\mu_i)$ and dispersion parameter σ^2, or in terms of a full *error model G*; see, for example, Jørgensen (2002). (Note that sometimes it is h rather than its inverse g which is termed the link function.)

The two levels of assumption correspond with the distinctions we have already drawn in respect of the linear model in our studies in Section 7.2.

As an example of a GLM consider, for example, a model where a response variable Y has a Poisson distribution, $P[\exp(\boldsymbol{\alpha}'\mathbf{x})]$, where the components of $\mathbf{x}$ are p explanatory (stimulus) variables, and those of $\boldsymbol{\alpha}$ are p parameters. Then $E(Y) = \exp(\boldsymbol{\alpha}'\mathbf{x})$, so that $\ln E(Y) = \boldsymbol{\alpha}'\mathbf{x}$ and we have a GLM with *logarithmic link function* and *Poisson error model*. Such a model is widely used in environmental studies, for instance, for plant or animal populations.

Such GLMs are powerful models and are readily applied and analysed, with extensive computer package support (e.g. Genstat or GLIM). See, for example, McCullagh and Nelder (1989).

We note that the explanatory (stimulus) variables enter only in terms of a linear combination of their values. The *probit* and *logistic (logit)* models we will be considering in Chapter 8 are further examples of GLMs. The latter has a *logit* link function $h(\mu) = \ln[\mu/(1-\mu)]$ and a *binomial error distribution*.

So how are we to analyse data to which we are fitting a GLM? We start with the minimal error structure assumptions of a variance function $V(\mu_i)$ and dispersion parameter σ^2. Parameter estimation in this mode usually addresses a slightly broader model where $\mathrm{Var}(Y_i) = \sigma^2 V(\mu_i)/w_i$, where the w_i are assumed known and reflect different weights which might be appropriate, for example, to replicate observations at 'level i'. The *deviance* (or *total deviance*) for the parameter set $\boldsymbol{\alpha}$ is then defined as the weighted sum of *unit deviances* $d(y_i, \mu_i)$ as

$$D(\boldsymbol{\alpha}) = \sum_{i=1}^{n} w_i d(y_i, \mu_i) \tag{7.30}$$

where

$$d(y_i, \mu_i) = 2 \int_{\mu_i}^{y_i} \frac{y_i - x}{V(x)} \, \mathrm{d}x$$

is essentially a measure of squared deviation of y_i and μ_i. As an specific example, when $V(\mu_i) = 1$ we have $d(y_i, \mu_i) = (y_i - \mu_i)^2$.

We then estimate $\boldsymbol{\alpha}$ by minimizing $D(\boldsymbol{\alpha})$; for example, by least squares in the special case when $V(\mu_i) = 1$. Of course, $\boldsymbol{\alpha}$ enters into the estimation process via $h^{-1}(\mu_i)$; see (7.28).

In the more structured case where an error model G is specified, more formal estimation methods can be used. Thus we might employ the *maximum likelihood approach*. This can be computationally rather intense, but can nowadays be readily handled either in full detail or in an approximating asymptotic version aided by such statistical software as GLIM or Genstat.

The asymptotic approach assumes that the log-likelihood for α takes the form

$$l(\alpha) = \text{constant} - \frac{D(\alpha)}{2\sigma^2}, \tag{7.31}$$

where the 'constant' involves only σ^2 and the data so that the estimator discussed above is indeed the maximum likelihood estimator; its solution will typically involve iterative methods.

Generalized linear model fitting can be approached by means of the *analysis of deviance* (Nelder and Wedderburn, 1972) which essentially generalizes the notion of the *analysis of variance* for linear models. This employs asymptotic arguments to test for different levels of dimensionality of α, usually through successive elimination of different components of α. The methods will differ somewhat, depending on whether or not we assume σ^2 to be known. Thus, for example, comparison of two possible linear predictors $\alpha_1' X_1$ and $\alpha_1' X_1 + \alpha_2' X_2$ (to examine the need for the parameters α_2) is achieved by examining the *scaled deviance*

$$S(1, 2) = -2 \ln (l_1/l_2), \tag{7.32}$$

where l_i is the maximized likelihood under model i (1 or 2). $S(1,2)$ will have an asymptotic χ_k^2 distribution where k is effectively the dimensionality of α_2.

Detailed discussion of applications, specification, fitting and estimation of relevant parameters in GLMs can be found in McCullagh and Nelder (1989), Dobson (1990) and McCulloch and Searle (2001).

Many other simple models can of course be interpreted as GLMs. These include (as we have already noted in some cases): analysis of variance for a designed experiment (normal error structure, identity link function, mean response linear in factor levels), multiple regression, log-linear models (Poisson error structure, logarithmic link function), multinomial data, probit and logit models, and logistic regression.

CHAPTER 8

Special Relationship Models, Including Quantal Response and Repeated Measures

...been together now for forty years.

A wide range of non-linear models can be accommodated under the label *generalized linear model* and handled in the way outlined in Section 7.3. However, certain such models have customarily been dealt with in their own right and not as part of a broader class. One such example is the range of models which may be used for **dose–response studies**, typically in environmental toxicology (see, for example, Bailer and Piegorsch, 2000; Bailer and Oris, 2002).

In this context, a particular level of exposure (a particular *dose*) of a stimulus may be sufficient to cause a specifically defined, even a *lethal*, effect in the response of an affected subject. The stimulus may be environmentally *encountered* (as in the level of sulphur dioxide in the air) or environmentally *administered* (as in the dose given to plants to kill off infestation or to a patient to ease a condition). The response may be quantitative, as in the effect on a biochemical blood measure of a patient, or qualitative (in healing the condition or killing the infecting insects). If it is qualitative, we may have what is known as a quantal response model. Such models are widely employed in environmental, epidemiological and toxicological studies.

Consider the special case where the response variable Y is discrete. Y may take values $0, 1, 2, \ldots, k$ corresponding to k different classified states $A_1, A_2, \ldots, A_k$, each of which describes a defined condition which can arise depending on the value of a stimulus variable X. The special case of *quantal response* is where $Y = 0$ or 1 (the response takes one of just two possible forms); for example, a dose X of an environmental pollutant may kill ($Y = 1$) or not kill ($Y = 0$) a subject exposed to it. Thus, for example, SO_2 in the atmosphere might at high enough levels destroy benign organisms such as lichens.

Example 8.1 Suppose we consider the effects of applying different dose levels (in appropriate units) of a trial insecticide for possible control of an environmentally undesirable form of infestation and obtain the following data on

Environmental Statistics V. Barnett
© 2004 John Wiley & Sons, Ltd ISBN: 0-471-48971-9 (HB)

numbers of insects treated and killed at different application levels of the insecticide.

Level of insecticide, x	1.6	3	4	5	6	7	8	9.6
Number of insects treated, n	10	15	12	15	13	8	17	10
Number killed, r	1	2	3	5	6	5	13	8
Proportion killed, r/n	0.100	0.133	0.250	0.333	0.462	0.625	0.765	0.800

Clearly the proportion killed tends to increase with the level (the dose) of the applied insecticide, but it seems to do so in a non-linear way (see Figure 8.1, which superposes an indicative form for the underlying relationship between dose and proportion killed).

To study such a relationship we will need to go beyond the linear model to explain what is going on, taking account of anything we know of the distributional behaviour of the response variable, Y. We will examine the details of this model in Section 8.2.

8.1 TOXICOLOGY CONCERNS

Dose–response relationships are of particular interest in **toxicology**, where we wish to examine how effective or toxic is the influence of certain stimuli on the behaviour of a response variable which describes the well-being or decline of an exposed subject. The subject may be a human, animal or plant, a community of

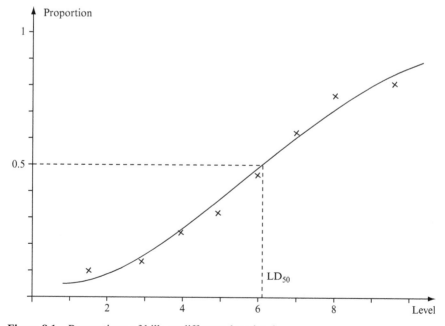

Figure 8.1 Proportions of kills at different dose levels.

such beings or even an entire ecosystem. Apart from widespread concern for environmental issues, the study of foods (including 'shelf-life' of products or effects of pesticide or fertilizer treatments in food development) and of drugs (newly developed or currently in use for disease control) is of major interest in toxicological terms. Bailer and Oris (2002) give an overview of environmental toxicology, and more detailed statistical treatment of toxicological methods can be found in Morgan (1992, 1996) and Piegorsch and Bailer (1997).

Dose–response effects are often summarized by certain features of the dose–response relationship (e.g. as illustrated in Figure 8.1). Commonly employed measures include those known as the LD_{50} *level*, the LD_5 *level* and the LD_{95}*level*: the 'lethal doses' that 'kill' (or have some critical effect on) 50%, 5% and 95% of the population, respectively. We will need to be able to estimate these. The LD_{50} is illustrated in Figure 8.1.

This is typical of the interest in the dose–response model as widely applied in biological, environmental, epidemiological and toxicological problems. Depending on application, the various measures LD_{50}, LD_5, LD_{95} may be of more or less interest. For example, if we are testing a new insecticide we will want it to be effective in controlling the offending agent (be it an insect or a particular form of biochemical infestation) and we will probably concentrate on the LD_{95} to reflect that the agent is under control. If we are concerned about the effect of an environmental pollutant on human health, we will want this effect to be low and will focus perhaps on the LD_5 or even lower effect measures such as the *NOEL* (the 'no obvious effect level') or the *NOAEL* (the 'no obvious adverse effect level'); see Bailer and Oris (2002) and Crump (2002). These purport to be levels below which no adverse affect has been statistically or biologically demonstrated. The NOEL and NOAEL pose serious problems in statistical terms especially since often no specific dose–reponse model is postulated. Even with such a model, the low levels of effect being studied are bound to lead to great difficulty in applying efficient inference procedures. We are effectively trying to predict variate values corresponding with (sometimes ill-defined) percentage points very low in the tail of the distribution and where no observations have been obtained; cf. extreme-value methods (Chapter 4). Further complications are spelt out by Crump (2002).

An alternative and related approach applied to dose–response data across environmental, epidemiological and general medical interests is that of *benchmark analysis* (Crump, 2002). The *benchmark dose* (BMD) or *benchmark level* is that dose or exposure level which *yields a specific increase in risk* compared with incidence in an *unexposed* population. Such increase is called the *benchmark risk* (BMR). Benchmark analysis proceeds by specifying the BMR and applying dose–response analysis to estimate the BMD in the form of a statistically inferred lower bound. Maximum likelihood methods have been proposed (Crump, 2002).

The assumed dose–response model often takes the form of a *binary quantal response* model – where one of two specific qualitative outcomes must arise at any dose for any individual – although extensions to quantitative response have been discussed. Again we have the problem of trying to infer behaviour in the extreme lower tail of a dose–response curve.

We have mentioned epidemiology in the context of dose–response relationships. This is inevitable. Epidemiology is concerned with the study of disease at large (in a population): its incidence and prevalence, how it is affected or modified by intervention in the form of infection, pollution, diet, financial policy, etc. All such concerns can involve the range of statistical considerations and methods we have pursued throughout this book, but particularly those we are discussing in this current chapter. Coverage of statistical methods and environmental interests in epidemiology is provided by, for example, the review by Wise (1982), the text by Clayton and Hills (1993) and the compendium volume by Elliott et al. (1992).

8.2   QUANTAL RESPONSE

Let us return to the basic inference problems for the quantal response model. Morgan (1992) discusses in detail the analysis of quantal response data. It is implicit that the response variable Y must be binomial, $Y \sim B[n, p(x)]$, with mean $np(x)$ depending on the level or dose, x.

We will need to assume that $p = F(x)$, where $F(x)$ is a distribution function (said to be the distribution function of the *tolerance distribution*). Thus $F(x)$ takes the general shape shown in Figure 8.2 which can be expressed mathematically in a variety of forms as long as they are monotone non-decreasing in x and range from 0 to 1. One form commonly used is the *normal model* or *probit model*,

$$F_1(x) = \Phi(\alpha + \beta x), \tag{8.1}$$

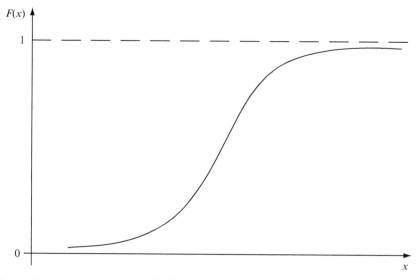

Figure 8.2   Form of tolerance distribution function.

where $\Phi(z)$ is the d.f. of the standard normal distribution (see Finney, 1971; Morgan, 1992). Another is the *logistic model*, where, possibly after a convenient transformation of x to, for example $\ln x$,

$$\begin{aligned} F_2(x) &= \frac{\exp(\alpha + \beta x)}{1 + \exp(\alpha + \beta x)} \\ &= [1 + \exp(-\alpha - \beta x)]^{-1} \end{aligned} \tag{8.2}$$

(see Hosmer and Lemeshow, 2000 on *logistic regression*).

For the logistic case (8.2), it is common to use the transform

$$\Psi(x) = \ln\left[\frac{p(x)}{1 - p(x)}\right] = \alpha + \beta x \tag{8.3}$$

This is the so-called *logit transformation* or *logit model*, which is linear in x (possibly in a suitably transformed space) and clearly indicates how an appropriate link function renders the model interpretable as a generalized linear model. Note the useful interpretation of $\Psi(x)$ in (8.3) as the *log-odds* measure of a 'kill' versus a 'non-kill'.

The probit model is correspondingly interpretable as a GLM.

The probit and logit models are widely applied and can be rather similar in form. The latter is easier to analyse and compute – but this is a relatively unimportant distinction with modern computing facilities. They also serve as general models for 'growth relationships' when we wish to model non-linear regression relationships. Other forms are also used. We have the Gompertz model and the Weibull model with wide application in biology, economics, medicine, reliability studies and other fields, and *compartment models* (involving systems of linear first-order differential equations especially in chemistry and pharmacokinetics). See, for instance, Bates and Watts (1988).

We will consider the logistic model in more detail. Suppose, for $i = 1, \ldots, k$, we take n_i independent observations at (log) dose x_i and observe r_i of the key responses ('kills'). Estimation of α and β can be conveniently carried out by *maximum likelihood* (or by other more *ad hoc* methods such as weighted least squares or minimum χ^2).

Consider the logistic model (8.2). For the form of data just described, the log-likelihood is

$$L_{\alpha,\beta}(\mathbf{r}, \mathbf{n}; \mathbf{x}) = \sum \{r_i \ln F_2(x_i) + (n_i - r_i) \ln[1 - F_2(x_i)]\}. \tag{8.4}$$

We can substitute $[1 + \exp(-\alpha - \beta x)]^{-1}$ for $F_2(x)$, differentiate with respect to α and β, equate each derivative to zero and seek to solve the resulting two equations for the maximum likelihood estimators $\hat{\alpha}$ and $\hat{\beta}$. This yields

$$\sum r_i = \sum n_i \left[1 + \exp(\hat{\alpha} + \hat{\beta} x_i)\right]^{-1} \tag{8.5}$$

and

$$\sum r_i x_i = \sum n_i x_i \left[1 + \exp(\hat{\alpha} + \hat{\beta} x_i)\right]^{-1} \tag{8.6}$$

which have no explicit solution and need to be solved iteratively for $\hat{\alpha}$ and $\hat{\beta}$ using Taylor series methods akin to those described in Section 7.2.3. Similarly, we will need iterative methods to obtain the approximate forms of the variances and covariance of $\hat{\alpha}$ and $\hat{\beta}$ from the information matrix. See, for example, Stuart *et al.* (1999, Sections 18.22 ff.).

Example 8.2   Returning to the trial insecticide data of Example 8.1, we can fit a logit model by maximum likelihood. We obtain

$$p(x_i) = [1 + \exp(4.32 - 0.554x_i)]^{-1}.$$

Note that we can estimate summary measures such as the LD_5, LD_{50} or LD_{95} in an obvious manner. For example, an estimate of the LD_5 can be obtained by solving

$$0.05 = \frac{\exp(\hat{\alpha} + \hat{\beta}x)}{1 + \exp(\hat{\alpha} + \hat{\beta}x)}.$$

Here we find an estimate of the LD_5 as $x = 0.86$.

We can readily extend these models to cases of several influencing variables. Thus x becomes a vector $\mathbf{x} = (x_1, x_2, \ldots, x_m)'$ and we observe r_i and $x_{1i}, x_{2i}, \ldots, x_{mi} (i = 1, 2, \ldots, k)$. The obvious multivariate generalization of the logistic model gives

$$\ln\left(\frac{p(\mathbf{x})}{1 - p(\mathbf{x})}\right) = \alpha + \mathbf{x}'\boldsymbol{\beta}. \tag{8.7}$$

This suggests a form of approximate modelling under which we adopt a *linear model*

$$y_i = \ln\left(\frac{r_i}{n_i - r_i}\right) = \alpha + \mathbf{x}'\boldsymbol{\beta} + \varepsilon_i. \tag{8.8}$$

Of course, the y_i will not be expected to have the same variance and we would thus need to conduct an appropriate *weighted least-squares* analysis.

8.3   BIOASSAY

Bioassay, or biological assay, is a body of methodology concerned with assessing the potency of a stimulus (e.g. a drug, a hormone, or radiation) in its effect on (usually) biological organisms (see Tsutakawa, 1982). It can take two basic forms: estimating stimulus response; and evaluating the potency of one stimulus (e.g. a new drug or a pollutant) relative to another (a 'standard' or familiar form). We have discussed the former at some length in Chapter 7 and the earlier sections of this chapter – particularly in our examination of quantal response in Section 8.2. Let us now consider the latter, the problem of *relative potency*, which was illustrated in the introductory examples of Chapter 1 (see Section 1.3.4).

Regression methods and maximum likelihood are used extensively in bio-assay. Consider two stimuli, T and S. They are said to have *similar potency* if the effect on a response variable Y of some level Z of the stimulus is such that there is a constant scale factor ρ such that

$$Y_T(Z) = Y_S(\rho Z). \tag{8.9}$$

This implies that the *stimulus–response relationship* is essentially the same for both of the stimuli; we have only to scale the dose level by a simple multiplicative factor to obtain identical relationships.

In many cases it proves reasonable to assume that Y has a linear regression relationship with $x = \ln z$. Then we can write

$$E[Y_T(z)] = \alpha + \beta x = \alpha + \beta \ln z. \tag{8.10}$$

And, if S *is similar in potency* to T, we have

$$E[Y_S(z)] = \alpha + \beta \ln \rho z = \alpha + \beta \ln \rho + \beta \ln z. \tag{8.11}$$

Thus we have two linear regression models of Y on $\ln z$ with the *same slope* β but *different intercepts* α and $\alpha + \beta \ln \rho$. So we could seek to *test if two stimuli have similar potency* in the more restricted sense of (8.10) and (8.11) by carrying out a hypothesis test for parallelism of the two regression lines; if this is accepted we can proceed to estimate ρ. This procedure is known as a *parallel-line assay*; it assumes that the regression lines are parallel and separated by a horizontal distance $\beta \ln \rho = \alpha_S - \alpha_T$, where we have $\alpha_T = \alpha$ and $\alpha_S = \alpha + \beta \ln \rho$.

Another form of regression-based assay is known as a *slope-ratio assay* – this assumes that the regressions are linear with respect to $x = z^\lambda$ for some appropriate power $\lambda \neq 0$.

Let us consider in detail how parallel-line assay operates. Suppose we have random samples of sizes n_1 and n_2 of stimulus and response for the two stimuli, T and S which we write in the forms

$$(y_{T1}, x_{T1}), \ldots, (y_{Tn_1}, x_{Tn_1}) \quad \text{and} \quad (y_{S1}, x_{S1}), \ldots, (y_{Sn_2}, x_{Sn_2}).$$

If the residuals are uncorrelated and the error distribution is normal with constant variance, the maximum likelihood estimators need to be chosen to minimize

$$\begin{aligned} R &= \sum_{i=1}^{n_1}(y_{Ti} - \alpha - \beta x_{Ti})^2 + \sum_{i=1}^{n_2}(y_{Si} - \alpha - \beta \ln \rho - \beta x_{Si})^2 \\ &= \sum_{i=1}^{n_1}(y_{Ti} - \alpha_T - \beta x_{Ti})^2 + \sum_{i=1}^{n_2}(y_{Si} - \alpha_S - \beta x_{Si})^2. \end{aligned} \tag{8.12}$$

Obviously, we have

$$\hat{\alpha}_T = \bar{y}_T - \beta \bar{x}_T \quad \text{and} \quad \hat{\alpha}_S = \bar{y}_S - \beta \bar{x}_S. \tag{8.13}$$

Now

$$\frac{\partial R}{\partial \beta} = -\sum_{i=1}^{n_1} y_{Ti}x_{Ti} + n_1\alpha_T\bar{x}_T + \beta \sum_{i=1}^{n_1} x_{Ti}^2 - \sum_{i=1}^{n_2} y_{Si}x_{Si} + n_2\alpha_S\bar{x}_S + \beta \sum_{i=1}^{n_2} x_{Si}^2.$$

So $\partial R/\partial \beta = 0$ yields the maximum likelihood equation

$$-\sum_{i=1}^{n_1+n_2} y_ix_i + \hat{\beta} \sum_{i=1}^{n_1+n_2} x_i^2 + n_1\bar{y}_T\bar{x}_T - n_1\hat{\beta}\bar{x}_S^2 + n_2\bar{y}_S\bar{x}_S - n_2\hat{\beta}\bar{x}_S^2 = 0$$

for $\hat{\beta}$. Thus we obtain the maximum likelihood estimator of the slope parameter as

$$\hat{\beta} = \frac{\displaystyle\sum_{i=1}^{n_1+n_2} y_ix_i - n_1\bar{y}_T\bar{x}_T - n_2\bar{y}_S\bar{x}_S}{\displaystyle\sum_{i=1}^{n_1+n_2} x_i^2 - n_1\bar{x}_S^2 - n_2\bar{x}_S^2}, \tag{8.14}$$

and hence we can then obtain $\hat{\alpha}_T$ and $\hat{\alpha}_S$.

So the maximum likelihood estimator of ρ is readily found to be $\exp\left(\frac{(\hat{\alpha}_T - \hat{\alpha}_S)}{\hat{\beta}}\right)$.

In the usual way, we can estimate σ^2 from the residual sum of squares $R_{\min}$ and can test for parallelism using the appropriate *likelihood ratio test*.

Example 8.3 In Section 1.3.4 we considered a bioassay problem concerning two treatment regimes applied to laboratory animal specimens. The data are shown in Table 1.1. We can readily seek to fit a parallel-line assay model to these data. In fact, it fits well and we obtain estimates $\hat{\alpha}_T = 67.71, \hat{\alpha}_S = 32.88$ and $\hat{\beta} = 50.32$ of the crucial parameters, where S denotes the standard treatment and T denotes the new treatment. These will provide an estimate of the scale factor; we obtain $\hat{\rho} = 2.00$. This appears intuitively reasonable when we plot the responses against the stimuli for the two treatments.

8.4   REPEATED MEASURES

There is a further class of situations which is proving to be highly relevant to many types of environmental problem where again it is not adequate to seek to model the relationship between a stimulus and a response in terms of the linear models discussed in this chapter and the last. Specifically, we might be interested in how a response variable Y relates to a predictor (or stimulus) variable X but we *cannot* obtain observation pairs $(Y_i, X_i), i = 1, 2, \ldots$, which are *independent*, nor readily model the nature of the dependence between Y and X.

Such situations are exemplified in the taking of successive observations over time on one individual or specimen. For example, we might be interested in the pattern of growth (Y) of a plant after a time (X) from planting, or the level (Y) of a physiological measure at different growth stages (X) of a person, and we want to investigate these interests by examining data on *different individuals* (plants, people) where for each individual we have a short series of measurements usually over time. We refer to such a data set and the corresponding model as *repeated measures data* and a *repeated measures model*.

The data from one individual are sometimes known as a *profile* and frequently take the form of a short time series (although the index set might be distance rather than time in some applications). Time-series models and methods generally are discussed in Chapter 10, but note that such methods are not usually appropriate for repeated measures studies due to the multiplicity of the profiles, the shortness of the time series and the non-stationarity of relationship within them. Kenward (2000) gives a concise introduction to this topic; Diggle *et al.* (1994) provide more detailed coverage in their treatment of the analysis of *longitudinal data* (observational studies over time). Koch *et al.* (1988) offer an extended review with emphasis on the wide range of different appropriate methods of analysis.

Data for each individual specimen (plant or person) will typically take the form $(y_1, x_1), \ldots, (y_n, x_n)$, satisfying some relationship

$$y_i = f(x_i) + \varepsilon_i,$$

where the 'errors' or residuals ε_i have an assumed structure (in terms of distribution and/or correlation) and $f(\cdot)$ needs to be specified or estimated. This is like a regression/linear (general linear) model problem, but standard methods will not readily apply because of the complex (and sometimes individual-specific) form of dependence between the observed values in a profile. Thus we may have somewhat different forms $f(\cdot)$ applying to different individuals.

Figure 8.3 illustrates typical repeated measures data in the form of just two profiles. They may have different numbers of observations (six and five here), show different relationships and typically exhibit less variation *within* individuals than *between*. It may be necessary to distinguish the one form of variation from another, denoting the error as ε_{ij} (where i refers to the individual and j to the treatment level, e.g. time).

In particular situations, the separate circumstances under which individuals are observed may be *qualitative* rather than *quantitative*. Thus, rather than, or as well as, being observed at different times, the individual may be observed after application of distinct and different successive treatments. For example, in a study of how patients react to a treatment regime for some environmentally related medical condition, they may be given one drug (A) and then another (B). Furthermore, these treatments A and B may be used at different dose levels A_i and B_j for any patient with all possible pairs of levels represented in the study. Here the 'levels' will be quantitative. But levels of treatment factors in general can be quantitative or qualitative. In the medical study a factor may have three 'levels': apply heat therapy, use ultrasound, and use neither heat nor ultrasound. Clearly the type of structure we are describing falls within the field of *designed experiments*; often a simple crossed-factor design would accommodate it (apart from the within-specimen temporal pattern). For

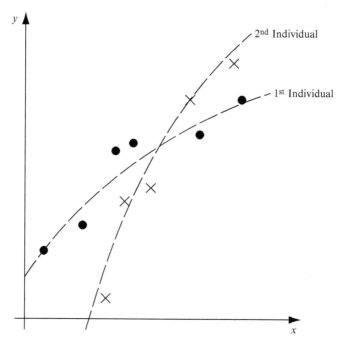

Figure 8.3   Two repeated measures profiles.

discussion of the design and analysis of experiments, see, for example, Cochran and Cox (1957), Box *et al.* (1978), John (1998) and Hinklemann and Kempthorne (1994).

In fact, there is a variety of prospects for analysing repeated measures data. The simplest cases will be where the data are *balanced* in a design of experiments context. This would be so, for example, if all profiles were measured at the same set of times, all other treatment factors had all their levels represented proportionately and if the data were complete (with no *missing values*). In such a case we would typically have crossed non-time factors and the time would be nested. See *Example 8.4* below.

Even with some *imbalance* in these respects it might be possible to assign a manageable (if complex) form of experimental design and to analyse the data accordingly using analysis of variance (ANOVA) methods (see Hocking, 1996, Chapters 15–17, for an overview of ANOVA techniques). We will discuss this approach further at a later stage in the chapter.

Sometimes specific or general linear models may be applicable – such as a particular GLM with appropriately designated error model (see below), or perhaps replicated logit models with declared internal correlation structure.

Finally, time-series models and methods may be considered, although their usefulness will be restricted by the fact that stationarity is unlikely to be a reasonable assumption and the time series of the profiles will typically be very short.

Example 8.4   Consider the data shown in Table 8.1 on plasma fluoride concentrations for litters of baby rats of different ages at different times after injection of different doses of a drug. This example is discussed by Koch *et al.* (1988).

This repeated measures example is balanced (in the sense that we have a full crossed age/dose design and the same three times from injection for all specimens); there are no missing data. It is conveniently set out in a particular way (as a *split-plot experimental design* – see for example Cochran and Cox, 1957, pp. 293 ff - frequently used for tractable repeated measures data and to which we will return). We see various characteristics of a repeated measures problem.

- Subjects (litters) are each measured at three different conditions (times after injections) and have been assigned (or are chosen to reflect) different levels of two other factors (*A*, age 6 or 11 days; and *B*, dose 0.10, 0.25 or 0.50μg).
- It can be assumed (typically) that we have independence between but not within subjects, and more variability between subjects than within.
- The post-injection times constitute a nested factor and apply to each subject. The factors age and dose are crossed – assigned in all combinations to the set of different subjects. The response can be thought of as a 3-vector from a multivariate distribution – it will, of course, have covariational structure. Subjects will typically be independent.

Table 8.1   Data on split-plot study of pairs of baby rats: average log plasma fluoride concentrations.

Age	Dose	Litter	Post-injection time (min.)		
(days)	(μg)	number	15	30	60
6	0.50	1	4.1	3.9	3.3
6	0.50	2	5.1	4.0	3.2
6	0.50	3	5.8	5.8	4.4
6	0.25	4	4.8	3.4	2.3
6	0.25	5	3.9	3.5	2.6
6	0.25	6	5.2	4.8	3.7
6	0.10	7	3.3	2.2	1.6
6	0.10	8	3.4	2.9	1.8
6	0.10	9	3.7	3.8	2.2
11	0.50	10	5.1	3.5	1.9
11	0.50	11	5.6	4.6	3.4
11	0.50	12	5.9	5.0	3.2
11	0.25	13	3.9	2.3	1.6
11	0.25	14	6.5	4.0	2.6
11	0.25	15	5.2	4.6	2.7
11	0.10	16	2.8	2.0	1.8
11	0.10	17	4.3	3.3	1.9
11	0.10	18	3.8	3.6	2.6

In such problems as that described in *Example 8.4* we could, if the number of levels is large enough to warrant it, carry out an internal multivariate analysis for each subject and then analyse across subjects. But this is not practicable for this specific example. Alternatively a *split-plot model* can be used, and we will consider this in more detail below for the data in Example 8.4.

In a split-plot designed experiment, *all but one* of the treatments are randomized over plots. Thus if we have treatments A and B which each are applied at four levels $(A_1, A_2, \ldots, A_4; B_1, B_2, \ldots, B_4)$ we randomly assign each of the 16 distinct $A_i B_j$ combinations to a separate experimental specimen or 'plot'. This can be represented diagrammatically as follows:

$A_1 B_1$	$A_4 B_3$	.	.
$A_3 B_2$	.	.	.

Then each plot is *split* to accommodate the different levels of another factor, C (e.g. into three parts if C has three levels). The whole design is replicated if necessary.

Example 8.5   Consider again the rat plasma fluoride study of Example 8.4. The 18 litters are the plots; age and dose are the primary factors A and B, and time from injection is the within-plot factor (C). The design is replicated three times. There are standard designed experiment ANOVA methods available to analyse such problems, and indeed they are readily handled by standard statistical packages such as Genstat, Minitab or S-PLUS. Clearly such a design supports the idea that the repeated measures within the plots (factor C) may be dependent and are likely to show less variability than is found between the plots (where the primary factors A and B are applied or replications occur).

If, again as in this example, the within-plot factor (here time) has quantitative levels, we can effectively carry out regression over those levels. This is done by means of dividing the within-plot variability into orthogonal contrasts representing constant effect, linear effect, quadratic effect, etc., in the standard way (see, for example, Draper and Smith, 1998, Section 6.3).

Consider the results. The means for different times (and orthogonal contrast summary values) for the 3×2 primary factor levels are as shown in Table 8.2. From these results we note that plasma fluoride concentration tends to decrease over time for each (age $\times$ dose) group and to increase with dose for each (age $\times$ time) combination. For the 11-day-old rats we find a greater decrease over time at the higher doses than for the 6-day-old rats. Such observations motivate

Table 8.2 Means and estimated standard errors for the logarithms of plasma fluoride concentrations and the orthonormal summary functions for the groups of baby rats in the split-plot study.

Age (days)	Dose (μg)	Minutes post-injection			Orthonormal summary functions		
		15	30	60	Average	Trend	Curvature
6	0.50	5.02	4.53	3.65	7.62	0.97	0.16
6	0.25	4.63	3.89	2.83	6.56	1.27	0.13
6	0.10	3.44	2.97	1.89	4.79	1.09	0.25
11	0.50	5.85	4.36	2.85	7.54	2.12	0.01
11	0.25	4.89	3.65	2.29	6.25	1.84	0.05
11	0.10	3.65	2.99	2.12	5.06	1.08	0.08
Estimated S.E. for all groups		0.41	0.53	0.39	0.70	0.20	0.20

considering three orthonormal summary functions of the data from each litter: the first reflects the average response for the three time periods; the second is the linear trend across log time; and the third is a measure of any lack of linearity (or curvature) in the trend.

An analysis of variance table can then be produced as follows:

		Separate functions		
Source of variation	Degrees of freedom	Average	Trend	Curvature
Overall mean	1	715.3[***]	35.04[***]	0.231
Age	1	0.008	1.460[***]	0.081
Dose	2	10.62[***]	0.431[*]	0.012
Age × dose	2	0.124	0.500[**]	0.003
Within-groups error	12	1.487	0.115	0.117

Asterisks denote significance at [*]5%, [**]1% and [***]0.1% level.

We see a significant effect of dose on the 'average response'; also a significant overall linear trend, and some an indication of an effect of dose level on the size of the linear trend. Also there is a significant age × dose interaction for the linear trend – with the differences between age effects increasing with dose. There seem to be no curvature effects. See Koch *et al.* (1988) for more details.

We have considered a single repeated measures example in some detail. Here a split-plot model was appropriate. We conclude by considering the more general case where a straightforward split-plot model cannot readily be applied. The broader prospect is to apply a generalized linear model. Often this can reasonably take the following form. Suppose we have k profiles of sizes

$n_i(i = 1, 2, \ldots, k)$. Starting with individual profiles we might postulate that the outcome set $\mathbf{Y}_i$ for profile i has the structure

$$E[g(\mathbf{Y}_i)] = \mathbf{X}_i \boldsymbol{\beta}_i \quad (i = 1, 2, \ldots, k)$$

for an individual $(n_i \times p)$ design matrix $\mathbf{X}_i$ where $\boldsymbol{\beta}_i$ is a $(p \times 1)$ vector of parameters. A crucial feature of the repeated measures model is dependence within the profile. Thus we would declare a variance-covariance model

$$V(\mathbf{Y}_i) = \boldsymbol{\Sigma}_i,$$

allowing for distinct covariational behaviour in and between each profile. Such a model is too broadly expressed to be reasonably specified – the distinct $\boldsymbol{\beta}_i$ could lead to unidentifiability of the parameters, and how are we to know the individual $\boldsymbol{\Sigma}_i$? As a partial remedy, it is not uncommon for a *linear model* to be used; taking $g(\cdot)$ as the identity function and with either a *common* within-profile slope parameter or an imposed (possibly time-series-based) constraining structure on the $\boldsymbol{\Sigma}_i$ (or both).

PART IV

Standards and regulations

CHAPTER 9

Environmental Standards

We have to keep up standards...

9.1 INTRODUCTION

Another important expression of the interests we have in the effect of a pollutant, contaminant or pharmacological agent on vulnerable subjects (particularly exposed human individuals) is to be found in the setting and maintenance of *environmental standards*. These can take a variety of forms, from general and often ill-defined principles to detailed specific quantitative, sometimes statistical, procedures.

Of the least formal type there are general aims such as those of trying to control the exposure levels to ensure that undesirable effects are *as low a reasonably achievable* (ALARA) or of using methods to control pollutants or production of potentially harmful agents that employ *best available technology not entailing excessive cost* (BATNEEC). The common appeal to the use of the so-called *precautionary principle* as the basis for safe production is often even less objectively defined in terms of its outcomes.

At the next level of formality we might seek to control the effects of a pollutant or contaminant by restricting its maximum levels or overall extent to be below some *critical level* or *critical load*. These are defined, respectively, as maximum levels or total quantities which would be likely to produce some specifically defined ill effect. However, such levels or quantities are seldom based on appropriate extensive empirical evidence or validated models. Instead, values are assigned by 'extrapolation' to the human condition from laboratory-based study of animals or animal organisms (see also the notions of NOEL and NOEAL in Section 8.1). Barnett and O'Hagan (1997) and Barnett and Bown (2002a) review these broad principles more comprehensively and give references to extended discussion; see also Cox *et al.* (1999).

More prescriptive controls are expressed sometimes in advisory terms but more often as legislative requirements. These may take the form of *upper* (or *lower*) *limit standards* or sometimes, less formally, as upper or lower limit *targets*. Many examples are given in Barnett and O'Hagan (1997) and Barnett

Environmental Statistics   V. Barnett
© 2004 John Wiley & Sons, Ltd   ISBN: 0-471-48971-9 (HB)

and Bown (2002a, 2002b). Examples include the requirements that (in different countries and in different contexts):

- the 24-hour running average level of particulate matter in the atmosphere for particles with a mass median aerodynamic diameter less than $10\,\mu m$ (known as PM_{10}) should be no more than $50\,\mu g/m^3$;
- the 95th percentile of the distribution of total nitrogen discharged into a river should not exceed 8 mgN/l;
- in a sample of six monthly readings of faecal coliforms in water, the median value must be below 150 colony-forming units (c.f.u.) and the two highest values below 160 c.f.u.;
- in a sample of 12 readings of lead pollution levels in a river, the mean should not exceed 12 ppm v/v.

Such standards, although well-intentioned, are seldom implementable in any formal statistical way which takes proper regard for prevailing uncertainties and variations. Indeed, in spite of the inevitable variational and probabilistic framework in which all environmental standards issues are encountered, there has been little attempt until very recently to take any account of uncertainty in setting, monitoring or implementing standards.

The four examples just described reflect two different forms of standard which have been termed *ideal* and *realizable* by Barnett and O'Hagan (1997). An *ideal standard* is set in terms of some prescribed characteristic of the *population* (or *distribution*) of all encountered pollution levels – for example, a limit is placed on a mean or on some specific percentile of the distribution of pollution levels (as with the particulates concentration, or the nitrogen levels, in the examples above). A *realizable standard*, in contrast, declares what we need to find in a particular (random) sample of observations if the standard is to be satisfied (as for the faecal coliforms, or lead pollution, levels in the examples above). Both forms are encountered in current standards specifications but are seriously deficient.

The ideal standard says what we want of the distribution, but specifies no inference procedure for testing compliance with this requirement. The realizable standard, in contrast, is unambiguous in its requirement of what we should find in data and readily implemented but embodies no statistical principles for inferring the characteristics of the overall distribution of pollution levels for reference beyond the immediate sample.

As remarked above, the range of different types of standard is discussed in some detail by Barnett and Bown (2000a). Barnett and O'Hagan (1997), in response to a request for relevant evidence by the UK Royal Commission on Environmental Pollution, presented a detailed review of prevailing approaches to environmental standards and, in particular, stressed the vital need for standards to be set in such a way that they take full regard of the uncertainty and variability of the environment. Specifically, they defined the prevailing ideal and realizable forms, but eschewed these as inadequate for the necessary handling of uncertainty and variability in standard-setting. They proposed that a standard should encompass both the ideal and the realizable emphases and should thus typically specify *both* a limit level for a pollutant or contaminant, etc. (in the form of a population statement, for example, about a mean or a percentile) *and* a degree of statistical assurance required (a variational or probabilistic statement) in demonstrating whether that limit level had been satisfied or breached. This approach was endorsed by the Royal Commission in its report to the UK government on setting environmental standards (Royal Commission on Environmental Standards, 1998). Examples of such a form of standard – termed a *statistically verifiable ideal standard* SVIS – are developed and applied in different contexts by Barnett and Bown (2002b, 2002c, 2002d).

9.2   THE STATISTICALLY VERIFIABLE IDEAL STANDARD

As we have just noted, the SVIS in principle links an ideal standard (which reflects uncertainty and variation in the population at large) with a compliance criterion to be used in practice to seek to verify the standard at some prescribed level of statistical assurance. Typical specific examples of SVIS might require, on the one hand, that (i) 'the 0.95 quantile of the pollutant distribution must not exceed level L, to be demonstrated with 95% statistical assurance' or, on the other hand, that (ii) 'the mean of the pollutant distribution must not exceed level M, to be demonstrated with 99% statistical assurance'. Characteristically, such standards are expressed in terms of 'levels', that is, acceptable limit values for appropriate summary statistics or parameters of a pollution or contamination distribution, whether of sulphur dioxide, *E. coli* or noise (see Example 4.1 in Chapter 4). It is both a strength and a flexibility of the approach that it does not specify the method by which statistical assurance is to be provided, allowing for developments in statistical methodology, technology and sampling technique.

Correspondingly, recent research on the setting of statistically meaningful environmental standards has concentrated on the Barnett and O'Hagan (1997) statistically verifiable ideal standard.

Barnett and Bown (2002b, 2002c, 2002d) have explored the effect of efficient sampling schemes on the SVIS verification procedures and have considered the use of quantile-based standards in a variety of environmental applications. Thompson *et al.* (2002) re-examine the US ozone air quality standard and seek to transform it into the SVIS format; they consider using distributional knowledge (or assumptions) to improve the 'non-parametric' approach of exceedance counting currently seen in many air quality standards – for examples, see USEPA (1997) and Department of the Environment, Transport and the Regions (2000); see also Cox *et al.* (1999).

We will consider how examples of an SVIS such as those described above might be implemented. We start with example (ii), where we aim for 99% assurance that the mean pollution level does not exceed a value M. We can express this more specifically as follows.

The expected value μ_X of the distribution of the pollution level X should not exceed M. If this is so, the probability of misclassification as in non-compliance (i.e. of concluding that $\mu_X > M$) should not exceed a specified small value α (0.01 in this case).

This is readily set up in terms of a test of significance. Suppose, for the moment, that we have $X \sim N(\mu_X, \sigma^2)$ with σ^2 known, and we have a random sample $x_1, x_2, \ldots, x_n$ of n pollution readings with sample mean $\bar{x}$. Then we need only conduct a level-α test of the hypothesis $H: \mu_X \leq M$ against $\overline{H}: \mu_X > M$, concluding that the standard is violated if

$$\bar{x} > M + \frac{z_\alpha \sigma}{\sqrt{n}}, \tag{9.1}$$

where z_α is the two-tailed α-point of the standardized normal distribution. For '99% assurance' as specified, we would use a 1% test and (9.1) becomes

$$\bar{x} > M + \frac{2.576\sigma}{\sqrt{n}}.$$

If σ^2 is not known, as is commonly the case, we would merely replace (9.1) with the form for the standard t-test.

It is clear to see, however, that standards expressed in terms of quantiles will often be more relevant than those expressed in terms of means; we frequently want to limit 'extreme' values which will be found in the tails of the distribution of X. Example (i) above will provide such a prospect. The most direct quantile-based SVIS is essentially 'non-parametric' in form in that it does not depend explicitly on the distribution of the pollution level, X.

Example (i) requires $p = P(X > L) \leq 0.05$. So if we take sample observations of pollution levels, each will be either greater than L (with probability p) or less than L (with probability $1 - p$). Thus if r out of n observations exceed L then r is an observation from a binomial distribution $B(n, p)$ and we have only to employ the standard 5% significance test of the hypothesis $H: p \leq 0.05$ to implement the SVIS.

Of course, we could base the SVIS in example (ii) more specifically on an assumed distribution for X, and we might expect to obtain a more efficient version in this more structured situation. Let us consider how this can be done (Barnett and Bown, 2002b, 2002d). We can state the standard as follows:

The $1 - \gamma$ quantile, $\xi_{1-\gamma}$, of the distribution of the pollution level X should not exceed L. If this is true, the probability of misclassification as in non-compliance (i.e. of concluding that $\xi_{1-\gamma} > L$) should be no more than some specified small value α.

This again provides a 'level of statistical assurance' of $1 - \alpha$, and such a SVIS can, as before, be expressed as a level-α test of significance, this time of the null hypothesis $H: \xi_{1-\gamma} \leq L$ against the alternative hypotheses $\overline{H}: \xi_{1-\gamma} > L$. Thus using the statistical machinery of significance testing, we would satisfy the requirement for statistical assurance at level $1 - \alpha$.

Barnett and Bown (2002d) discuss best linear unbiased quantile estimators based on random samples (and also, indeed, on *ranked-set samples*) as the basis for carrying out such a SVIS in the case of normally distributed pollution levels. The methods are illustrated for a standard on levels of ammonia (as nitrogen) in river water.

Let us briefly examine how the quantile-based SVIS would work in the case of $N(\mu_X, \sigma^2)$ pollution levels X, where σ^2 is assumed known. Notice that the quantile $\xi_{1-\gamma}$ is defined by $P(X < \xi_{1-\gamma}) = 1 - \gamma$ so that, in this situation, we have

$$\xi_{1-\gamma} = \mu_X + z_{z\gamma}\sigma. \tag{9.2}$$

Thus the null hypothesis $H{:}\xi_{1-\gamma} \leq L$ is equivalent to $H{:}\mu_X \leq L - z_{z\gamma}\sigma$ so that the significance test is again just the standard test of a normal mean.

Of course, σ^2 is unlikely to be known in practice and no such simple approach will be feasible. Barnett and Bown (2002b, 2002d) discuss this issue, comparing different quantile estimators based on moments or order statistics and the significance tests relevant to them; they also consider the implications of using *ranked-set sampling* rather than random sampling and demonstrate impressive efficiency gains from ranked-set sampling.

We note that this significance test (and the earlier described tests) assigns the benefit of the doubt to the polluter (i.e. the polluter could meet the standard even with a sample estimate of the quantile that is in excess of L). Such standards could of course be set the other way round, with the benefit of the doubt assigned to the regulator. We have only to interchange the null and alternative hypotheses. Let us examine this benefit-of-the-doubt principle in more detail.

There may be good reason for such imbalance; on the one hand, it might have been enshrined in original negotiation or, on the other, perhaps it was imposed by legislation. But the imbalance cannot be regarded as entirely satisfactory. It requires one party to operate on different conditions than the other. For example, with the quantile standard as described, sample levels of the estimated quantile can be well in excess of L before the regulator can claim infringement. With the hypotheses reversed, the potential 'polluter' will need to demonstrate sample values well below L to be 'in the clear'. We will return to this issue in Section 9.3 below.

Methods for implementing a SVIS when, as is likely, pollution distributions are non-normal have also been considered. Often pollution levels will have positively skewed distributions; the gamma family may provide a reasonable model for such cases. Bown (2002) derives various appropriate forms of SVIS for gamma-distributed pollution levels and discusses their application.

9.2.1 Other sampling methods

The SVIS just discussed was based on a random sample of pollution observations. It is interesting when illustrating the nature of the SVIS further to consider a situation in which *composite sampling* is employed in a standards context. (Composite sampling was discussed in Section 5.3.) Composite sampling confers general benefits of cost reduction when sample testing and measurement are expensive and some standards (e.g. for contaminated land in Australia) advocate (or at least do not inhibit) the use of composite samples.

We will consider a new SVIS for pollutants in land contamination issues (Barnett and Bown, 2002c). This approach using composite sampling improves on the conventionally used and highly conservative 'divide-by-n' approach which we will define below.

In Australia, appropriate sample values for land contamination obtained using specific sampling protocols dictated by standard AS4482.1-1997 (Standards Australia, 1997) must be compared with so-called environmental investigation limits (EILs) which define upper limits on the concentrations of various contaminants (ANZECC and NHMRC, 1992). Of special interest, the Australian approach to contaminated land investigation (based on standard AS4482.1-1997) does indeed consider the use of composite sampling for compliance assessment.

Let us examine the underlying ideas. Suppose we take a random sample of n soil specimens (with potential but *unobserved* pollution levels $x_i, i = 1, 2, \ldots, n$) but measure (observe) only the single *composite* sample level of contamination in all soil samples put together. If, as is often reasonable, we assume that the level of the composite is the sum of the levels in the individual specimens, we will then obtain a single measured value $t = \sum x_i$ and we can estimate the mean pollution level for the sampled specimens as $\bar{x} = t/n$.

Statisticians and environmentalists in Australia (and around the world) are keen to use composite sampling in contaminated site assessment based on EILs. It offers important measurement cost savings. The approach, however, encounters some criticism; standard AS4482.1-1997, whilst claiming composite sampling to be 'generally unsuitable for the definitive assessment of site contamination' (Standards Australia, 1997), does not rule it out but restricts the number of specimens in any one composite sample to at most four. No statistical justification is offered for this restriction.

The issue behind the controversy of composite sampling in standards verification concerns the adjustment of the standard limit originally set for a single observation x_i to an appropriate limit for the composite observation. Suppose that the usual (*realizable*) standard requires that $x_i \leq x_0$ for $i = 1, \ldots, n$. In such a situation x_0 becomes the EIL for the case in hand. Then under the commonly employed 'divide-by-n rule' we would be required to conclude that the standard is *not* breached *only if*

$$\bar{x} < x_0' = \frac{x_0}{n}. \tag{9.3}$$

Note the effect of this! In a most extreme and pivotal case, when $n - 1$ of the (unobserved) values from a set of n specimens happen to be *zero* and one takes the limit value x_0, we would declare the site to be 'in violation'. This is nonsense. All but one of the sample members have no contamination at all –

one just reaches (but does not exceed) the EIL. Thus the principle behind the 'divide-by-n' rule is highly conservative; it protects against breaches of the standard at the expense of falsely declaring non-compliance for composite samples of observations that are well below the limit, since such declaration requires only that the average value of the individual observations should reach or exceed x_0/n.

Consider an even more extreme form of this. If all (unobserved) values x_i were x_0/n (i.e. all are *well below* the EIL) we would have $t = x_0/n$ and would conclude that the standard was breached.

Example 9.1 Consider the Australian EIL for copper (ANZECC and NHMRC, 1992). This standard requires to user to set $x_0 = 60\,mg/kg$. So, when $n = 4, x_0'$ becomes 15. However, in the case where each observation has a contamination level of just half the standard limit (i.e. $x_i = 30; i = 1, \ldots, 4$), the composite observation will be $t/n = 30 > x_0'$. So the standard will be failed, even though we know it has been clearly met by each observation. Standard AS4482.1-1997 allows the use of composite sampling for compliance assessment only when the 'divide-by-n' rule is used, judging that 'no other adjustment [to the original standard level] is acceptable in health and environmental risk assessments' (Standards Australia, 1997).

We will show that we can usefully employ composite sampling in standards assessment but can do much better than to use a realizable standard and to adopt the 'divide-by-n' rule for the adjustment of the EIL. Employing a hypothesis-testing framework, as described by Barnett and Bown (2002c), we will set up an SVIS that builds on the current practice of using EILs under the ANZECC and NHMRC (1992) guidelines. In most cases this leads to 'adjusted limits' sensibly higher than under the 'divide-by-n' rule.

We start by assuming independent and identically normally distributed contaminant levels, $X \sim N(\mu, \sigma^2), i = 1, \ldots, n$. Distributions of contaminant levels often demonstrate positive skewness as mentioned above, but are frequently amenable to normal transformation. Again we need to make a *probabilistic statement* about the level of contamination that can be allowed, and we will require that some high percentile, $\xi_{1-\gamma}$, for a small value $\gamma > 0$ (e.g. the upper 1% point), of the contaminant distribution should be at most L.

Thus, assuming σ^2 known for the moment, we require $P(X_i > L) < \gamma$ or

$$\frac{L - \mu_X}{\sigma} > \Phi^{-1}(1 - \gamma), \tag{9.4}$$

where $\Phi(z)$ is the distribution function of the standardized normal distribution $N(0,1)$. In this way, L is thus specified in terms of the contaminant distribution. If $\xi_{1-\gamma} > L$, or equivalently if

$$\mu_X > L - \sigma\Phi^{-1}(1 - \gamma) = \mu_0 \tag{9.5}$$

(say), it is concluded that contamination is at an unacceptable level and that the standard has been breached.

However, these statements are expressed in terms of parameters of the distribution of contaminant levels, and we need to construct a SVIS as an inference procedure based upon sample data. Again we take the general form used above in the random sampling context, but modify it to give the benefit of the doubt to the regulator. Thus we require the following form of SVIS:

> The $1 - \gamma$ quantile, $\xi_{1-\gamma}$, of the distribution of pollution level X should not exceed L (the EIL). If it does exceed L, the probability of misclassification as in compliance (i.e. of concluding that $\xi_{1-\gamma} < L$) should be no more than some specified small value α.

However, we will now choose to test a sample for compliance with the standard by using composite sampling.

Consider a composite sample of n observations, yielding a single composite observation X. Again we take $\bar{X} = X/n$, assuming observations are independent and that concentration levels are additive. Clearly, $\bar{X} \sim N(\mu, \sigma^2/n)$. We thus need a level-$\alpha$ hypothesis test of

$$H{:}\mu_X \geq \mu_0 \quad \text{against} \quad \bar{H}{:}\mu < \mu_0$$

based on the composite sample data.

The implied *benefit of the doubt* to the regulator is also the spirit of the 'divide-by-n' rule, where we conclude compliance only if we obtain $\bar{x} < x_0/n$ for measured composite mean $\bar{x}$ (we saw how demanding this can be in the numerical example above).

It is interesting to compare the effects of the 'divide-by-n' rule with the results we would obtain from the significance-test approach, where rejection of H – for classification as 'compliant' – can be shown to require

$$\bar{x} < L - \varepsilon,$$

where $\varepsilon > 0$ takes a known form expressed in terms of σ, γ and α. Specifically, from (9.4) and (9.5) above, we need

$$\bar{x} < L - \sigma[\Phi^{-1}(1 - \alpha)/\sqrt{n} + \Phi^{-1}(1 - \gamma)]. \tag{9.6}$$

Thus if we equate the quantile limit L for the SVIS to the EIL x_0 of the 'divide-by-n' approach, $x_0 - \varepsilon$ will become the *adjusted limit* for the composite sample mean (compared with x_0/n for the 'divide-by-n' rule).

Example 9.2 Barnett and Bown (2002c) discuss an example on lead contamination which employs the maximum allowed size of composite samples ($n = 4$) under the AS4482.1-1997 guidelines. The EIL is assumed to be $x_0 = 300$ mg/kg and for comparative illustration they set the standard upon the 0.98 quantile of the lead concentration distribution, with $L = 300$ and $\gamma = 0.02$. Assuming a normal distribution with mean μ and standard deviation $\sigma = 60$, we have, from (9.5),

$$x_0 = 300 - 60 \times 2.0537 = 176.778 \, \text{mg/kg}.$$

The critical value for the SVIS above at the 5% significance level is, from (9.6),

$$L - \varepsilon = 176.778 - 1.6449 \times 60/2 = 127.431 \, \text{mg/kg}.$$

Thus, using the SVIS, the proposed adjusted limit greatly exceeds the limit of 75 mg/kg that would be implemented under the 'divide-by-n' rule; instead of requiring $\bar{x} < 75 \, \text{mg/kg}$ for compliance, we only need to show that $\bar{x} < 127.4 \, \text{mg/kg}$.

If σ is not known, as is often the case, we need to modify the procedure just described and illustrated; see Barnett and Bown (2002c). The use of ranked-set sampling (see Section 5.3) has also been considered in the setting of standards and demonstrates impressive efficiency advantages; see Barnett and Bown (2002d) and Bown (2002).

9.3   GUARD POINT STANDARDS

We have already discussed a crucial point in relation to environmental standard-setting using hypothesis-testing principles (e.g. as a basis for an SVIS): namely, that there is implicit assignment of 'benefit of the doubt' either to the regulator (as in Section 9.2.1) or, for complementary hypotheses, to the producer of the potential pollution (the 'complier').

Let us review this matter in the case where the aim of the environmental standard is to limit the effects of a pollutant by imposing a constraint on the *expected value*, μ, of the distribution of the pollution level, X. We will assume that the upper level of concern for μ is $\mu_0 (= 120$, say). With the form of SVIS described above, the test either assumes the default status of the population to be 'compliance' and the regulator will then need strong evidence (i.e. a high observed mean pollution level) in order to reject this, or assumes 'non-compliance' ('violation') at the outset and the complier will need to produce at a level far below μ_0 in order to escape the risk of classification as 'non-compliant'.

This leads to inevitable conflicts of interest and cannot be regarded as ideal; it seems to imply an intrinsic unfairness to one party or the other. Barnett and O'Hagan (1997, Section 5.2.2) suggested a basic means of overcoming this problem, and Bown (2002) has developed a corresponding compromise procedure where it is not necessary for one party or the other to take a rigid 'benefit of the doubt' stance. This she describes as a *guard point standard*, and it operates on the following principles.

Suppose that both parties (the regulator and the complier) are prepared to compromise. Instead of setting the single standard level, $\mu_0 = 120$, the regulator is prepared to set an *upper guard point* above μ_0, at $\mu_2 = 130$ say, provided there is assurance that pollution at this level will be detected with probability $1 - \alpha$ (for α small, e.g. $\alpha = 0.05$). Correspondingly, the complier will accept a *lower guard point* below μ_0, at $\mu_1 = 110$, say, if assured that when pollution levels are as low as this there will probability $1 - \beta$ (for β small) of correct classification as compliant.

We suppose that in any application, $\mu_1 < \mu_0 < \mu_2$ and that α and β can be appropriately chosen to be regarded as 'fair' to both parties. Then if $\mu > \mu_2$, the probability of failing the standard is at least $1 - \alpha$, whilst if $\mu < \mu_1$, the probability of satisfying the standard is at least $1 - \beta$. Mean pollution levels between μ_1 and μ_2 are regarded as relatively unimportant: the guard point compromise principle defines such situations as 'uncertain' (or 'too close to call'). Thus there are now three possible outcomes: *compliance*, *violation* and *uncertainty* (where more information may need to be sought).

Bown (2002) suggests the following *statistically verifiable ideal guard point standard* (SVIGPS), for specified $\mu_1 < \mu_0 < \mu_2, \alpha$ and β:

> if a population has μ in excess of μ_2, the probability that the population is declared as non-compliant should be no less than $1 - \alpha$; if μ does not exceed μ_1, the probability that the population is declared as compliant should be no less than $1 - \beta$.

Bown (2002) develops a testing procedure for the SVIGPS case of normally distributed X, and adapts it for setting a SVIGPS for a population quantile rather than mean. It is further developed to apply to an asymmetrically distributed pollutant level, specifically from the gamma family. Application of such a standard is illustrated for some real-life pollution problems.

9.4 STANDARDS ALONG THE CAUSE–EFFECT CHAIN

Consider a situation in which an industrial plant lies alongside a waterway such as a river. When not fully under control, the plant might discharge effluent into the river from where it flows downstream, possibly flooding farmland, running past a school in a neighbouring village and flowing eventually into a lake which is home to fish and wildfowl. There are various pollution risks and corresponding prospects for standards to be operated. For example, we might impose a standard on the possible *cause* of trouble: the level of pollution initially released in the factory discharge (we call this the 'entry stage'). Various so-called 'pollution contact' *effects* are possible: damage to crops in the flooded field is a hazard, as are skin complaints among the schoolchildren, even death of fish or birds in the lake.

Thus we consider the situation typified by Figure 9.1 (see Barnett and Bown, 2002e). This is a typical example of the *cause–effect chain* of environmental risk (see, for example, the discussion of the *cost–benefit chain* in Barnett and O'Hagan, 1997). Standards might be set at various points in the chain: on the levels of source pollution (at the entry stage), on its levels as it travels downstream, or on its *effects* (on schoolchildren, on fish or on birds).

Let us distinguish two distinct emphases. The first concerns pollution levels (the *causes*), whether at *entry* or *in contact* (possibly later) with vulnerable subjects. The methods described above using the SVIS form of standard are relevant to this interest. Note, however, that we now have a new concern.

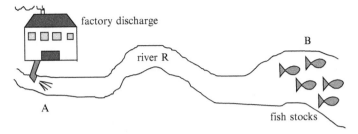

Figure 9.1   The cause–effect chain.

Surely any standard set at the *entry stage* should be mutually compatible with any standard set at a *contact stage*: it should represent the same or equivalent levels of concern and control. So we need *harmonization of standards* over distance (space) and time.

The second emphasis concerns *effects*, such as the probability that a particular form of fish is killed – again possibly at different distances from the industrial plant. So once more we need to strive for *harmonization* over distance and time, but now we have the added difficulty of determining what form of standard it is appropriate to apply to the probability of death of a fish. We might need to develop such standards in either or both of the *absolute* and *conditional* forms, for example for the 'marginal probability of death' or for the 'probability of death when we have pollution level $X = x$'. Such standards will be different in form from those described earlier, although the principle of the SVIS will still be appropriate.

We have explained the need for harmonization (mutual consistency) of standards at different *points* (in distance and time) *of the cause–effect chain*. But a new consideration now enters. We also need to *harmonize* (to seek mutual consistency for) standards set on different functions, for example, on pollution level (a *cause*) or on 'probability of death' (an *effect*). In general terms, what we are saying is that any requirements we might set on the levels of discharge at the factory must surely be in harmony with requirements on resulting levels downstream and with what we require of the protection of crops, children, fish or fowl.

Let us formalize some of these ideas. Referring to Figure 9.1, suppose that pollution levels at the factory (A) are represented by a variable X; those at a fishery (B) by a random variable Y, distributed differently from X but induced by X. Setting a SVIS on X and monitoring it at A is relatively straightforward (based on the distribution of X or using a 'non-parametric' approach as described in Section 9.2). Now let D be some appropriately defined *distressing outcome* for some vulnerable individual – for example, the pollution-related death of a trout. To examine consistency of standards we will need to know about not just the distribution of X but also about the corresponding absolute or conditional pollution measures (*effects*)

$$P_A(D) = P(D \text{ at } A)$$

and

$$P_A(D \mid X) = P(D \text{ at A} \mid X)$$

involving, almost inevitably, an appropriate dose–response model (see Section 8.1).

These three quantities, X, $P_A(D)$ and $P_A(D|X)$, are all measured at point A. We also need to examine corresponding transferred levels and effects at the point B. Various prospects for harmonization arise: harmonization of standards between functions and between sites. There are $3 \times 3 = 9$ such possibilities for any two sites A and B.

One possible concern (from the nine prospects) is to seek to harmonize standards on X (at A) and the corresponding $P_B(D|Y)$: the probability of a 'distressing outcome' at B given the pollution level at B. In this example we seek to harmonize over distance and over function. Clearly, this will involve two forms of modelled relationship; firstly, that of how the pollution level distribution transfers from that of X at A to that of Y at B, and then of the level/effect (dose–response) relationship $P_B(D|Y)$.

The easier version of this is to consider dose–response modelling for mutually consistent standard-setting between different pollution measures *within* the site, A, focusing on the pollution level X and the probability of the distressing outcome given X, $P_A(D|X)$.

As we have remarked, dose–response (level/effect) modelling is becoming central to many current environmental standard-setting procedures. An example is found in the New Zealand Ministry for the Environment use of an exponential model linking the bacteriological indicator *enterococci* and the probability of illness as a result of bathing in polluted water. Clearly not all standards will be expressed (as most of our examples so far) in terms of *levels of contact* with pollutants. Sometimes the standard is set on the *effect* of the pollutant. An example is found in a standard limiting the number of swimming-associated illnesses to 19 per 1000 swimmers (Barnett and Bown, 2002e).

It is then possible in principle to extend the approach (via a *pollution transfer model*) to examine standard-setting for a pollution level X at A and for that of Y at B – proceeding then to the harmonization of the effect measure $P_B(D|Y)$ as proposed above.

Barnett and Bown (2002e) develop such consistent standards within and between sites at least in simple cases where we have clear information about the transfer from X to Y and about relevant dose–response models. Theory and methods are illustrated by practical applications, such as standard-setting for the rivers investigated in the detailed hydrological study of Wilson *et al.* (1997).

For full implementation much remains to be done in terms of allowing for the inevitable uncertainties of knowledge about the transfer relationship and the dose–response patterns – some fascinating challenges for statistical inference.

PART V

A many-dimensional environment: Spatial and temporal processes

CHAPTER 10

Time-Series Methods

There's a time and a place for everything!

10.1 SPACE AND TIME EFFECTS

We have seen many examples of environmental processes whose outcomes are influenced by time or by location in space. Temporal and spatial variation are of the very essence of the intrinsic variability we encounter in the environment. For example, pollution levels in the atmosphere, or in water, are bound to vary with time in response to the random stimuli that yield them (discharges of contaminants) and their interaction with climatic conditions (mitigated by heavy rain in some cases, enhanced by high humidity in others), and to vary over space (i.e. from one location to another in response to local environmental characteristics).

Thus *when* we observe the environment and *where* we observe it can in many cases strongly influence the observed levels. This must be true, for example, of social noise from clubs and discos, of nitrogen dioxide levels in different towns (Smith, 1999), or of *E. coli* found in recreational and bathing water. Figure 10.1, from Goudey and Laslett (1999), shows sites of *E. coli* monitoring in Port Phillip Bay, Melbourne, over a 35-year period.

In much of the methodology we have considered so far we have not taken specific notice for the effects of time and space. We have assumed that environmental variables (pollution levels, say) behave as random variables – that is, there is some probability distribution which describes their overall uncertainty and variability. Seldom have we used time or physical location as a direct covariate. They have just been part of the indeterminism in the generating process which contributes to the uncertainty and variability.

But this is not always good enough. There are bound to be situations where we want time or location specifically to be part of the explanatory process (we saw this with the climatological examples in Sections 1.3.2 and 1.3.3. Many environmental networks are set up precisely to examine how pollution levels vary at different places within the network or at different times (of the day or of the year). Thus environmental problems inevitably involve observation of

Environmental Statistics V. Barnett
© 2004 John Wiley & Sons, Ltd ISBN: 0-471-48971-9 (HB)

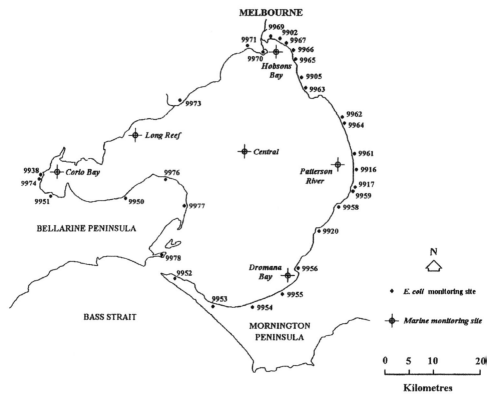

Figure 10.1 Sites of *E. coli* monitoring in Port Phillip Bay, Melbourne (Goudey and Laslett, 1999).

measures or effects which vary over time (*t*) or over (two- or three-dimensional) space (**s**), or both.

Some typical more detailed examples might include the following:

- Weed development over a cultivated field. In the early 1990s spatial modelling of weed growth in crop land was used as a basis for attempting to develop a locally sensitive automated weed treatment regime, where herbicide was applied to the land precisely as needed in terms of the spatial deployment of the weeds (see Wilson and Brain, 1991; Brain and Marshall, 1999).

- Ozone levels over a city area. Plots of hour-by-hour variation in modelled and observed city ozone levels at three city sites (relative to the prevailing ozone standard) are shown in Figure 10.2, taken from the operational evaluation of air quality models by Sampson and Guttorp (1999), which stresses the importance of such *time-series plots* as one of the components of model assessment.

- Nitrogen dioxide measures month by month in different neighbouring communities. Figure 10.3, from Smith (1999), shows monthly NO_2 levels

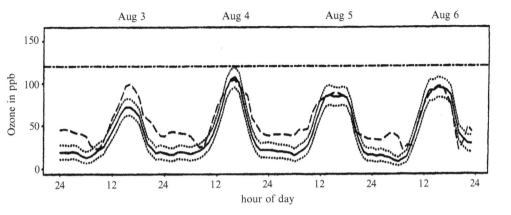

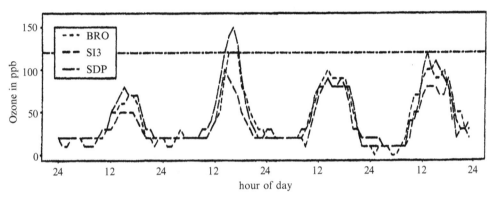

Figure 10.2 Modelled and observed ozone levels at different city sites over 4 days (Meiring *et al.*, with kind permission of Kluwer Academic Publishers).

for six different towns in Yorkshire over one year, in a review of the use of air quality statistics in environmental policy-making.

All of these examples involve temporal and spatial variation: the weeds develop *over the growing period* (*t*) to different extents *at different parts of the field* (**s**); ozone levels characteristically *differ from time to time* (*t*) throughout the day (higher in the afternoon, as we note in Figure 10.2) and are more serious in some places (**s**) than others; the NO_2 levels (Figure 10.3) show large variations from month to month (*t*) throughout the year and from one town to another (**s**).

Basic variables (weed development or air quality measures, in our examples) may be univariate or multivariate, they may vary significantly in terms of the temporal or spatial covariates (e.g. ozone level with time of day or NO_2 levels in different towns). Alternatively, they may be relatively unaffected by time or location, with variations reflecting only random fluctuation around some prevailing level (possible for the spatial variation in weed crops). Sometimes, however, rather than regarding the time or space variation as mere random

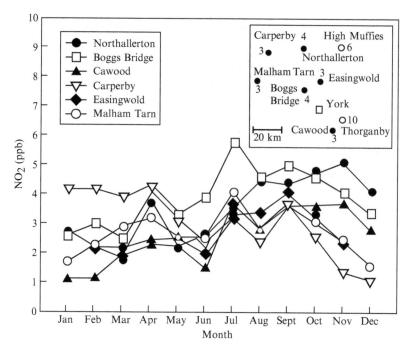

Figure 10.3 Monthly NO$_2$ levels at six different Yorkshire towns and their relative locations (from Smith, 1999).

fluctuation, we may need, as an essential part of an investigation, to *model* the space/time influence, to *predict* variables from covariates or to *interpolate* values at unobserved times or positions.

A substantial example of this latter interest has already been illustrated in Section 1.3.3, which described an agro-environmental study needing climatological variables (temperature, precipitation, etc.) to be interpolated in time and space at new locations using the recorded values at 212 dispersed UK meteorological stations (as described by Landau and Barnett, 1996).

Some of our earlier methodology has been able to take some account of space or time effects as in analysing extremes (Chapter 4), in using linear or generalized linear models (Chapter 7) or examining repeated measures structures (Chapter 8). But we must now move on to more specific models and methods for representing spatial or temporal effects. The full conjoint *spatial-temporal* situation is of major importance in environmental study but it has proved difficult to represent, model and analyse, and very much remains to be done.

10.2   TIME SERIES

We will start our examination of the effects of time and space by considering the time effect alone, using the models and methods of time-series analysis. Later, in Chapter 11, we examine spatial models and methods and review what has been achieved in conjoint space/time (spatial-temporal) representations.

An interesting starting point for studying temporal effects is to be found in some of the classical long time series of environmentally related measures. We will be referring to three particular time series. The first is an extended form of the well-known set of sunspot data from 1700 to 1960 based on the Wolf index of the mid-1900s; see the review of Wolf's work by Waldemeier (1961) and the history of sunspot indices by Feehrer (2000). A smoothed time-series plot of the sunspot data showing the sunspot number (the average annual sunspot index) from 1700 to 2000 is shown in Figure 10.4 in the form used in Bloomfield (2000) – it exhibits a marked *periodic* effect.

The second time series is the set of annual wheat price index data from 1500 to 1869 discussed by Beveridge (1921); see the plot in Figure 10.5 which again uses the representation employed by Bloomfield (2000) – we see a gradual (if irregular) upward *trend* in the mean level of the index over the 250 years or so.

Our third example relates to a set of time series data on the yields of leas (grasses, etc.) from 1856 to 1990 from the untreated plots of the Rothamsted Classical Experiment on Park Grass; Figure 10.6 is from Jenkinson *et al.* (1994), a study of the possible effects of global warming – in particular, increases in atmospheric CO_2 – on crop yield. It shows the total yield (continuous line) and the residual yield (dotted line) after fitting a regression model against time, for three untreated plots ((a),(b),(c)). No conspicuous trend is evident.

Bloomfield (2000) provides internet references for accessing the sunspot and wheat price data sets as used in Figures 10.4 and 10.5.

The *cyclic variation* (otherwise termed *periodicity* or *seasonality*) exhibited in the sunspot data of Figure 10.4 is one of the principal characteristics of time-varying phenomena. We see it also in the hourly ozone levels of Figure 10.2, but

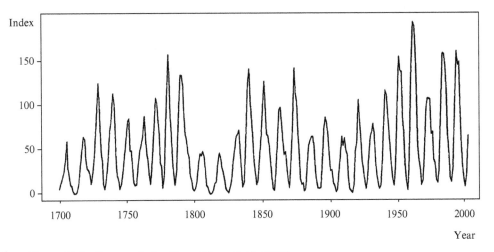

Figure 10.4 Sunspot index (from Bloomfield, 2000).

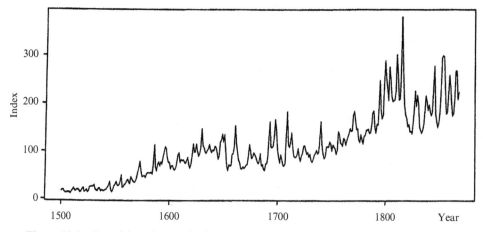

Figure 10.5   Beveridge wheat price index (from Bloomfield, 2000).

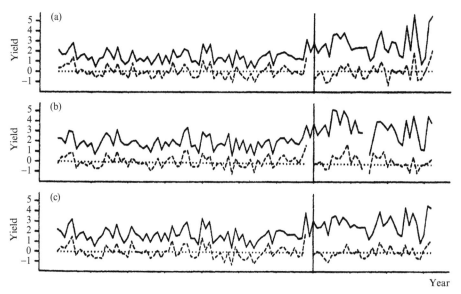

Figure 10.6   Yields of lees on Park Grass, Rothamsted (from Jenkinson *et al.*, 1994).

here the period of the cyclic variation is the 24-hour day (with maximum values in the early afternoon) compared with about 11 years for the sunspot data. Commonly, we experience seasonality over an annual cycle; whether for climatologically affected environmental variables, for economic measures or for commercial sales data (see, for example, Chatfield, 1996, Figure 1.3).

Another principal characteristic of time-series data is revealed in the Beveridge wheat index data of Figure 10.5, where over the 250 years we see a tendency for the index to rise from about 20 to about 200 but with large perturbations around this increasing *trend*. Such trends are commonly encountered. They may be linear in form as time progresses or may be more complex.

Perturbations around seasonal or trend effects, possibly in the nature of *random variation*, constitute a third principal characteristic of a time series. This may be the only feature evident in the harvest yields from the Rothamsted Park Grass experiment shown in Figure 10.6.

10.3   BASIC ISSUES

Let us examine some fundamental issues. Consider an outcome X at time t. This can be thought of as a *stochastic process* and represented as a family of random variables $\{X(t); t \in \tau\}$ indexed by the variable t. When the index variable is *time*, $X(t)$ is also known as a *time-series process* (although we will qualify this definition a little later) and t might be discrete (when we write X_t) or continuous (when we write $X(t)$) – for example, annual total rainfall X_t for the particular year t, or ozone levels $X(t)$ measured at arbitrary times t of the day. At a specific time t, we might observe $X(t) = x(t)$; the set $\{x(t); t \in \tau\}$ is known as a realization of the time series.

We recall that we have distinguished three basic characteristics or features of a time series:

- *secular trend*, where the mean value $E[X(t)]$ of the process seems to be changing (possibly increasing) with time (we may also find time variation in the variance of $X(t)$);
- *cyclic or seasonal variation*, where there is clear indication of a cyclic or periodic pattern of variability;
- *random error*, where there is obviously further random variation ('noise') around any secular trend or cyclic variation – such random changes may or may not be independent.

The last of these suggests consideration of a fourth crucial characteristic of a time series, namely:

- *serial correlation*, where, in many applications, we expect successive X_t or $X(t)$ to show *correlation* at the adjacent time points, possibly stronger the closer the time points are to each other – this feature must also be represented in any time-series model we seek to use.

We will be examining time series in terms of these four basic characteristics. Before proceeding, however, to more formal study of time series it is important to note or expand on some further distinctions.

The random variable $X(t)$ or X_t can take various forms – whether continuous or discrete. In its most primitive form it is *binary*, representing just the presence

or absence of some effect. Two particular versions are the point process and the binary process.

In the *point process*, $X(t) = 0$ except at a discrete set of time points $t_1, t_2, \ldots$ where $X(t) = 1$; here 0 and 1 represent absense or presence of some effect: For example, environmental disasters may occur at times $t_1, t_2, \ldots$, or a system may fail at these times. Such processes need special methods of investigation – see Cox and Lewis (1966) for detailed study, and Section 10.7 below for introductory developmental comments.

In a *binary process* $X(t)$ is continuously in states 0 or 1, 'switching' between them. For example, an environmental control system may be either on or off at any instant in time:

Throughout this chapter we will consider more general forms than these, but we will be restricting attention in the main to a discrete time series X_t ($t = 0, 1, 2, \ldots$). After all, this is usually what we observe (the outcome values x_t at discrete time points) even if the underlying process is in fact continuous. Continuous analogue output from (often electronic) measurement devices is sometimes available but may then need to be discretized for numerical processing.

Our interests can range widely in sophistication from pictorial representation and *descriptive summary* (Figures 10.2–10.6 all provide substantial intuitive understanding of what is going on in these processes), through *prediction* ('forecasting') to complex *model fitting* for in-depth understanding of the time effects. We will be following (necessarily briefly) this route of increasing sophistication employing the four-part characterization of a time series comprising possible trend, cyclic variation, random error and serial correlation which prompts distinct methodological approaches.

At a descriptive level, we might seek to use (more or less) informal methods to try to estimate the trend and cyclic components or to eliminate them to reveal more clearly the correlated error structure. We consider this in Section 10.4.

More detailed study operates in one or other of *two distinct modes* and concentrates on time series with no trend or cyclic components and usually time-independent variance structure (possibly preprocessed to achieve these characteristics); this leads us to consider what are called *stationary time series* or (in more limited form) *covariance-stationary time series* (see Section 10.5).

In the first mode, *time-domain* models and methods are concerned with representing changes from one time instant to another with emphasis on the serial correlation (*autocorrelation*) as a major defining element. In the second mode, *frequency-domain* models and methods (with corresponding *harmonic analysis*) regard the time series as a superposition of cyclic (harmonic) components which combine to generate the cyclic and random-variational form of the process as we observe it. Even a trend has been described as a combination of cyclic components whose periods exceed the length of the observed time series (Granger, 1966). Central to this approach is the notion of the *spectrum* of the process and of its *spectral decomposition*.

In both approaches all aspects of inference arise: model identification and fitting, estimation, testing, prediction (forecasting). We examine time-domain and frequency-domain aspects in Sections 10.5 and 10.6, respectively.

Of course, the time series may be *multivariate*, simultaneously representing a number of variables at any time (see Hannan, 1970; Bloomfield, 2000). Even the 'time' variable may not be scalar under special circumstances! We will only consider the univariate case here.

The literature on time series and the broader field of stochastic processes is wide-ranging, reflecting the extensive developments in these important fields since the pivotal text by Bartlett (1955). Some key texts on aspects of stochastic processes include Cox and Miller (1965), Feller (1966) Karlin and Taylor (1981) and Parzen (1962). For time series specifically, see Anderson (1994), Bloomfield (2000), Box *et al.* (1994), Brillinger (1975), Brockwell and Davis (1987), Fuller (1995), Hannan (1960, 1970) and Priestley (1981). The thorny issue of outliers in time series is discussed by Barnett and Lewis (1994, Chapter 10); see Section 10.6.1 below for further comment. A discussion of ecological time series is given by Solow (2002).

10.4   DESCRIPTIVE METHODS

We will concentrate on a discrete time series $\{X_t; t = 0, 1, 2, \ldots\}$ and its realization $\{x_t; t = 0, 1, 2, \ldots\}$. Consider first of all any possible *trend* in the process. Suppose we have

$$X_t = \mu_t + \varepsilon_t, \tag{10.1}$$

where μ_t describes how $E(X_t)$ varies with t (it is the *trend function*) and ε_t describes random variation ('noise') around the mean or trend. We would expect that $E(\varepsilon_t) = 0$ and normally assume that $Var(\varepsilon_t) = \sigma^2$, that is, that variability around the trend is independent of t. How are we to estimate μ_t or to transform the process to eliminate it, leaving a trend-free process?

10.4.1   Estimating or eliminating trend

If in (10.1) the trend may be assumed to be linear, even very simple descriptive measures can be and are applied. For example, we have the so-called *semi-average*, which provides a quick method of estimating the linear trend function. We simply divide the series into two equal parts and join the means of each part by a straight line (this has been used quite widely in economics).

More specifically, we can use linear model and regression methods, even generalized linear model methods (see Section 7.3), to estimate the trend whether or not we assume it to be linear. Linear (or, more usually, polynomial) regression methods may be used to estimate straight-line or curvilinear trends. However, if there is any periodicity or serial correlation the residual errors around the fitted model will reflect them and use of linear or generalized linear models is then at best an informal method. Of course, if we are able to assume no periodicity and that the ε_t are independent and identically distributed, we restore the properties of the methods as described in Chapter 7.

More complex trend functions for the mean μ_t may be needed and fitted using GLM principles. For example, if we anticipate an ogival form reaching a constant asymptotic value as $t \to \infty$, we might use the logistic model (see Section 8.2), expressed perhaps in the form

$$\mu_t = \alpha(1 + e^{-\beta t})^{-1}.$$

Another widely used form is the Gompertz growth-curve model, which can be written

$$\mu_t = \alpha - \beta \gamma^t, \quad \text{with } 0 < \gamma < 1.$$

For elimination of trend (rather than estimation) it can be useful to adopt a process of *differencing*. Consider first of all a differencing process of order 1. Here we transform the time series x_t to a new one y_t where

$$y_t = \Delta x_t = x_t - x_{t-1} \tag{10.2}$$

(here Δ is known as the *difference operator*). If X_t has linear trend, that is,

$$X_t = \alpha + \beta t + \varepsilon_t \tag{10.3}$$

where the ε_t are independent random errors with constant variance σ^2 (so-called 'white noise'), then

$$Y_t = X_t - X_{t-1} = \beta + \varepsilon_t - \varepsilon_{t-1} = \beta + \eta_t,$$

and we have eliminated the linear trend. However, we have paid a price for this advantage since it turns out that the first-order differencing enhances the variance and induces serial correlation. With Var $(\varepsilon_t) = \sigma^2$ we clearly have Var$(\eta_t) = 2\sigma^2$. Also

$$\text{Cov}(Y_t, Y_{t-1}) = \text{E}[(\varepsilon_t - \varepsilon_{t-1})(\varepsilon_{t-1} - \varepsilon_{t-2})] = -\text{E}(\varepsilon_{t-1}\varepsilon_{t-1}) = -\sigma^2.$$

So the first-order differencing has removed the linear trend but doubled the residual variance and introduced first-order serial correlation $\text{Corr}(Y_t, Y_{t-1}) = -\frac{1}{2}$.

Example 10.1   What happens if we use second-order differencing under similar assumptions of independent errors with constant variance? It can readily be seen that this will eliminate quadratic trend but the error variance is now $6\sigma^2$, and again we induce serial correlation. Thus, with second-order differencing we transform the time series x_t to

$$y_t = \Delta^2 x_t = x_t - 2x_{t-1} + x_{t-2}, \tag{10.4}$$

and if $X_t = \alpha + \beta t + \gamma t^2 + \varepsilon_t$ we find that $E(Y_t) = 2\gamma$ and $\text{Var}(Y_t) = 6\sigma^2$.

We notice two things about differencing. The process of differencing where we change one time series into another by means of a linear transformation is an illustration of applying what is called a *linear filter* to the process we are studying. This will prove to be a most important principle – it underlies, for example, the Box–Jenkins approach which we will be describing later.

Further, higher-order differences can be regarded as repeatedly applied first-order differences. Thus second-order differencing in Example 10.1 can be thought of as first-order differencing of the first-order differences of x_t. Again, this is an important principle we will see illustrated in the Box–Jenkins approach.

In general, pth-order differencing eliminates polynomial trend of order p but increases the residual error variance to $\binom{2p}{p}\sigma^2$. We should note that the method is useful only for polynomial trend. For example, a differenced sine wave is a sine wave!

As well as removing trend, linear filters can 'smooth' the variability in time series and provide 'local estimates' of trend through the time points for the purposes of prediction or forecasting. Consider, for example, the linearly transformed process

$$y_t = \frac{1}{3}(x_{t-1} + x_t + x_{t+1})$$

for all t. Suppose again that we have a linear trend with

$$X_t = \alpha + \beta t + \varepsilon_t.$$

Then x_{t-1}, x_t and x_{t+1} must lie effectively on a line (apart from random errors). We find that $E(Y_t) = \alpha + \beta t + \bar{\varepsilon}_t$, where $\bar{\varepsilon}_t = (\varepsilon_{t-1} + \varepsilon_t + \varepsilon_{t+1})/3$. So the transformed process conserves the trend, $\alpha + \beta t$, but with reduced residual variance since, if the ε_t are independent random errors with constant variance σ^2, Y_t will have variance $\sigma^2/3$.

Around any time point t we can, of course, estimate the scale and intercept parameters α and β locally from x_{t-1}, x_t and x_{t+1}. The local mean is $\alpha + \beta t$ and is, of course, estimated by y_t more accurately than by x_t. We use the notation

$\frac{1}{3}(1, 1, 1)$ to refer to this linear filter, and it has the effect of smoothing the process by fitting a linear form through each three successive time points.

Example 10.2   The same effect can be achieved with quadratic trend by using the linear filter $\frac{1}{35}(-3, 12, 17, 12, -3)$. This fits a quadratic form through five successive time-series values, and provides a smoothed version of the time series which conserves the quadratic trend but with reduced residual variance. In this case it is reduced to about $\sigma^2/2$.

The linear filters $\frac{1}{3}(1, 1, 1)$ and $\frac{1}{35}(-3, 12, 17, 12, -3)$ are examples of what is called a *moving average* process. These are specific examples of the general form

$$Y_t = \sum_{i=-r}^{s} \alpha_{t+i} X_{t+i} \tag{10.5}$$

which, if $\sum_{i=-r}^{s} \alpha_{t+i} = 1$, will have the effect of conserving the trend and smoothing out the local variation. Often we use the symmetric form having $r = s = q$, say, and $\alpha_i = \alpha_{-i}$ so that $\alpha_i = 1/(2q+1)$. Thus

$$W_t = (2q+1)^{-1} \sum_{j=-q}^{q} X_{t+j} = (2q+1)^{-1} \sum_{j=-q}^{q} \mu_{t+j} + (2q+1)^{-1} \sum_{j=-q}^{q} \varepsilon_{t+j} \sim \mu_t + \eta_t$$

if μ_t is approximately linear over $(t-q, t+q)$. By letting $\hat{\mu}_t = W_t$, updated estimates of the trend μ_t can be obtained as the process progresses, leading to a smoother process.

Another particular form of linear filter with $q = 7$ is the 15-point moving average $(-3, -6, -5, 3, 21, 46, 67, 74, 67, 46, 21, 3, -5, -6, -3)/320$, which conserves *cubic trend*.

One-sided moving averages can also be used for smoothing, particularly when we are interested in predicting or forecasting later values in the time process. For example, the predictor

$$\hat{\mu}_t = aX_t + (1-a)\hat{\mu}_{t-1}, \tag{10.6}$$

with $\hat{\mu}_1 = x_1$, is an *exponentially weighted moving average*, where

$$\hat{\mu}_t = \sum_{j=0}^{t-2} a(1-a)^j x_{t-j} + (1-a)^{t-1} x_1$$

$$= ax_t + a(1-a)x_{t-1} + a(1-a)^2 x_{t-2} \ldots + (1-a)^{t-1} x_1, \tag{10.7}$$

and we have a one-sided linear filter in which the earlier observations are more and more discounted or downweighted as we go further from the current time.

We should note that it has been observed that moving average methods tend to induce periodicity in the transformed error terms – this is the so-called Slutzsky–Yule effect.

10.4.2   Periodicities

We noted that another specific characteristic of a time series is the presence of possible cyclic components or periodicities. There are descriptive methods to handle this behavioural aspect of the process. Suppose the time-series realization looks like Figure 10.7, with periodic variation superimposed on a (rising) trend. We can again seek to estimate the periodic components or to eliminate them. If the period, p, of the seasonality is known, *ad hoc* approaches include the following:

1. We could *estimate* the periodic components in a trend-free time series (or one with almost no trend) by simply averaging successive observations of time separation p.
2. We could use the linear filter $y_t = x_t - x_{t-p}$ to seek to *remove seasonality* and *linear* trend. For example, $\Delta_{12} = x_t - x_{t-12}$ will remove annual seasonal effects in monthly data; higher-order differences (at constant separation) can again remove higher-order trends. These considerations underlie the widely used more formal Box–Jenkins method (see below).
3. We could carry out smoothing by (centred) moving average methods of the same extent as the seasonality. For example, for quarterly data, the seasonal effect might be removed by using $y_t = (x_{t-2}/2 + x_{t-1} + x_t + x_{t+1} + x_{t+2}/2)/4$.

Typical time-series models reflecting trend and periodicities can be expressed in the generic form

$$X_t = \mu_t + s_t + \varepsilon_t,$$

where μ_t is the trend, s_t is the periodic or seasonal effect and ε_t is the random-variational element. An alternative and sometimes more realistic representation is one

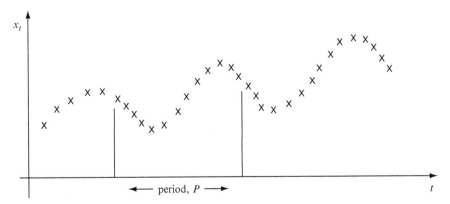

Figure 10.7   Time series with periodic component and trend.

in which the seasonal component is proportional in magnitude to the trend.

Having eliminated trend, a simple estimate of the seasonal component can be obtained (as in approach 1 above) merely by averaging over each 'season' or period over which it is reasonable to expect to find cyclic variation.

For a review of periodic time series, see Lund and Seymour (2002).

10.4.3 Stationary time series

Time series with trends and seasonal effects *develop in time*. Typically, the mean function μ_t (possibly also the variance funtion, $\sigma_t^2 = \text{Var}(X_t)$, and the covariances) *depend on t*. Sometimes, however, the process ε_t may be assumed to be a 'white noise' process – that is, to be made up of independent random quantities of mean 0 and constant variance. This does not develop in time. Such processes are important when we seek deeper understanding of time-series, and an important approach to time series analysis involves seeking to transform the process to this form before more detailed study. To this end, we are thus lead to define and to study what is termed a *stationary time series*.

Most formal study of time series models and methods is indeed concerned with stationary time series. The discrete time series $X_t(t = 1, 2, \ldots)$ is said to be *strictly stationary* if $(X_{t_1}, \ldots, X_{t_i})$ and $(X_{t_1+k}, \ldots, X_{t_i+k})$ have identical distributions for all $t_1, t_2, \ldots, t_i$ and k.

This is highly restrictive, often unnecessarily so, and difficult to confirm or justify – for practical use, *second-order stationarity*, or *covariance stationarity*, usually suffices. This requires merely that μ_t, σ_t^2 and $\text{Cov}(X_t, X_{t+s})$, $s = 1, 2, \ldots$, should be independent of t.

If we start with a time series which we know or suspect to have trend and seasonality, the usual approach is to seek to eliminate the trend and the seasonal effects and thus to produce a smoothed time series to which we hope to fit one of the many *second-order* stationary time-series models. These fall into two groups depending on the mode of study, as we remarked in Section 10.3. Firstly, we have *time-domain* models and methods which represent and examine how the correlational structure develops in time. An example would be what is termed a *first-order autoregressive process*, AR(1), defined by the relationship

$$X_t = \alpha X_{t-1} + \varepsilon_t \tag{10.8}$$

where the ε_t are independent and have mean 0 and constant variance σ^2. The *pth-order autoregressive process*, AR(p), is defined as

$$X_t = \alpha_1 X_{t-1} + \alpha_2 X_{t-2} \ldots \alpha_p X_{t-p} + \varepsilon_t.$$

Another example is known as a *pth-order moving average process*, MA(p), represented as

$$X_t = a_0 Z_t + a_1 Z_{t-1} + \ldots + a_p Z_{t-p}$$

and defined in terms of random variables Z_t which are independent with zero mean and constant variance, σ_Z^2.

Secondly, we have *frequency-domain* models and methods which stress the harmonic nature of the process and seek to represent it as a combination of superimposed sinusoidal terms, for example

$$X_t = \sum_{i=1}^{k} b_i \cos\left(\frac{2\pi i t}{\lambda} + \alpha_i\right) + \varepsilon_t.$$

Both types of models, and methods to infer their properties from data, are widely and successfully used. We start by examining the time-domain approach in more detail.

10.5   TIME-DOMAIN MODELS AND METHODS

The fundamental time-domain measures of a covariance stationary process X_t are: the *mean function*,

$$\mu_X = E(X_t);$$

the *variance function*,

$$\sigma_X^2 = \text{Var}(X_t),$$

and the *autocovariance function*

$$\gamma_s = \text{Cov}(X_t, X_{t+s}) \text{ with } \gamma_0 = \sigma^2,$$

all of which will be independent of t. The related *serial correlation of lag s,*

$$\rho_s = \frac{\gamma_s}{\gamma_0},$$

is also known as the *autocorrelation function*. We will have $\rho_0 = 1$ and $\rho_s = \rho_{-s}$. This provides the key characterization of the process in time-domain terms, and it is important to be able to interpret different qualitative forms for ρ_s. Let us examine some basic models.

We defined the AR(1) process in Section 10.4. We can successively substitute for $X_{t-1}, X_{t-2}, \ldots$ in (10.8) to give

$$X_t = \alpha X_{t-1} + \varepsilon_t$$
$$= \alpha(\alpha X_{t-2} + \varepsilon_{t-1}) + \varepsilon_t$$
$$= \cdots$$
$$= \sum_{i=0}^{\infty} \alpha^i \varepsilon_{t-i}$$

which shows that this process takes the form of an exponentially weighted sum of an infinite number of independent increments. (Note that the moving average process is a linear combination of a *finite* number of such increments.)

So what can we say of the AR(1) process? Firstly, if we have $E(\varepsilon_t) = 0$ and $\text{Var}(\varepsilon_t) = \sigma^2$, we find that $E(X_t) = 0$ and that

$$\text{Var}(X_t) = \sigma^2 \sum_{i=0}^{\infty} \alpha^{2i} = \frac{\sigma^2}{1 - \alpha^2}. \tag{10.9}$$

Note that (10.9) is independent of t. So we need $|\alpha| < 1$ for (10.8) to be a valid representation.

Now

$$E(X_{t+s}, X_t) = \sigma^2 \sum_{i=0}^{\infty} \alpha^{s+i} \alpha^i = \alpha^s \text{Var}(X_t).$$

So the autocorrelation function is readily confirmed to be $\rho_s = \alpha^s$ (see Figure 10.8).

The *second-order autoregressive process*, AR(2), defined by

$$X_t = \alpha X_{t-1} + \beta X_{t-2} + \varepsilon_t \tag{10.10}$$

is a particularly flexible model.

Example 10.3 Consider an AR(2) process. It is revealing (as an exercise for the reader) to determine its autocorrelation function and examine the various prospects that can arise for different values of the parameters α and β. Note that (10.10) is an example of the general pth-order form AR(p) defined in Section 10.4 with $p = 2$, $\alpha_1 = \alpha$ and $\alpha_2 = \beta$.

Consider now the *moving average process*. We observed that the AR(1) process can be expressed as an *infinite* moving average process with weights $a_i = a^i$. Finite moving average processes can also be useful models. Suppose

$$X_t = a_0 Z_t + a_1 Z_{t-1} + \ldots + a_p Z_{t-p} \tag{10.11}$$

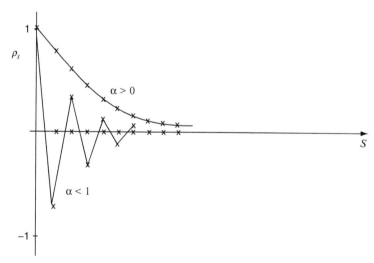

Figure 10.8   AR(1) autocorrelation function.

in which we usually take $a_0 = 1$. This is the moving average process of order p mentioned in Section 10.4; it has $E(X_t) = 0$ if the Z_i are independent with zero means and constant variances σ_Z^2. Furthermore,

$$\text{Var}(X_t) = \left(\sum_{i=0}^{p} a_i^2 \right) \sigma_Z^2,$$

and we readily confirm that

$$\gamma_s \begin{cases} = \left(\sum_{j=0}^{p-s} a_j a_{j+s} \right) \sigma_Z^2 & s = 0, 1, \ldots, p \\ = 0 & s = p = 1, p+2, \ldots \end{cases}$$

So what is this process like?

Example 10.4   Consider a moving average process of order p with

$$a_j = \frac{1}{p+1}.$$

This process is the simple average of $p + 1$ independent increments. We obtain

$$\rho_s = \begin{cases} 1 - \frac{s}{p+1} & s = 0, 1, \ldots, p \\ 0 & s = p+1, p+2, \ldots \end{cases}$$

so that we have a triangular autocorrelation function (Figure 10.9).

We should note a general form for representing autoregressive and moving average processes. If B is the backward shift operator satisfying

$$B^j X_t = X_{t-j}, \qquad j = 0, \pm 1, \pm 2, \ldots, \tag{10.12}$$

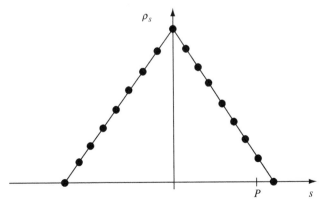

Figure 10.9   MA(p) autocorrelation function.

and $\phi(z)$ and $\theta(z)$ are pth- and qth-degree polynomials

$$\phi(z) = 1 - \alpha_1 z - \ldots - \alpha_p z^p$$

and

$$\theta(z) = 1 + a_1 z \ldots + a_q z^q$$

then the AR(p) and MA(q) processes can be represented as

$$\phi(B)X_t = Z_t \quad \text{and} \quad X_t = \theta(B)Z_t,$$

respectively, where Z_t is a trendless, independent-increment white noise process with variance σ^2.

So far we have just considered models in the time domain, specifically of autoregressive or moving average form. More sophisticated study of time series is not restricted to the simple autoregressive or moving average models of the underlying processes. What are used in detailed methodology (e.g. in the Box–Jenkins approach) are hybrid models described as autoregressive moving average (ARMA) or autoregressive integrated moving average (ARIMA) processes. These are again defined in terms of linear difference equations with constant coefficients. Let us consider how these are represented.

The *autoregressive moving average process* of orders p and q, denoted by ARMA(p, q), is defined by

$$\phi(B)X_t = \theta(B)Z_t. \tag{10.13}$$

An even more structured model representation is that which is termed an *autoregressive integrated moving average process* of orders p, d and q, denoted by ARIMA(p, d, q). This is defined by declaring that a transformed version of X_t in the form $Y_t = (1 - B^d)X_t$ has an ARMA (p, q) form.

In practice, we are unlikely to know what is a reasonable model for the process we are examining and, as with all statistical inquiry, we need to analyse the data to hand to infer properties of the underlying process. This might involve initially assessing the stationarity assumption, which in turn needs us to estimate the mean, variance and covariance function and to fit a reasonable model.

In the *Box–Jenkins approach* (Box et al., 1994; Anderson, 1982) we aim to fit (possibly with iteration) a plausible parametric model to a realization of length n of a discrete time series and to assess the adequacy of the fit. The fitted model is then used to express the form of the underlying process. A wide class of parametric families of models is entertained, all having the forms of linear transformations of a white noise process, Z_t, where the components are independent with zero mean and constant variance.

Essentially the aim is first to remove trend and periodic components and then to fit an ARMA(p,q) to the modified process. For example, the basic times series X_t might be 'pre-whitened' to remove annual seasonals and complex trend structure by transforming it to $Y_t = \Delta_{12} \Delta^m X_t$ for a suitable choice of m, and then modelled in the ARMA(p,q) form $\phi(B)Y_t = \theta(B)Z_t$ with the parameters estimated from the realization to hand.

In this brief review we will not be taking a detailed look at time-series inference, and the reader is referred to the various texts mentioned in Section 10.3 above. However, there is one estimator which we should examine: an estimator of the autocorrelation function. Thus if we have a realization $x_1, x_2, x_3, \ldots, x_n$ we can seek to estimate the autocorrelation function ρ_s in terms of an estimate of the autocovariance function and the variance function. There are several intuitively appealing moment estimates of these but perhaps the commonest form of *sample covariance function* is

$$g_s = \sum (x_t - \bar{x})(x_{t+s} - \bar{x})/n \qquad (10.14)$$

where $\bar{x} = \sum x_t/n$. Correspondingly, we estimate the autocorrelation function by $r_s = g_s/g_0$ and a plot of r_s against s is called the *correlogram*. It provides a useful pictorial representation of the observed time series. Some interesting illustrations are given by Chatfield (1996, pp. 22–24) showing typical correlograms for times-series realizations reflecting:

- random (time-independent) data, where the correlogram also shows random values typically between $\pm 2/\sqrt{n}$;
- short-term correlation, where the correlogram rapidly decays to insignificant values;
- oscillatory behaviour, where the correlogram alternates between positive and negative values of decaying magnitude;
- seasonal fluctuations.

Example 10.5 Let us examine the sunspot data in more detail. Brockwell and Davis (1987) present values of the sunspot index from 1770 to 1869 (see Table 10.1); their graphical representation (Figure 10.10) shows again the clear

Table 10.1 Sunspot numbers for 100 years from 1770 (as quoted in Brockwell and Davis, 1987).

1770	101	1788	131	1806	28	1824	8	1842	24	1860	96
1771	82	1789	118	1807	10	1825	17	1843	11	1861	77
1772	66	1790	90	1808	8	1826	36	1844	15	1862	59
1773	35	1791	67	1809	2	1827	50	1845	40	1863	44
1774	31	1792	60	1810	0	1828	62	1846	62	1864	47
1775	7	1793	47	1811	1	1829	67	1847	98	1865	30
1776	20	1794	41	1812	5	1830	71	1848	124	1866	16
1777	92	1795	21	1813	12	1831	48	1849	96	1867	7
1778	154	1796	16	1814	14	1832	28	1850	66	1868	37
1779	125	1797	6	1815	35	1833	8	1851	64	1869	74
1780	85	1798	4	1816	46	1834	13	1852	54		
1781	68	1799	7	1817	41	1835	57	1853	39		
1782	38	1800	14	1818	30	1836	122	1854	21		
1783	23	1801	34	1819	24	1837	138	1855	7		
1784	10	1802	45	1820	16	1838	103	1856	4		
1785	24	1803	43	1821	7	1839	86	1857	23		
1786	83	1804	48	1822	4	1840	63	1858	55		
1787	132	1805	42	1823	2	1841	37	1859	94		

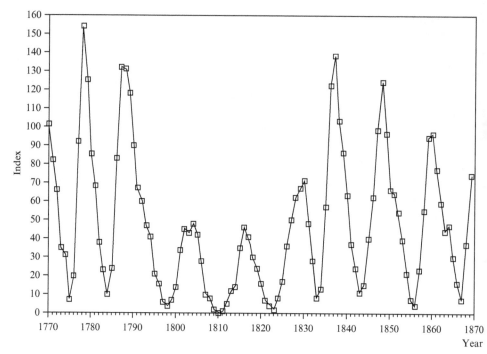

Figure 10.10   Plot of sunspot data of Table 10.1 (from Brockwell and Davis, 1987)

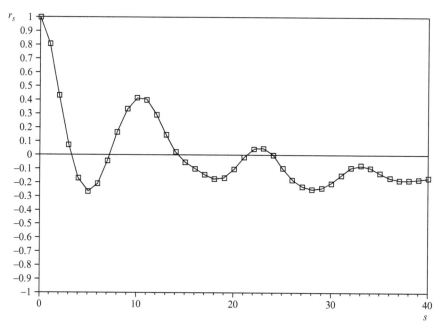

Figure 10.11   Correlogram of sunspot data of Table 10.1 (from Brockwell and Davis, 1987).

indication of periodicity. They calculate the correlogram, and it takes the interesting form shown in Figure 10.11.

We noted above that 'oscillatory' or periodically varying data would be expected to produce a cyclic form of correlogram of decreasing amplitude just as we have confirmed for the sunspot data.

How are we to interpret the characteristics of the correlogram? There are clear serial correlations in the data which we see even from examining the successive values in Table 10.1, and the correlogram shows a peak at about 10–11 years, but clearly there is much more structure to be investigated by more detailed analyses (see Example 10.6).

10.6 FREQUENCY-DOMAIN MODELS AND METHODS

We now consider some features of the dual frequency-domain representation of a time series. In the time-domain approach we represented a second-order stationary process X_t in terms of uncorrelated elements, ε_t, and serial correlation features. Fundamental frequency-domain measures centre on what is defined to be the *spectrum* of the process.

Specifically, in the frequency domain we represent a detrended and stationary process X_t in terms of periodic components of Fourier type:

$$X_t = \int_{-\pi}^{\pi} e^{i\omega t} dS(\omega) \tag{10.15}$$

where $S(\omega)$ is a stochastic process with zero mean and where $S(\omega)$ and $S(\omega')$, $\omega \neq \omega'$, are uncorrelated. This is known as the *spectral* (or Cramér) *representation*.

10.6.1 Properties of the spectral representation

The integral form (10.15) can be roughly interpreted as

$$X_t = \frac{2\pi}{m} \sum_{i=0}^{m-1} \left[A_i \sin\left(\frac{2\pi it}{m}\right) + B_i \cos\left(\frac{2\pi it}{m}\right) \right] \tag{10.16}$$

for large m and where the A_i and the B_i are random variables with zero means and are uncorrelated (i.e. as a linear combination of sine and cosine elements with random weights).

The measure $\mathrm{Var}[S(\omega)] = G(\omega)$ (say) will be a non-decreasing function with a maximum value $G(\pi)$ which is just $\mathrm{Var}(X_t) = \sigma_X^2$.

The *spectral distribution function* (or *integrated spectrum*)

$$F(\omega) = \frac{G(\omega)}{\sigma_X^2} \tag{10.17}$$

is typically monotone increasing from 0 to 1 and gives a full description of X_t; it represents the proportion of variability contributed by frequencies up to ω.

Also, as far as autocorrelation is concerned, we find that

$$\rho_t = \int_{-\pi}^{\pi} e^{i\omega t} dF(\omega)$$

(this is just the *characteristic function* of the distribution with distribution function $F(\omega)$).

The first derivative process $f(\omega) = F'(\omega)$ is called the *spectral density function* or the *spectrum*. Any discrete components in the spectrum (i.e. in $dF(\omega)$) can be interpreted as genuine periodicities in X_t; any continuous part is a measure of a range of 'short-term periodicities'.

Let us consider some examples. A spike in the spectrum, or a localized peak, will correspond respectively with a genuine periodicity or a 'quasi-periodicity'; see Figure 10.12.

To progress further with this approach we would need to consider detailed inference procedures. Estimation of the spectrum and derived quantities is difficult and needs much care to avoid anomalies (see Bloomfield, 2000; Brockwell and Davis, 1987).

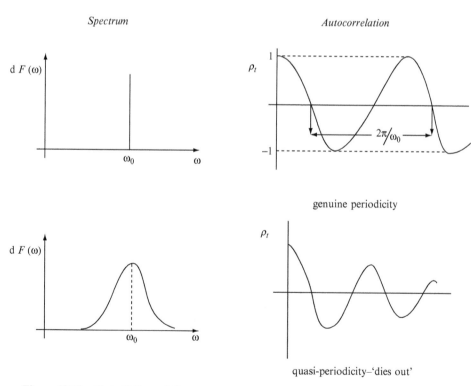

Figure 10.12 Periodicity and the spectrum.

However, one estimator we should consider here is the so-called period-ogram, which provides a sample estimate of the spectral density or spectrum $f(\omega)$. This takes the form

$$I_n(\omega) = \frac{1}{2\pi} \left| \sum_{t=1}^{n} (x_t - \bar{x}) \exp(-i\omega t) \right|^2 = \frac{1}{2\pi} \sum_{k=-n+1}^{n-1} g_k \exp(-i\omega k), \qquad (10.18)$$

with $\bar{x} = \sum x_t / n$ and where g_s is the sample covariance estimate $\sum (x_t - \bar{x})(x_{t+s} - \bar{x})/n$ as discussed above.

This is clearly an overparameterized statistic and will not be a consistent estimator. But it is asymptotically unbiased and provides asymptotically uncorrelated estimates at different frequencies $\omega_j = 2\pi j/n$. A consistent estimator can be constructed by taking an appropriate smoothed form of (10.18) with normed symmetric weights (see Fuller, 1995, Chapter 7; Brockwell and Davis, 1987).

Example 10.6 Let us return to the sunspot data we considered earlier. Bloomfield (2000, Section 6.4) discusses the harmonic analysis of the sunspot data and presents (as Figure 6.10) the periodogram of the data (see also Figure 5.10 of Bloomfield, 1976). As is often the case, the periodogram shows relatively little structure and suggests no clear evidence of periodicity. Nonetheless, there is a region of larger values which peaks at a period of just over 11 years. It is this sort of relatively weak evidence which underlies the belief that there is an 11-year period in heightened sunspot activity.

Bloomfield (2000, pp. 77–78) qualifies such conclusions by remarking that the sequence of peaks and troughs in the basic data are by no means regularly spaced, noting a gap of 15 years from 1787 to 1802 with no peaks and, in contrast, successive peaks in 1761 and 1769. Thus it would be misleading to name any single frequency as dominant. Further, the amplitudes of the peaks vary widely, as we can clearly see in the subset of the data presented in Table 10.1, and the patterns of the oscillations are irregular. Bloomfield concludes that it is thus not surprising that the harmonic analysis does not give a clear picture.

10.6.2 Outliers in time series

Analysis of outliers in time series is an important but complex issue. It is discussed in detail in Chapter 10 of Barnett and Lewis (1994) centring on the distinction between two fundamentally distinct types of contamination or outlier manifestation, *additive outliers* (AO) and *innovations outliers* (IO), corresponding with *isolated extreme values* superimposed on the time series and *inherent contamination* which transmits through the correlation structure.

Time-domain and frequency-domain methods have been developed and are concerned with tests of discordancy for outliers and outlier-robust (accommodation) procedures. In the latter context, the robust M-estimators (maximum likelihood type estimators; see, for example, Barnett and Lewis, 1994, pp. 148–151) have been developed for AR, MA, ARMA and ARIMA processes distinguishing between the AO and IO manifestations. In frequency-domain developments the emphasis is also on accommodation – on seeking to construct outlier-robust versions of the smoothed periodogram to estimate the spectrum. Outliers can seriously affect estimation of the spectrum if not properly taken account of. Additive outliers can cause the greatest problems. Some Bayesian methods for dealing with outliers (e.g. Abraham and Box, 1979) also feature the AO–IO distinction.

Although we will not be discussing time-series outliers in detail, we conclude this section by demonstrating some of the surprises that we can encounter in this field.

Example 10.7 Barnett and Lewis (1994, Chapter 10) illustrate how the spectrum of a time series can be markedly changed by the presence of just one or two mild outliers whilst, in contrast, a time series which apparently has many extreme outliers is in fact uncontaminated but also non-stationary.

Kleiner et al. (1979) contribute some significant proposals for frequency-domain robust estimation in the face of AO and IO outliers. They also discuss a practical problem with 1000 measurements of minute distortions in copper-coated steel ducting for telephone cables. Figure 10.13 (from Kleiner et al.,1979) shows the raw data and highlights two apparently modest outliers. What is surprising is the marked effect these can have when seeking to estimate the spectrum.

Figure 10.14 shows a non-robust and an outlier-robust estimate where it seems that the non-robust version (the dotted curve) is essentially exhibiting just the effect of the two outliers (above a frequency of 0.1).

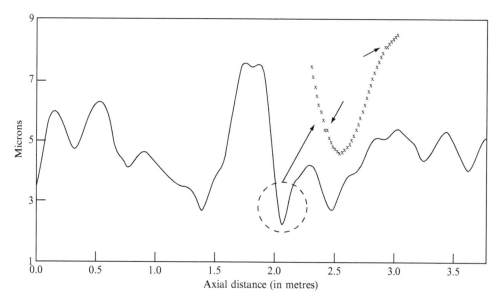

Figure 10.13 Distortions in copper-coated steel ducting (from Barnett and Lewis, 1994, and Kleiner et al., 1979).

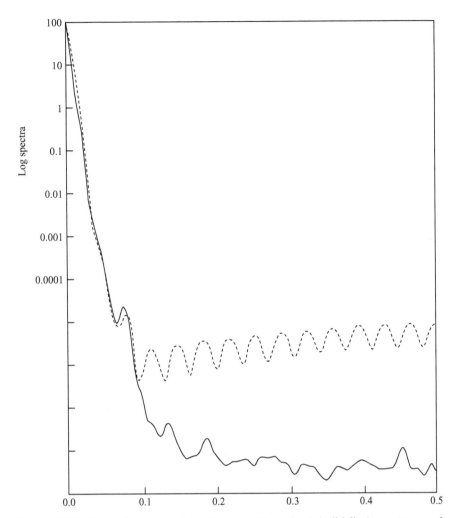

Figure 10.14 Non-robust (dotted line) and outlier robust (solid line) spectrum estimates of ducting data (Barnett and Lewis, 1994; Kleiner *et al.*, 1979).

Another anomaly is seen in the time-series data shown in Figure 10.15 (taken from Barnett and Lewis, 1994, attributed to Subba Rao, 1979). These data seem to have very clear and substantial (additive) outliers. But this is not so. The data are a simulated sequence of uncontaminated observations from the non-stationary process defined by

$$X_t = 0.4X_{t-1} + 0.8X_{t-1}Z_{t-1} + Z_t$$

where the Z_t are independent $N(0, 1)$. It looks stationary and seems to be contaminated with outliers, but neither of these features is present.

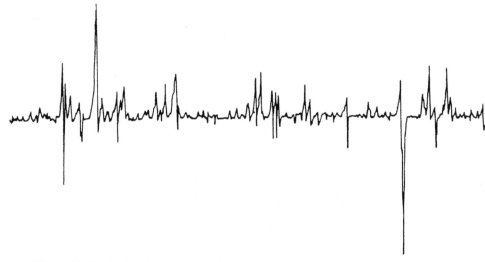

Figure 10.15 A simulated non-stationary process (from Barnett and Lewis, 1994; Subba Rao, 1979).

10.7 POINT PROCESSES

We noted earlier (in Section 10.1) a form of temporal process in which events of interest occur at specific time points $t_1, t_2, t_3 \ldots$. Such a process is a *point process* and is one of a class of processes reviewed by Cox and Lewis (1966) under the heading 'series of events' defined as 'point events occurring in a haphazard way in time or space'; see also Cox and Isham (1980). We will consider the univariate temporal form of such processes – this will provide relevant motivation for the discussion of spatial point processes in Section 11.1. Cox and Lewis (1966) introduce the topic by means of illustrations involving coal-mining disasters, faults in aircraft air-conditioning and failures of computers. The relevance to environmental problems is clear.

10.7.1 The Poisson process

Let us start with the simplest point process, which is known as a *Poisson process*. The Poisson process has the following characteristics:

- the number of events occurring in any interval of time of duration s has a Poisson distribution with mean λs, where λ is the *rate* of the process;
- this number of events is independent of the number of events occurring in any other *non-overlapping* interval of time of duration s.

Thus if $N_{t,s}(s > 0)$ is the number of events in $(t, t + s)$ we have

$$P(n) = P(N_{t,s} = n) = \frac{(\lambda s)^n e^{-\lambda s}}{n!} \quad (s = 0, 1, 2, \ldots .) \tag{10.19}$$

and $N_{t,s}$ is independent of $N_{u,s}$ for $u > t + s$. This process arises if the probability of an event occurring in an infinitesimally small interval $(t, t + \delta t)$ is *constant* at $\lambda \delta t$, independent of the occurrence of an event in any other distinct infinitesimally small interval and if at most one event can occur in any such infinitesimally small interval.

We notice that there is no development in time – no trend; λ is independent of t, as is the random variable $N_{t,s}$. The Poisson process has been found to be a good model for a wide range of time-varying occurrences of a specific event where randomness of occurrence in time is the principal feature, for example emissions from a radioactive source (over a time period which is small in relation to the half-life) or breakdowns in electronic monitoring equipment.

What can we say about the time intervals of consecutive events? If we denote by X the time to the first event, we have

$$P(X > x) = P(N_{0,x} = 0) = e^{-\lambda x}, \tag{10.20}$$

so that X has an *exponential distribution* with parameter λ (or mean $1/\lambda$).

Because of the time-independence features of the Poisson process, we can conclude that the distribution of the time from the time origin to the first event is the same as the distribution of the time from any event to the next one, or from an arbitrary time instant to the next event. Also the conditional distribution of $X - x_0$, given that $X > x_0$, has the same exponential distribution as X; this is known as the 'lack-of-memory property'.

There is another characteristic of the Poisson process which has important implications. Given that n events have occurred in a time period $(0, t)$, it is readily shown that the time points of the events are uniformly distributed in $(0, t)$. This has implications for inference. In particular, we can simulate a Poisson process by choosing at random an observation from the appropriate Poisson distribution (10.19) and then distributing that observed number of events uniformly (in probabilistic terms) over the time interval in question.

Example 10.8   Barnett and Lewis (1994, Example 6.2) discuss traffic flow data and consider times on one day at which every fourth vehicle travelling in a specific direction on a particular road in Hull passes an observer. In all just 36 vehicles were observed giving the eight observations of the times (in seconds) between four vehicles:

$$17 \quad 6 \quad 18 \quad 12 \quad 40 \quad 8 \quad 8 \quad 21$$

They ask whether, if the flow is random, the observation 40 is a discordant outlier (e.g. due to some adventitious hold-up in the traffic flow.) With random flow we have a Poisson process so that the sample would be made up of independent observations of a gamma variate with shape parameter $r = 4$.

A test for an upper outlier in a gamma sample (the Barnett and Lewis test Ga1) uses a test statistic

$$\frac{x_{(n)}}{n\bar{x}} = \frac{40}{130} = 0.3077.$$

The upper 5% point for this statistic is 0.3043. So we conclude that there is some ground for concern about whether the outlier 40 is indeed a genuine observation from an assumed Poisson process.

10.7.2   Other point processes

We may often need to depart from or modify the Poisson process to properly represent different practical situations. For example, a more general type of point process, in which we assume merely that intervals between events are independent and identically distributed, may be appropriate in some circumstances. Such a process is termed a *renewal process* (see Cox, 1962, for basic results in this field).

Alternatively, we might start with the Poisson process but remove the time-independence assumption by letting λ be a function of time: $\lambda = \lambda(t) = \lambda_0 + \lambda_1 t$. This introduces a trend in the process. The straight linear form just mentioned is not usually chosen in view of the restrictions required on λ_0 and λ_1 to ensure that the *rate* remains positive. Instead, a more common and useful form is $\ln(\lambda_0 + \lambda_1 t)$. This is of course just a generalized linear model (see Section 7.3). Its probabilistic properties are readily established, and inferences from sample data are straightforward.

A more sophisticated variation is one where the rate function $\lambda(t)$ is a stationary stochastic process, or time series, in continuous time. Cox and Lewis (1966) call such a process a *doubly stochastic Poisson process*.

Two other forms of process which should be mentioned for their potential importance in environmental problems are (a, b) *processes* and *superimposed processes*. In the former, we have a process where events occur fairly regularly with time intervals a but with perturbations b around the regular pattern, where b is a random variable of mean zero and standard deviation which is typically small relative to a. This has been applied to the supply of crude oil by tanker from distant oil fields.

In superimposed processes we jointly consider a series of point processes as a single process. For example, we may have a system made up of many subsystems each subject to failures from time to time. The failure characteristics of the overall system are then described by the superimposed process. The subprocesses may or may not be identical (e.g. some components may be more prone to failure than others). There is an important *limit law* which declares that under fairly wide conditions on the subprocesses themselves *the superimposition of a large number of subprocesses behaves essentially like a Poisson process* (see Cox and Lewis, 1966, Chapter 8).

Important related areas of study with widespread relevance to environmental issues are those of reliability and survival analysis where internal probabilistic features and overall lifetime characteristics of systems are modelled and analysed. See, for example, Mann et al. (1974) and Lee (1992). There is special relevance in areas of monitoring, testing equipment, sensors, response of humans to debilitating environmental effects, etc. Interesting coverage of survival methods in epidemiology with clear environmental relevance is provided by Newman (2001).

Example 10.9 Barnett and Kenward (1996, 1998) propose and examine an interesting environmental point process model to describe the social nuisance phenomenon of the occurrence of false alarms on security systems. One interest is in whether routine maintenance of the security systems is in any way statistically related to the incidence of false alarms. To this end they consider the sequences of false alarms which occur between successive maintenance checks. In modelling this situation, we have, essentially, a hierarchical point process for the maintenance check times $x_1, x_2, \ldots$ and, within this process, for the false alarms $y_{i1}, y_{i2}, \ldots, y_{in_i}$ which occur during the ith intermaintenance period (x_i, x_{i+1}).

Many questions arise, such as whether the y_{ij} are distributed at random in (x_i, x_{i+1}), and whether maintenance affects the false alarm process. A specific form of this latter question is to examine whether $Y_{i1} - X_i$ (the time from maintenance to the next false alarm) might be stochastically larger than $X_i - Y_{i-1, n_{i-1}}$ (the time from the previous false alarm to maintenance). Such matters are considered in detail, with various models fitted to extensive real-life data.

It is common for there to be regulatory standards for maintenance of security systems. In the UK, these typically require two maintenance checks each year. Do these help? If false alarms occur at random in time, independent of maintenance checks, then no timing of such checks can be useful in reducing their incidence due to the familiar lack-of-memory property of the exponential distribution. The time to a false alarm from any maintenance time will follow the same exponential distribution as that of the inter-false-alarm process.

But the false alarms may be affected by the maintenance check process! Barnett and Kenward (1998) examine this prospect specifically with a surprising conclusion, namely, that the expected value of $Y_{i1} - X_i$ is smaller than the expected value of $X_i - Y_{i-1, n_{i-1}}$, concluding that the maintenance check seems to *accelerate* the occurrence of a false alarm!

CHAPTER 11

Spatial Methods for Environmental Processes

...about the space of two hours...

We have just considered point processes (Section 10.7) where events occur in time in some random or indeterministic way. For example, in a Poisson process events occur purely at random in time so that the times between events are independent and follow an exponential distribution. Such a model might apply, for example, to times of failure of some environmental monitoring device or even to the times of occurrence of environmental and social catastrophes such as earthquakes, volcano eruptions or accidents; see Carn (2002) and Cox and Lewis (1966), who analyse data on intervals in time between coal mining disasters from Maguire *et al.* (1952). Such a process developing in time and starting from a time base defined to be zero time can be represented thus:

with $t_0 = 0$.

Suppose instead that we are progressing along a transect line (see Section 6.3) through a forest and marking the distances d from the starting point $d_0 = 0$ at which we observe a particular rare mammal and that we see specimens at d_1, d_2, d_3 and d_4. We can represent this process in terms of distance (in *space*) in precisely the same way as the *time* process above, replacing the symbol t with the symbol d. So our *spatial* process looks just like our time process. Indeed, Cox and Lewis (1966), in reviewing models and methods for the study of 'series of events', are at pains to point out their interest in 'point events occurring in a haphazard way in space or time'. So we can start our study of spatial processes precisely in terms of such an analogy. Why should spatial processes be any different from temporal processes?

There are some obvious qualitative differences: time is ordered (we progress from one time to a later time), *space* is not. Further, on any normal definition, time is unidimensional whereas space often needs to be considered in two- or three-dimensional form. Such qualitative differences are material to our interests but do not prevent useful carry-over from the study of time processes to

Environmental Statistics   V. Barnett
© 2004 John Wiley & Sons, Ltd   ISBN: 0-471-48971-9 (HB)

that of spatial processes. We will start with this emphasis on duality (see also Section 11.7 for more graphic illustration of this duality with a three- or four-dimensional representation of space and time simultaneously).

11.1 SPATIAL POINT PROCESS MODELS AND METHODS

The general spatial model is one which describes the characteristics of a random variable $Y(\mathbf{x})$, where Y is a random outcome variable occurring at a position $\mathbf{x}$ in n-space. Often Y is scalar and $\mathbf{x}$ is two-dimensional, representing a planar region.

Consider, first, a data set giving locations and girths of pine trees in a 200 m × 200 m region of a forest (as discussed by Cressie, 1993, p. 580). These data are presented graphically in Figure 11.1 (from Cressie, 1993, who tabulates the complete underlying data set of locations and breast-height diameters for 584 longleaf pine trees over a 4-hectare study region). Here Y is a quantitative variable, the girth of the tree, and $\mathbf{x}$ is its location. We should note that in this example the $\mathbf{x}$-values are not prescribed (as we might wish them to be in seeking to describe any *relationship* between girth and location); we have not decided beforehand at what set of values of $\mathbf{x}$ we will observe Y. Instead, the $\mathbf{x}$-values *occur at random in some sense*; they are where the trees happen to be found.

This is a *spatial point process*. In its most basic form we might just be interested in *occurrence*: in the locations $\mathbf{x}$ at which an event occurs, for example where a longleaf pine tree is growing. In other contexts, we might have a spatial point process where $Y(\mathbf{x})$ represents the occurrence of a childhood leukaemia case over the geographic region D or the presence of a specific astronomical feature in space, S. In such problems, we have $Y = 0$ over most of D (or S), but at isolated points we will put $Y(\mathbf{x}) = 1$ to signify an observed case of the phenomenon of interest at location $\mathbf{x}$. This is a direct analogue of the point process in time; see the discussion in Section 10.7.

It is important to consider what models can be used to describe where the values $Y(\mathbf{x}) = 1$ will be encountered, that is, where the event of concern occurs. Natural interest centres on how we are to represent, in probabilistic terms, different possible forms of outcome, such as randomness, clumpiness, clustering, regularity, and pattern. Indeed, certain notions of complete randomness, of regularity and of clustering are well understood and readily modelled probabilistically. Corresponding statistical methods exist to analyse them in the context of observed data. Such matters are included in the detailed coverage of spatial processes by Cressie (1993) and Ripley (1981). See also Upton and Fingleton (1985, 1989).

The homogenous Poisson process and the Cox process are two examples of models for randomness; the Poisson cluster process allows us to express 'clumpiness'. Various more regularly patterned forms can also be modelled and analysed, usually in terms of perturbations around a deterministically defined pattern such as a lattice or regular tessellation.

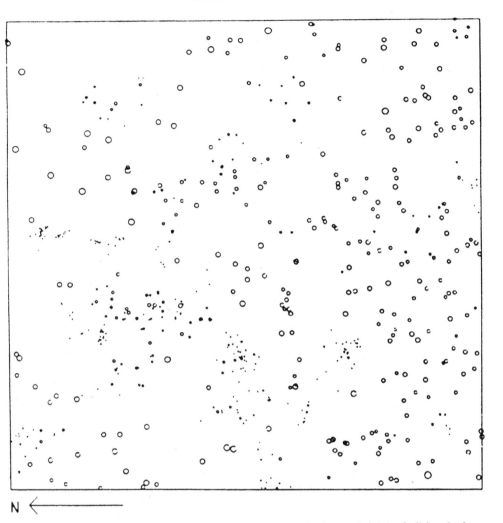

N ←――――――――――

Figure 11.1 Map of locations and relative diameters (at breast height) of all longleaf pines in the 200 m × 200 m (4 ha) study region in 1979 (from Cressie, 1993, Figure 8.1).

Let us start with the simplest model, that of the *Poisson process*. It mirrors the one-dimensional Poisson process in time discussed in Section 10.7. Suppose $\mathbf{x} = (x_1, x_2)$. Here the probability of encountering an event in a small rectangular region where x_1 goes from x_{10} to $x_{10} + \delta x_1$ and x_2 goes from x_{20} to $x_{20} + \delta x_2$ is assumed to be proportional to the area of the region. Thus

$$P(x_{10} < X_1 < x_{10} + \delta x_1, x_{20} < X_2 < x_{20} + \delta x_2) = \lambda \delta x_1 \delta x_2, \qquad (11.1)$$

where λ is the *intensity* of the process. Events in distinct small rectangular regions are assumed independent. Again it is readily shown that the distribution of the number of events in a finite region $\{x_{10} < X_1 < x_{10} + a, x_{20} < X_2 < x_{20} + b\}$ is $P(\lambda ab)$, that is, Poisson with mean λab.

This model is widely used as a 'null model' – as an expression of no structure, or of *complete spatial randomness* against which more structured forms can be compared and contrasted, and statistically tested if data are available. For example, let us return to the tree-girth data of Figure 11.1, ignoring girths and considering only the locations of the trees as a realization of a spatial point process. If we fit a Poisson process model to the locations of the tree, we find that it is not sustainable. Cressie (1993) samples the location data by means of 100 random quadrats of diameter 6 m and shows that the resulting frequency distribution of numbers of trees in the quadrats cannot reasonably be assumed to follow a Poisson distribution (this prospect is rejected by a χ^2 gooodness-of-fit test).

This is hardly surprising. Presumably the pine trees were originally planted according to some more or less regular pattern of equally spaced planting positions, and some semblance of this pattern still remains. Obviously, the clear regularity of the original planting design may well be much distorted by losses of trees, and in-fill sapling growth, over the years. Could it be that there is statistical evidence of regularity of planting or, in contrast, of clustering of trees in small groups rather than at random? Before considering *models* to represent such departures from complete spatial randomness, it is interesting to note that more qualitative measures (or indexes) have been proposed to make such distinctions. Cressie (1993) reviews many of these, including such intuitively natural choices as the 'relative variance', $I = S^2/\bar{r}$, which is the ratio of the sample variance and the mean of the number of events in a sample of regions of fixed size, and other measures based on it. For a Poisson process, we obviously expect $I = 1$, with values greater or less than one being indicative of clustering and of regularity, respectively. For the pine tree data we have $I = 2.34$, suggesting a degree of clustering – but see below for further discussion.

For more detailed study, we could try to fit more complex models. An obvious extension of the Poisson process (11.1) is to set $\lambda = \lambda(x_1, x_2)$, allowing the 'rate' of the process to vary with the location, **x**. This is then known as an *inhomogeneous Poisson process*, with *intensity function* $\lambda = \lambda(x_1, x_2)$. In any application, the intensity function will need to be delimited – by, for example, expressing it in some parametric form – and will need to be estimated from the data. Lawson (1988) illustrates this for data on cancer incidence.

An alternative development of the Poisson process is one where λ is assumed to be a *random variable* rather than a constant. This is known as a *Cox process* (Cox, 1955; Lundberg, 1940). When λ is degenerate, taking the fixed value λ with probability one, we revert to the homogeneous Poisson process as a special case. Many levels of Cox process can be considered, ranging from those which specify a fixed spatially homogeneous distribution for λ to those which allow spatial dependence, and other factors such as Markovian linking, into the distribution of λ.

An important class of models for clustering starts with the so-called Neyman–Scott process; see Neyman (1939) and, in the context of positions of stars and galaxies, Neyman and Scott (1958).

Here, we start with a homogeneous Poisson process to produce the locations of a first generation of outcomes. We then assume that at each realized location a random number N of offspring is produced independently according to some fixed non-negative distribution with probability function $p(n)$. These offspring are then assumed to be located at positions relative to the parent according to a homogeneous probability distribution $f(\mathbf{z})$ centred on the location $\mathbf{z}$ of the parent. The process is made up of the offspring. Bartlett (1975) has shown that certain Neyman–Scott processes and certain Cox processes are formally equivalent.

Example 11.1 Consider again the data on pine tree girths discussed in Cressie (1993, Chapter 8). We rejected the Poisson process as an adequate description of the underlying process, informally preferring a clustering alternative. Cressie (1993, pp. 667–669) fits a Neyman–Scott process with Poisson numbers of offspring at circular normal locations. The analysis again supports a degree of clustering but the fitted Neyman–Scott model is also not adequate. The data are reluctant to reveal their structure!

We have considered models for randomness and for clustering. What about *regularity*? Lattice structures are possible. Consider again the pine trees. Suppose we postulate that the pine trees were originally planted on a rectangular lattice with distances α and β in the X_1 (north–south) and X_2 (east–west) directions, respectively. We would (apart from edge effects in small regions) have had equal numbers in each distinct non-overlapping region of fixed area. Losses over the years would somewhat reduce the original numbers, so we might expect to find numbers which were fairly constant but with perturbations of small coefficient of variation – much less than the value of $\sqrt{(\lambda ab)^{-1}}$ we expect for a Poisson process or equivalently, as remarked above, with *variance ratio* less than one in value. (Sapling infill might cloud the regularity and, in contrast, suggest a degree of clustering.) We can check these prospects through such informal statistical measures as described above – noting, in fact, an inclination towards clustering – or we might again seek to set up (and fit) a more structured model to represent regularity.

Cressie (1993) proposes a class of *inhibition models* which have an in-built propensity to restrict the prospect of outcomes being less than a certain distance apart. For example, we might have a homogeneous Poisson process but delete all pairs of outcomes in any realization which are less than a distance δ apart. This model was proposed by Matérn (1960). An alternative proposal by Matérn involves labelling any outcome of the homogeneous Poisson process with a value of a random variable Z. Then any original outcome is deleted if there is another within distance δ with a smaller Z-value.

Various generalizations have been advanced; see, for example, Bartlett (1975), Diggle *et al.* (1976) and Stoyan (1988). Ripley and Kelley (1977) propose a different form of inhibition model based on Markov point processes. See also Diggle (1975).

Both Ripley (1981) and Cressie (1993) offer extensive discussion of models and methods for spatial point processes. In particular, Cressie (1993, Chapter 8) presents, for each of a wide range of models, information and discussion on *simulation, model fitting, parametric* and *non-parametric estimation* of relevant characteristics and *diagnostics*. Cressie (1993, Section 8.6) discusses broader issues of modelling and analysing *multivariate* spatial point processes.

11.2   THE GENERAL SPATIAL PROCESS

Suppose $Y(\mathbf{x})$ is the random outcome of observing a variable Y at position $\mathbf{x}$ ($\mathbf{x}$ may be a point on a line, in a plane, in space, etc.). We will continue to concentrate on the two-dimensional problem where $\mathbf{x}$ is a point in a plane. In Section 11.1, $Y(\mathbf{x})$ took a very special form represented as predominantly 0 except at a set of distinct points $\mathbf{x}$ at which it took the value 1 to reflect the presence of some outcome of interest. We called such processes *spatial point processes*.

For the pine tree data of Figure 11.1 we considered only the locations of the trees as the spatial variable, ignoring the quantitative outcome representing the girth of the tree. Instead $Y(\mathbf{x})$ could have been that girth measure. Again it would be non-zero only at points where trees were found but at those points it would take values from some non-negative continuous distribution rather than being restricted to 0 or 1. Even more generally, $Y(\mathbf{x})$ might be discrete or continuous, univariate or multivariate and take positive or negative values at *any* point $\mathbf{x}$. We will consider only the univariate case – a single quantity is observed (or is observable) at a point $\mathbf{x}$ usually in the plane, i.e. in two-dimensional space.

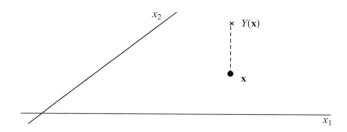

There are many possible interests. Suppose we observe $Y(\mathbf{x})$ at a finite number of points $\mathbf{x}_1, \mathbf{x}_2, \ldots, \mathbf{x}_n$. We might want to use the observed values $y(\mathbf{x}_1), y(\mathbf{x}_2), \ldots, y(\mathbf{x}_n)$:

- to model $Y(\mathbf{x})$ and estimate relevant parameters;
- to fit a relationships between $Y(\mathbf{x})$ and a covariate z (e.g. between rainfall Y at location $\mathbf{x}$ and altitude z);

- to interpolate or predict $Y(\mathbf{x}_0)$ at an unobserved point $\mathbf{x}_0$ within or beyond the region in which we have taken the observations.

We will start with the third of these interests.

11.2.1   Predication, interpolation and kriging

Suppose $Y(\mathbf{x})$ is observed at n sites, giving observations $Y(\mathbf{x}_1)$, $Y(\mathbf{x}_2)$, $\ldots$, $Y(\mathbf{x}_n)$, and our interest is in predicting (estimating, interpolating) some function $g[Y(\mathbf{x})]$. For example, for Cressie's (1993) pine tree data (Figure 11.1) we might wish to estimate the mean tree girth ('diameter at breast height') over the study region or the proportion of the trees with girth in excess of 50 cm.

It may be that we expect there to be no trend in $Y(\mathbf{x})$ over a study region D (or that the trend has been removed) but we anticipate spatial correlation from one location to another. A possible model in this situation declares that

$$Y(\mathbf{x}) = \mu + \varepsilon(\mathbf{x}), \tag{11.2}$$

in which μ is an unknown constant and $\varepsilon(\mathbf{x})$ is an *intrinsically stationary* random process over D. We need to define this concept of intrinsic stationarity: it requires that

$$E[\varepsilon(\mathbf{x} + \mathbf{h}) - \varepsilon(\mathbf{x})] = 0$$

and

$$Var[\varepsilon(\mathbf{x} + \mathbf{h}) - \varepsilon(\mathbf{x})] = 2\gamma(\mathbf{h}) \tag{11.3}$$

for two points a distance $\mathbf{h}$ apart; $2\gamma(\mathbf{h})$ is known as the *variogram*, which will turn out to be a pivotal measure of the spatial process in terms of our current studies ($\gamma(\mathbf{h})$ is termed the *semivariogram*).

Suppose now that we want to predict or to estimate some known functional $g(\{Y(\mathbf{x}); \mathbf{x} \in D\})$ of the random process $Y(\mathbf{x})$. More simply, we will refer to the quantity to be predicted as $g[Y(\mathbf{x})]$ or $g[Y]$.

Let the *predictor* of $g[Y(\mathbf{x})]$ be $p(Y, g)$. We will restrict attention to predictors taking the *linear form*

$$p(Y, g) = \sum_{i=1}^{n} \lambda_i g[Y(\mathbf{x}_i)] \tag{11.4}$$

subject to $\sum_{i=1}^{n} \lambda_i = 1$ to ensure unbiasedness. We will aim to obtain optimal predictors (estimators) in the sense that, within the class (11.4) and subject to $\sum_{i=1}^{n} \lambda_i = 1$, they minimize the expected squared error $E\left[\{g[Y(\mathbf{x})] - p(Y, g)\}^2\right]$.

Before proceeding to explore the prediction process we need to examine some possible forms for $g[Y(\mathbf{x})]$.

Forms of $g[Y(\mathbf{x})]$

A typical problem is to predict the average value of Y over some block B within D. Here we would have

$$g[Y] = \int_B Y(\mathbf{x})d\mathbf{x}/|B| = Y(B), \qquad (11.5)$$

say, where $|B|$ is the area of B. At an opposite extreme, but with obvious practical importance, we might want to predict $Y(\mathbf{x}_0)$, the value of $Y(\mathbf{x})$ at some specific new (unobserved) point $\mathbf{x}_0$. In this situation $g[Y] = Y(\mathbf{x}_0)$. This is just a special limiting case of (11.5) where $B = \{\mathbf{x}_0\}$.

The condition $\sum_{i=1}^{n} \lambda_i = 1$ implies *uniform unbiasedness* in the sense that

$$E[p(Y, g)] = E[p(Y:B)]$$
$$= \mu$$
$$= E[Y(B)].$$

Optimal prediction will be approached using a method known as *kriging*, or more specifically *ordinary kriging* (we will consider other forms later), where (as remarked above) the aim is to seek to minimize

$$\sigma_E^2 = E\{[Y(B) - p(Y, g)]^2\}$$

for the predictor

$$p(Y:B) = \sum_{i=1}^{n} \lambda_i g[Y(\mathbf{x}_i)]$$

subject to $\sum_{i=1}^{n} \lambda_i = 1$. This particular approach, which is proving very useful in environmental studies, illustrates the successful transference to the environmental domain of a body of techniques originally devised for a different and rather exotic field of investigation, exemplified by mining exploration for gold.

The broader methods of *geostatistics* for the local exploration of spatial aspects of the surface structure of the earth were pioneered by the work of Georges Matheron and his associates in the French schools of mining at Fontainebleau; see, for example, Matheron (1963, 1965, 1969). The development of specific ideas for prediction from empirical data stems from the proposals of Krige (1966) after whom the particular method of *kriging* is named although the general approach echoes work by Matérn in the early 1960s and approaches by Bartlett and by Whittle at a similar time. Introductory treatments of the topic within the geostatistics context are provided by Webster and Oliver (2001) who outline basic principles and methods and illustrate them with interesting agricultural examples. A detailed and extensive mathematical treatment is provided by Cressie (1993). See also the fine review of model-based geostatistics by Diggle *et al.* (1998).

In the special case $B = \{\mathbf{x}_0\}$, ordinary kriging allows us to range over different $\mathbf{x}_0$ and to 'map' $Y(\mathbf{x})$ over some region of interest. We will use the *variogram*, defined in (11.3) above as

$$2\gamma(\mathbf{h}) = \text{Var}[Y(\mathbf{x}) - Y(\mathbf{x} + \mathbf{h})] = E\{[\varepsilon(\mathbf{x}) - \varepsilon(\mathbf{x} + \mathbf{h})]\}^2$$

as the measure of proximity, or relationship, of $Y(\mathbf{x})$ and $Y(\mathbf{x} + \mathbf{h})$ at different distances $\mathbf{h}$. We will need to assume at this stage that $\text{Var}[Y(\mathbf{x}) - Y(\mathbf{x} + \mathbf{h})]$ depends only on the separation h and is independent of the *direction* of the separation. Such a model is said to be **isotropic**. Of course $\gamma(\mathbf{h})$ will not be known in any practical situation and will need to be estimated. In the process we will also need to explore from empirical evidence whether or not our two assumptions are plausible.

Before pursuing such practical matters further, let us examine the optimization process for an assumed form of isotropic variogram. We consider the special case of $B = \{\mathbf{x}_0\}$ to start with. Using the method of Lagrange multipliers, we need to minimize

$$\mathrm{E}\left[\left(Y(\mathbf{x}_0) - \sum_{i=1}^{n} \lambda_i Y(\mathbf{x}_i)\right)^2\right] - 2\phi\left(\sum_{i=1}^{n} \lambda_i - 1\right) \tag{11.6}$$

subject to

$$\sum_{i=1}^{n} \lambda_i = 1.$$

This is made easier by noting that

$$\left(Y(\mathbf{x}_0) - \sum_{i=1}^{n} \lambda_i Y(\mathbf{x}_i)\right)^2 = Y^2(\mathbf{x}_0) - 2Y(\mathbf{x}_0)\sum_{i=1}^{n} \lambda_i Y(\mathbf{x}_i) + \left\{\sum_{i=1}^{n} \lambda_i Y(\mathbf{x}_i)\right\}^2$$

$$= Y^2(\mathbf{x}_0) - 2Y(\mathbf{x}_0)\sum_{i=1}^{n} \lambda_i Y(\mathbf{x}_i) - \sum_{i=1}^{n} \lambda_i Y^2(\mathbf{x}_i)$$

$$+ \sum_{i=1}^{n}\sum_{j=1}^{n} \lambda_i \lambda_j Y(\mathbf{x}_i) Y(\mathbf{x}_j) + \sum_{i=1}^{n} \lambda_i Y^2(\mathbf{x}_i)$$

$$= \sum_{i=1}^{n} \lambda_i (Y(\mathbf{x}_0) - Y(\mathbf{x}_i))^2 - \frac{1}{2}\sum_{i=1}^{n}\sum_{j=1}^{n} \lambda_i \lambda_j (Y(\mathbf{x}_i) - Y(\mathbf{x}_j))^2. \tag{11.7}$$

So, under the model $Y(\mathbf{x}) = \mu + \varepsilon(\mathbf{x})$, we have to minimize, from (11.6),

$$2\sum_{i=1}^{n} \lambda_i \gamma(\mathbf{x}_0 - \mathbf{x}_i) - \sum_{i=1}^{n}\sum_{j=1}^{n} \lambda_i \lambda_j \gamma(\mathbf{x}_i - \mathbf{x}_j) - 2\phi\left(\sum_{i=1}^{n} \lambda_i - 1\right). \tag{11.8}$$

Differentiating with respect to $\lambda_1, \lambda_2, \ldots, \lambda_n$ and ϕ, and equating to zero, we must therefore solve the system of equations

$$-\sum_{i=1}^{n} \lambda_i \gamma(\mathbf{x}_i - \mathbf{x}_j) + \gamma(\mathbf{x}_0 - \mathbf{x}_i) - \phi = 0, \quad i = 1, 2, \ldots, n,$$

$$\sum_{i=1}^{n} \lambda_i = 1,$$

and we find that the optimal $\lambda_1, \lambda_2, \ldots, \lambda_n$ thus satisfy the matrix equation

$$\boldsymbol{\lambda}_0 = \mathbf{M}^{-1}\boldsymbol{\gamma}_0, \tag{11.9}$$

where

$$\boldsymbol{\lambda}_0' = (\lambda_1, \lambda_2, \ldots, \lambda_n, \phi)$$
$$\boldsymbol{\gamma}_0' = (\gamma(\mathbf{x}_0 - \mathbf{x}_1), \gamma(\mathbf{x}_0 - \mathbf{x}_2), \ldots, \gamma(\mathbf{x}_0 - \mathbf{x}_n), 1)$$

and

$$M_{ij} = \begin{cases} \gamma(\mathbf{x}_i - \mathbf{x}_j) & i, j = 1, 2, ..., n \\ 1 & i = n + 1, j = 1, 2, ..., n \text{ and } j = n + 1, i = 1, 2, ..., n \\ 0 & i = n + 1, j = n + 1 \end{cases}$$

$$\tag{11.10}.$$

So $\mathbf{M}$ is a symmetric $(n + 1) \times (n + 1)$ matrix that has the partitioned form (11.10) for which we can obtain its inverse using, for example, the general result in Rao (1973, p. 49),

$$\begin{pmatrix} \mathbf{A} & \mathbf{B} \\ \mathbf{B}' & \mathbf{D} \end{pmatrix}^{-1} = \begin{pmatrix} \mathbf{A}^{-1} + \mathbf{FE}^{-1}\mathbf{F}' & -\mathbf{FE}^{-1} \\ -\mathbf{E}^{-1}\mathbf{F} & \mathbf{E}^{-1} \end{pmatrix},$$

where $\mathbf{E} = \mathbf{D} - \mathbf{B}'\mathbf{A}^{-1}\mathbf{B}$ and $\mathbf{F} = \mathbf{A}^{-1}\mathbf{B}$. Thus we find $\boldsymbol{\lambda} = (\lambda_1, \lambda_2, \ldots, \lambda_n)$ from (11.9) and (11.10) in the form

$$\boldsymbol{\lambda}' = \left(\boldsymbol{\gamma}_0 + \mathbf{1}\left[\frac{1 - \mathbf{1}'\mathbf{G}^{-1}\boldsymbol{\gamma}_0}{\mathbf{1}'\mathbf{G}^{-1}\mathbf{1}} \right] \right)' \mathbf{G}^{-1} \tag{11.11}$$

and

$$\phi = -\left(\frac{1 - \mathbf{1}'\mathbf{G}^{-1}\boldsymbol{\gamma}_0}{\mathbf{1}'\mathbf{G}^{-1}\mathbf{1}} \right), \tag{11.12}$$

where $\mathbf{G}$ is just the $n \times n$ matrix, $\mathbf{G} = \{\gamma(\mathbf{x}_i - \mathbf{x}_j)\}$.

This is the optimal prediction process which is known as *ordinary kriging*. The solution involves a reasonable amount of computational manipulation but is manageable. Of course, it depends on *knowing* the semivariogram $\gamma(\mathbf{h})$ and on it being *isotropic*. In practice, we can only examine these prospects empirically, and in practice we will need to use an *estimate* of the semivariogram.

The optimal (ordinary kriging) predictor is denoted $\hat{p}(Y; \mathbf{x}_0)$, or it might be expressed as $\hat{Y}(\mathbf{x}_0)$, where

$$\hat{Y}(\mathbf{x}_0) = \sum_{i=1}^{n} \lambda_i Y(\mathbf{x}_i) \tag{11.13}$$

where the λ_i satisfy (11.11) above.

The minimized mean-square error of prediction, from (11.6), is known as the *kriging variance* or *prediction variance* and takes the form

$$\sigma_0^2(\mathbf{x}_0) = \boldsymbol{\lambda}_0' \boldsymbol{\gamma}_0$$

$$= \sum_{i=1}^{n} \lambda_i \gamma(\mathbf{x}_0 - \mathbf{x}_i) + \phi$$

$$= \boldsymbol{\gamma}_0' \mathbf{G}^{-1} \boldsymbol{\gamma}_0 - \frac{\left(\mathbf{1}'\mathbf{G}^{-1}\boldsymbol{\gamma}_0 - 1\right)^2}{\mathbf{1}'\mathbf{G}^{-1}\mathbf{1}}. \qquad (11.14)$$

Equivalently, it can be expressed as

$$\sigma_0^2(\mathbf{x}_0) = 2 \sum_{i=1}^{n} \lambda_i \gamma(\mathbf{x}_0 - \mathbf{x}_i) - \sum_{i=1}^{n} \sum_{j=1}^{n} \lambda_i \lambda_j \gamma(\mathbf{x}_i - \mathbf{x}_j).$$

Example 11.2   Observations of fish catches (in tonnes) at two sites 6 km apart in a commercial fishing field were 55.26 and 14.17, respectively. Kriging has been regularly used for prediction over this fishing field and a well-based empirical estimate of the semivariogram is $\gamma(x) = 5.1[1 - \exp(-0.49x)]$, where x is the distance in kilometres between two sites in an assumed isotropic situation. We want to predict the likely catch at a site 1 km from the former, and 5.4 km from the latter, reference sites.

The coefficients in the optimal ordinary kriging predictor, from (11.9), are

$$\begin{pmatrix} \lambda_1 \\ \lambda_2 \\ \phi \end{pmatrix} = \begin{pmatrix} 0 & 4.83 & 1 \\ 4.83 & 0 & 1 \\ 1 & 1 & 0 \end{pmatrix}^{-1} \begin{pmatrix} 1.976 \\ 4.738 \\ 1 \end{pmatrix}$$

$$= \begin{pmatrix} -0.1035 & 0.1035 & 0.5 \\ 0.1035 & -0.1035 & 0.5 \\ 0.5 & 0.5 & -2.415 \end{pmatrix}^{-1} \begin{pmatrix} 1.976 \\ 4.738 \\ 1 \end{pmatrix} = \begin{pmatrix} 0.786 \\ 0.214 \\ 0.942 \end{pmatrix}.$$

So the predicted catch at the reference point will be

$$\hat{Y}(\mathbf{x}_0) = \sum_{i=1}^{n} \lambda_i Y(\mathbf{x}_i) = 0.786 \times 55.26 + 0.214 \times 14.17 = 46.47 \text{ tonnes.}$$

11.2.2   Estimation of the variogram

The approach above assumes that $\gamma(\mathbf{h})$ is known. In practice, we will need to estimate it from sample data; see Cressie (1993, Section 2.4) and the detailed discussion in Webster and Oliver (2001, Chapter 5). Model-based estimation has been widely examined; see Webster and Oliver (2001, Chapter 6) for an extensive review. The effect of estimation of $\gamma(\mathbf{h})$ on the kriging variance is not very well understood: the kriging variance is likely to exceed $\sigma_0^2(\mathbf{x}_0)$ due to the added uncertainty arising from having to estimate $\gamma(\mathbf{h})$. Nonetheless, practical applications of kriging usually just replace $\gamma(\mathbf{h})$ with a sample estimate.

Estimation of $\gamma(\mathbf{h})$ also usually assumes *isotropy*, so that for empirical estimation we can plot sample estimates $\hat{\gamma}$ merely against radial distance r to form estimates $\hat{\gamma}(r)$ of the semivariogram. *Anisotropic* situations will, of course, require quite different and more complex procedures.

The observed data often suggest some characteristics of $\gamma(r)$ which require interpretation (Figure 11.2). It may be that

$$\gamma(0) = C_0 \neq 0.$$

But $\gamma(0) = \lim_{h \to 0} [\gamma(\mathbf{x} + \mathbf{h}) - \gamma(\mathbf{x})]$, which should in principle be zero. The interesting fact is that *in practice* $\hat{\gamma}(0) = C_0 \neq 0$. This is likely to reflect measurement errors, and C_0 (from its mining origins) is called the *nugget effect*, *nugget* or *nugget variance*, the term arising from Matheron (1962), who explained the effect as due to microscale variation ('small nuggets'). The role of the nugget in modelling $\gamma(r)$ is discussed by Cressie (1993, Section 2.3.1).

Another characteristic that is observed, and modelled, is the existence of a *constant* asymptotic semivariogram value. This is called the *sill* (s in Figure 11.2) and it represents the *maximum possible amount of variability* between $Y(\mathbf{x})$ values at distinct points.

A further characterizing feature is the radial distance between the origin and the point at which the asymptotic maximum value sets in: this is known as the *range* and takes the value a in Figure 11.2.

Much attention has been paid to what forms of variogram (what models) make sense both in practical terms (e.g. in respect to observed characteristics of soils or rocks) and in fundamental mathematical terms; again see Webster and Oliver (2001, Chapter 6).

There has also been much corresponding study of how to estimate the variogram from empirical data, with particular stress on the challenges in a two- (or higher-)dimensional framework. An obvious basic form of estimator of $\gamma(r)$ *for sufficient data* is the sample average squared difference between all observed $Y(\mathbf{x})$ values which are at a distance r apart; the so-called *classical estimator*. Alternatively, robust estimators have been proposed (Cressie, 1993,

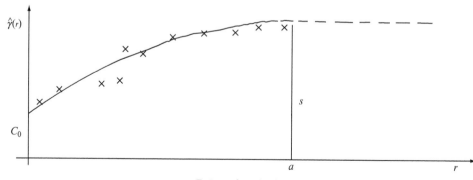

Estimated semivariogram

Figure 11.2   Estimated semivariogram.

Section 2.4.3) or specific forms are adopted to model the variogram, and these models are fitted to the sample data. Certain features must be common to all models. We would obviously expect that $\gamma(\mathbf{h})$ should increase with $|\mathbf{h}|$. In turn, this means that closer points must have more influence on our estimation of $Y(\mathbf{x}_0)$ than those further away (reflected in larger weights λ_i). Further, the variogram will need to be monotone non-increasing as $r \to 0$, with a minimum value $\gamma(0)$ which may or may not be zero depending on whether we anticipate and model a nugget effect. A wide variety of forms for $\gamma(\mathbf{h})$ have been proposed, fitted and generally studied (see Webster and Oliver, 2001, Chapters 5 and 6).

Example 11.3   Webster and Oliver (2001) present and discuss many practical examples of kriging. It is interesting to consider one of the examples that they discuss in some detail, concerning the potassium content of the soil at Broom's Barn research station (a sister site to the renowned agricultural research station at Rothamsted, which is about 100 km away). Broom's Barn is an approximately rectangular flat site of about 80 ha in area with a road entering in the middle of the long side and running to the centre of the site. A number of test sites (435 in all) indicated direction isotropy and provided an empirical estimate (the 'classical estimate') of $\gamma(\mathbf{h})$ in the expected general form – as shown in Figure 11.3 (note the nugget effect and the sill).

A number of different models were fitted to the data to seek to refine the estimated variogram. Ordinary kriging can be applied to any accepted form of the variogram to obtain the optimal predictor of potassium level at any new site and to map the levels of potassium over the whole Broom's Barn farm.

Figure 11.4 (from Webster and Oliver, 2001) shows for a particular assumed variogram the resulting weights which would be applied to different observed values, depending on their distance from the point at which we wish to predict

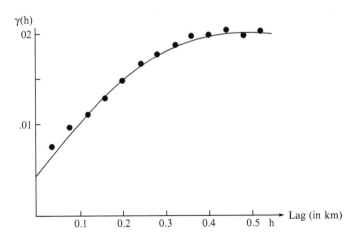

Figure 11.3   Estimated variogram for Broom's Barn data (from Webster and Oliver, 1990).

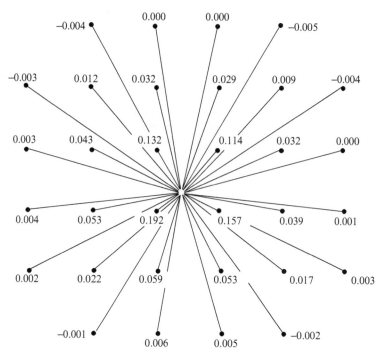

Figure 11.4   Kriging weights for Broom's Barn data (from Webster and Oliver, 1990).

the potassium level. We see that the nearest sample point attracts about 20% of the weight; the four nearest, over 60%.

It is inevitable that we should want to go beyond local estimation, to the mapping of a whole region. This can be done by producing local estimates at all points on a network or grid covering the region. We can then fit a surface to the set of discrete estimates. Alternatively, the estimates can be displayed (e.g. via computer graphics) either in raw or in a smoothed or interpolated form (e.g. as a shaded figure or perspective drawing).

There are many sophisticated computer software systems available for this purpose. Some impressive illustrations of the output are given by Webster and Oliver (2001) for the Broom's Barn data; Figures 11.5 and 11.6 show the kriging estimates and their estimated variances, respectively, of potassium content over the whole farm based on 50 m × 50 m block estimates and 435 observations overall.

11.2.3   Other forms of kriging

There is much potential in the kriging method, both for environmental studies in general as well as in particular disciplines such as agriculture, geology, mining, and soil science. Developments continue apace and comparisons are made with other possible approaches; for example, Voltz and Webster (1990)

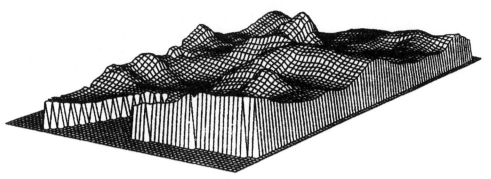

Figure 11.5 Estimates of potassium levels at Broom's barn using block kriging (from Webster and Oliver, 2001, Figure 8.17).

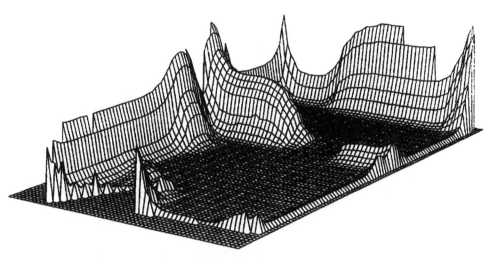

Figure 11.6 Estimation variances of potassium at Broom's Barn in block kriging (from Webster and Oliver, 2001, Figure 8.21).

compare kriging, cubic splines and broader classification methods in soil studies.

Ordinary kriging as described above is only the most basic approach. Many extensions of the model, interpretation and method are possible – see, for example, Cressie (1993) and Webster and Oliver (2001). There are specific alternative kriging methods for univariate responses $Y(\mathbf{x})$ which we will briefly describe below; details can be found in the references.

Block kriging. In any approach to kriging we can think of 'target' sites: those where we wish to predict or estimate $Y(\mathbf{x})$. The target sites may be specific point locations or blocks, B, of such locations. In the latter case we can seek to predict, say, the expected value of $Y(\mathbf{x})$ over B; see (11.5) above. Kriging for expected values over blocks is termed *block kriging*. The illustration of kriging for potassium levels at Broom's Barn (see Example 11.3 above) actually uses block kriging over 50 m × 50 m blocks. Punctual kriging (see below) could be used here instead, and Webster and Oliver (2001) compare this approach – see, for example, their Figure 8.15.

Punctual kriging. This is kriging where, in contrast to block kriging, we aim to predict $Y(\mathbf{x})$ at a specific point location $\mathbf{x}$. Ordinary kriging in the form described and developed in detail above is an example of this.

Indicator kriging. There are situations where we are interested in a response variable $Y(\mathbf{x})$ which takes a particularly simple form: it is either 0 or 1 at any point $\mathbf{x}$. This would be the case, for example, if we classified locations in terms of whether or not some environmental standard was met. Thus if the ozone level exceeded the local standard the point might be coded 1, otherwise 0. We can use *indicator kriging* in this situation to predict whether a new point location should be coded 1 or 0, that is, declared in violation of the standard or satisfying it. The method in fact estimates the expected value (at a point or over a block) of the indicator variable. The expected value is just $P(\mathbf{Y}(\mathbf{x}) = 1)$: the probability of violating the standard.

Disjunctive kriging. As for indicator kriging, where we considered the question of standards violation as an illustration, we may be concerned not with a step-function classification of a site as in violation or complying with the standard (i.e. $Y(\mathbf{x}) = 1$ or 0, respectively) but with the probability $P(Y(\mathbf{x}) = 1)$ of violation of the standard. If violation is equivalent to, say, $Y(\mathbf{x}) > y_0$, a form of non-linear kriging of the basic $Y(\mathbf{x})$ (not its indicator equivalent) known as *disjunctive kriging* can handle such problems with advantages over indicator kriging. These include the modelling and estimation of fewer variograms and the solution of a less extensive set of kriging equations if certain distributional assumptions hold; see Cressie (1993, pp. 278–281) or Webster and Oliver (2001, Chapter 10).

Robust kriging. This uses robust estimators of variograms in the ordinary kriging context.

Universal kriging. This replaces the stationary model (11.2) with one in which the mean is no longer constant but varies with the location $\mathbf{x}$ as an unknown linear combination of a set of k specific functions. So

$$Y(\mathbf{x}) = \mu(\mathbf{x}) + \varepsilon(\mathbf{x}),$$

where $\mu(\mathbf{x}) = \sum_{i=1}^{i=k} \alpha_i f_i(\mathbf{x})$ for specified functions $f_i(\mathbf{x})$.

A major development would be to extend the kriging approach to the case of a multivariate response $\mathbf{Y}(\mathbf{x})$ – for example potassium and sodium concentrations and pH on the Broom's Barn farm site. Progress has been made on this important topic (see McBratney and Webster, 1983).

One particular approach is known as *cokriging*. Suppose we have data as $k \times 1$ vectors $\mathbf{Y}(\mathbf{x}_1), \mathbf{Y}(\mathbf{x}_2), \ldots, \mathbf{Y}(\mathbf{x}_n)$ where $\mathbf{Y}$ has components $Y_1, Y_2, \ldots, Y_k$. We might wish to predict $Y_1(\mathbf{x}_0)$ but to do so taking account of $Y_2(\mathbf{x}_i), Y_3(\mathbf{x}_i), \ldots, Y_k(\mathbf{x}_i)$ for $\mathbf{x}_i, i = 1, 2, \ldots, n$. That is, we want to take advantage of *covariates* $Y_2, Y_3, \ldots, Y_k$ observed at the same sample points as the basic variable Y_1.

For example, a pollutant Y_1 tends to occur with other pollutants Y_2 and Y_3. We want to predict Y_1, but taking account of the association between levels of Y_1 and those of Y_2 and Y_3. The method of cokriging allows this to be done by predicting $Y_1(\mathbf{x}_0)$ in terms of a linear combination of *all* observed $\mathbf{Y}(\mathbf{x})$ values (i.e. smoothing over $Y_1, Y_2, \ldots, Y_k$ *and* $\mathbf{x}_1, \mathbf{x}_2, \ldots, \mathbf{x}_n$). Note, however, that more basic methods such as *regression* could have been applied in such situations. With cokriging, no form of relationship between Y_1 and $Y_2, Y_3, \ldots, Y_k$ is assumed or implied.

Example 11.4   Bown (2002) discusses an interesting application of kriging in the context of setting standards over a region. Suppose we are concerned with air quality throughout a metropolitan area, and levels are monitored at a set of n networked sites. We could apply local standards at each of the network sites when readings are taken, perhaps each month. To conform with the standard it is necessary for the level of the 'pollutant' to be no more than y_0, *say*. Then a violation would occur if in any month any of the sites was showing a reading in excess of the standard level y_0. We have here the classical multiple-comparisons problem. If exceedance is assessed in terms of some test of significance, then the larger the value of n (and hence the more impressive the level of monitoring of the air quality in the metropolitan area), the larger the number of spurious 'exceedances' which will be encountered. We could try to deal with this problem using standard statistical measures, but there is much to be said for setting some truly regional standard based on the accumulated data rather than attributing conformity or violation to each specific location separately.

One possibility is to set what Bown (2002) terms a *reference point standard*. This takes the form of a requirement on the relevant air quality measure at some single specified reference point to be assessed by using the network data from all sites. The reference point is not necessarily a network site; indeed, it is unlikely to be so.

A possible approach which is explored in some detail is to predict the pollution level at the reference site by kriging using the complete set of network data. Thus, given observations $Y(\mathbf{x}_1), Y(\mathbf{x}_2), \ldots, Y(\mathbf{x}_n)$, we obtain the kriged value $Y(\mathbf{x}_0)$ for the reference point $\mathbf{x}_0$ and conclude that the standard is met or is violated depending on whether $Y(\mathbf{x}_0) < y_0$ or $Y(\mathbf{x}_0) > y_0$, respectively, for an appropriate choice of the value y_0.

What is 'appropriate' needs to be assessed in terms of the principles outlined in the discussion of setting standards in Chapter 9 above; it will vary depending on what is the population characteristic (e.g. mean or percentile) for which we wish to obtain specified statistical assurance through application of the SVIS principle – see Section 9.2.

Thus, suppose we set a standard on the mean μ of the distribution of a pollutant level X by saying that we require $\mu < \mu_0$. Then a SVIS might ask for statistical assurance at a 95% level, and this can be implemented by conducting a 5% test of significance of the hypothesis

$$H: \mu < \mu_0$$

against the alternative

$$\overline{H}: \mu < \mu_0.$$

This will need to be operated in terms of the null distribution of the appropriate statistic which is, for a reference point standard, the kriged reference point predictor $\hat{Y}(x_0)$. So all depends on being able to determine this null distribution, which is not necessarily a simple matter. Bown (2002) discusses this issue in some detail.

11.3 MORE ABOUT STANDARDS OVER SPACE AND TIME

We have just considered a situation where levels of some environmental agent are monitored over a network of sites and at different times. We saw in Example 11.4 how kriging might help in assessing whether levels were meeting some prescribed standard. The notion of the 'level' of a pollutant (say) is a somewhat nebulous one. We often need to distinguish locally and environmentally generated (additional) inputs of the pollutant from the levels at which it occurs naturally at the site of interest. That is, we have two sources: *generated* and *background*. Depending on circumstances, we may decide to set standards on the total pollution levels (background and generated), on the grounds that it is this total effect which must be controlled; or we might set standards on just the excess (generated) pollution levels, arguing that the 'polluter' can hardly be responsible for the natural (background) contamination already present. These distinctions are difficult to maintain and represent since often we can only measure the *total amount* of a polluting agent.

Yet another manifestation of the notion of pollution levels is found in monitoring networks – large or small. If we measure the pollution levels at the network sites at different times we may find that isolated unusually high levels are encountered from time to time or for the odd site. Consider a landfill facility with a set of monitoring sites. A particular site, due to geological characteristics or dumping patterns, may be 'hotter' than other sites, or it may turn out that at some specific time much of the region, or a site or two, may 'blow hot'. Such *hotspots* (over space or time) are important but in a different way from the overall distributionally representative pollution levels.

Thus we might set a standard on the overall pollution levels, possibly implemented using a reference point standard as described in Example 11.4, with the standard level μ_0 determined by general considerations of public safety and comfort. On the other hand, we might have to accept the *rare* occurrence of hotspot behaviour which might not violate some overall standard even if instantaneously excessive. But it is likely that in concern for the operation of the landfill facility rather than the surrounding region (i.e. its internal ambience and safety of staff employed there) we also need to maintain a check on hotspot levels by operating a relevant *hotspot standard*.

Bown (2002, Chapter 9) explores this very problem, showing how to detect hotspots of contamination activity in a region, estimating their levels and setting up an appropriate hotspot standard. Following a proposal by Barnett, a procedure based on outlier methods is developed in terms of composite sampling (thus avoiding the need for multiple testing at sites or times), for both normal and gamma-distributed contamination levels where hotspots are interpreted as *outliers* or *contaminants* on a location-slippage model (see Section 3.3 above) in respect of the underlying distribution of pollution levels. Three types of hotspot activity are discussed and parameter estimators provided, given that a hotspot has been identified. These consist of cases where there is hotspot behaviour over the facility at an isolated time, or in an isolated site over various times or at a single site and single time. The methods are illustrated for an example on hotspot behaviour on land in Australia contaminated by arsenic.

11.4   RELATIONSHIP

We have considered an approach called *cokriging* in Section 11.2.3 where we wished to infer characteristics of one component of a multivariate spatial process in terms of the covariational import of the other components. Specifically, we considered data in the form of $k \times 1$ vectors $\mathbf{Y}(\mathbf{x}_1), \mathbf{Y}(\mathbf{x}_2), \ldots, \mathbf{Y}(\mathbf{x}_n)$ where $\mathbf{Y}$ has components $Y_1, Y_2, \ldots, Y_k$. We aimed to predict $Y_1(\mathbf{x}_0)$ at some new point $\mathbf{x}_0$ by taking account of $Y_2(\mathbf{x}_i), Y_3(\mathbf{x}_i), \ldots, Y_k(\mathbf{x}_i)$ for $\mathbf{x}_i, i = 1, 2, \ldots, n$. The *covariates* $Y_2, Y_3, \ldots, Y_k$ are, of course, observed at the same sample points as the basic variable Y_1.

A typical example concerns the prediction of a key environmental variable Y_1 (e.g. levels of some air pollutant) when we also have available in that situation the corresponding levels of two other pollutants Y_2 and Y_3 which we expect to be correlated with those of Y_1. Thus if we want to predict Y_1 taking account of the association between the levels of Y_1, and those of Y_2 and Y_3, we can use the *cokriging* method. We noted that with cokriging no specific form of relationship between Y_1 and Y_2 and Y_3 needs to be adopted.

But, in the spirit of the relationship models discussed in Chapter 7, we could use more basic methods such as *regression* or *linear model* analyses instead of cokriging. Thus we could regress Y_1 on $Y_2, Y_3, \ldots, Y_k$, assuming

$$E(Y_1) = h(y_2, y_3, \ldots, y_k) \tag{11.15}$$

for observed (or fixed) values $y_2, y_3, \ldots, y_k$. We would need to specify a particular form for the model (11.15). A simple case would be the *linear model*

$$E(Y_1) = \alpha + \sum_{i=2}^{k} \alpha_i y_i$$

to allow for an intercept term. We would then use standard methods (e.g. least squares) to fit this linear model to n data points $y_2(\mathbf{x}_j), y_3(\mathbf{x}_j), \ldots, y_k(\mathbf{x}_j)$ for $j = 1, 2, \ldots, n$.

For a new point, x_0, a number of possibilities arise. If $y_2(x_0)$, $y_3(x_0), \ldots, y_k(x_0)$ are given or observed, the least-squares fit enables us to obtain an immediate estimate of $E[Y_1(x_0)]$, say $\tilde{Y}_1(x_0)$. For example, Y_1 may be minimum temperature over a winter period, Y_2 and Y_3 may be latitude and longitude, and Y_4 altitude. This is the situation described in Section 1.3.3 where we wish to interpolate minimum temperatures at sites which are *not* meteorological stations in terms of their locations and altitudes. For a new location Y_2, Y_3 and Y_4 will be known.

Alternatively, we might, in the case where $y_2(x_0), y_3(x_0), \ldots, y_k(x_0)$ are not known, *use the model* to predict $Y_1(x)$ at all $x_i (i = 1, 2, \ldots, n)$ and then do *simple kriging* of the $\hat{Y}_1(x_i)$ to predict $Y_1(x_0)$.

Both approaches assume a well-structured model. If this is not felt appropriate, we can revert to *cokriging*.

Example 11.5   Let us return to an example introduced in Section 1.3.3. The results by Landau *et al.* (1998) on the winter wheat models and climatic conditions encountered at wheat trial sites illustrate different ways of interpolating (predicting) outcomes over space (our current interest) and time (which we considered in the previous chapter).

We recall that a major data set was assembled of wheat yields for about 1000 sites over many years throughout the UK where wheat was grown under controlled and well-documented conditions. These data constituted the 'observed yields' which were to be compared, for validation purposes, with the corresponding yields which the wheat models would predict. However, crucial inputs for the models included the daily maximum and minimum temperatures and the rainfall and radiation levels.

Such climatological measures were available for all relevant days and for 212 meteorological stations throughout the relevant parts of UK. However, these meteorological stations were not of course located at the sites at which the wheat yields were measured. So it was necessary to carry out a major interpolation exercise for the meteorological variables. This was successfully achieved as described by Landau and Barnett (1996); see also Section 1.3.3 above.

This weather interpolation example provides an illustration of a major interest in spatial analysis as developed in this chapter; spatial interpolation and the use of kriging. But it also moves us on to a more complex consideration!

Spatial interpolation was carried out (Landau and Barnett, 1996) by various methods to examine which of them proved most efficient. These included: simple pointwise numerical interpolation with no concern for statistical variability; point-by-point kriging; and regression analyses using spatial coordinates and altitudes as covariates with kriging of residuals. However, there was need to conduct such studies for a series of different years, thus introducing also a temporal component. This was achieved by applying the three methods just described separately for the different years.

These methods were complemented by a joint spatial-temporal approach. For each of the meteorological stations a second-order autoregressive time-series model was fitted over the years and its two parameters were estimated. These estimates were then spatially interpolated for the different wheat trial sites using regression and kriging. The joint spatial/temporal approach was the one finally adopted since it produced interpolation estimates which were at least as good as the separate spatial estimates for each year separately and with much more parsimonious parameterization.

Example 11.5 provides our first illustration of how it might be possible to examine *influences of space and time simultaneously*. This is of major concern for environmental study, and although modelling and methods for such *conjoint* interest are far less well developed that others we have examined, they represent the current frontier of concern. We will briefly consider what progress has been made in this difficult area in Section 11.7.

11.5 MORE ABOUT SPATIAL MODELS

We discussed point process models in Section 11.1. These related to situations where we were concerned about whether or not a specific condition applied at particular spatial points. The response variable $Y(\mathbf{x})$ expressed merely the presence or absence of the condition.

In Section 11.2, our discussion of kriging related to a quantitative response variable $Y(\mathbf{x})$ expressed in the main by a parsimoniously specified stationary model, with constant mean response and modest second-order variational assumptions, described in terms of the variogram and under the assumption of isotropy. More structure was introduced in Section 11.4, where it was assumed that the mean response for any component in a multivariate spatial response $Y(\mathbf{x})$ depended in covariational form on the other components of Y.

Of course, progressively more structured models of the spatial response variable, even for the univariate response $Y(\mathbf{x})$, might be contemplated. A basic case is where we might model the mean response in terms of just the location variables, $\mathbf{x}$. Perhaps we might expect Y to vary in a systematic way with $\mathbf{x}$ and would postulate that $E[Y(\mathbf{x})] = h(\mathbf{x})$.

For $h(\mathbf{x}) = \alpha + \beta_1 x_1 + \beta_2 x_2$ we have just a simple linear regression model. The methodology for such linear models was developed in Chapter 7 and is readily applicable if we incorporate the inevitable spatial correlation structure for the response which we would expect to encounter at least at small separations.

Universal kriging (Section 11.2.3) is one way of approaching this, but the spatial correlation could also be represented within the more usual regression (linear) model. An example of this was described in Section 1.3.3 where, in a study of wheat growth, we sought to interpolate minimum temperatures at sites which are *not* meteorological stations, in terms of their locations. An extended form was also encountered there in the respect that, for a new location, we also knew its altitude. Thus instead of regressing Y on just x_1 and x_2 (the latitude and longitude, the basic spatial coordinates), we were also able to include altitude x_3 as a further regressor variable. We revisited this problem also in *Example 11.5*.

Thus in regression models (or even in non-linear relationship models) for spatial processes the associated regressor variables need not be restricted to those describing location but can include other covariates as well. The notion of *response surfaces* and *response designs* also provides an approach to regression-based study of spatially distributed data. See, for example, Draper and Smith (1998) and further comments in Section 11.6 below.

Thus, we have a range of models in which the mean response $E[Y(\mathbf{x})]$ for a set of spatial data is expressed as a specific function of a range of influential (regressor) variables which may or may not include, but need not be restricted to, the location $\mathbf{x}$ at which the response is observed. Thus typically we adopt a model in the form

$$E[Y(\mathbf{x})|\mathbf{z}] = h(\mathbf{x}, \mathbf{z}), \qquad (11.16)$$

where $\mathbf{x}$ is location and $\mathbf{z}$ is a set of 'drive variables' or covariates which are related to the mean response in the way expressed by the specific model. See again the climatological example of Section 1.3.3 and *Example 11.5*, where location $\mathbf{x}$ enters directly in a linear model which also includes altitude (z) as a single further covariate.

Of course, $h(\mathbf{x}, \mathbf{z})$ can be linear in $\mathbf{x}$ and $\mathbf{z}$ in the sense described in Chapter 7, or may be non-linear. A *generalized linear model* formulation may be more appropriate. Where $\mathbf{x}$ enters the model for the mean response, spatial correlation structure must also be included appropriately as part of the model. An assumption of independence of responses (as in our study of relationship models in Chapter 7) will be untenable. When $\mathbf{x}$ does not feature explicitly in the model (11.16), we essentially leave the spatial domain and can revert to the range of models and methods discussed in Chapter 7.

11.5.1 Types of spatial model

We can essentially categorize spatial models into various classes. Firstly, there are the point process type models which we examined in Section 11.1.

General stochastic models which progress beyond binary response to quantitative outcomes include *random fields*, by which we mean stochastic processes where the indexing one-dimensional 'time' parameter is extended to higher-dimensional form – in particular, to two-dimensional *space*. See, for example, Adler (1985). Special cases relevant to spatial study include *Markov random fields* (again see Adler, 1985) in which the customarily 'directional' Markovian property is extended to two or more dimensions to encompass spatial interests, and *Gaussian random fields* (Adler, 1985; Worsley *et al.*, 1998), where the properties of the process can be fully represented in terms of the *covariance structure*.

The range of covariance-motivated spatial models and methods is reviewed by Sampson (2002); the stationary form underlies all the work on kriging. For non-stationary covariance-based models we could seek to apply locally stationary models subsequently smoothed over extended space using *kernel-based methods*; see the review by Guttorp and Sampson (1994); see also Oehlert (1993) and Loader and Switzer (1992) for relevant environmental applications.

A range of methods for non-stationary spatial processes uses the *empirical orthogonal function decomposition* of the spatial data set. This goes back to the 1960s but has seen a revival of interest recently in the work of, for example, Nychka *et al.* (2000). As well as discussing such so-called *basis function models*, Sampson (2002) reviews two other categories of model: *process-convolution models*, using convolutions of a Gaussian 'white noise' spatial process convolved with a smoothly spatially varying kernel function, and *spatial deformation models* (Guttorp and Sampson, 1994; Meiring *et al.*, 1997).

11.5.2   Harmonic analysis of spatial processes

A novel approach to spatial modelling and analysis of spatial structure is to be found in Mugglestone and Renshaw (1998, 2001) and related earlier publications. Prompted by the powerful influence of frequency-domain (harmonic) representation of a time series, they transfer this approach to the spatial context and propose the use of a two-dimensional form of spectral analysis.

In the form that most closely echoes the time-series methods, Mugglestone and Renshaw (1998) suggest the use of the *periodogram* (or *sample spectrum*) for two-dimensional lattice data, that is, where we have quantitative measurements at the nodes of a regular lattice.

Specifically (for pixelated data $X_{s,t}$, on an $m \times n$ rectangular array, normed for the mean) Mugglestone and Renshaw (1998) propose using the estimator

$$I_{p,q} = mn(a_{p,q}^2 + b_{p,q}^2), \tag{11.17}$$

where

$$a_{p,q} = (mn)^{-1} \sum_{s=1}^{m} \sum_{t=1}^{n} X_{s,t} \cos\left[2\pi(ps/m + qt/n)\right]$$

and

$$b_{p,q} = (mn)^{-1} \sum_{s=1}^{m} \sum_{t=1}^{n} X_{s,t} \sin\left[2\pi(ps/m + qt/n)\right],$$

and apply the approach to look for lineations in glacial landforms with some impressive results.

Mugglestone and Renshaw (2001) adopt the approach to the construction of tests of spatial randomness for arbitrary points in the plane. They show that their tests, particularly those based on what they term the 'scaled cumulative R-spectrum' and on the Anderson–Darling fit test statistic, for testing for complete spatial randomness are more powerful than space-domain tests. The R-spectrum $f_R(r)$ is defined for data $y_{r,s}$ ($r = 1, 2, \ldots, m$; $s = 1, 2, \ldots, n$) at mn lattice nodes centred to the form $x_{r,s} = y_{r,s} - \bar{y}$ (where $\bar{y}$ is the sample mean) as the average periodogram for periodic elements ('ordinates') with similar values of $r = \sqrt{(p^2 + q^2)}$. Another summary form known as the Θ-spectrum, $f_\Theta(\theta)$, averages the periodogram over ordinates with similar values of $\theta = \tan^{-1}(p/q)$. The R-spectrum and the Θ-spectrum can be used to examine 'scales of pattern' and 'directional features', respectively

Such methods have been in development for some time; early definitive work is found in Bartlett (1964) and in Renshaw and Ford (1983, 1984), the latter setting the framework for forming one-dimensional summaries of the periodogram for examining spatial structure. See also Mugglestone and Renshaw (1996a, 1996b).

An extension of these ideas has been used to develop a novel approach to modelling and analysing outliers in spatial lattice data. Nirel *et al.* (1998) show how to obtain outlier-robust estimates of autocorrelation functions and spectral density functions by appropriately detecting and replacing any outliers (contaminants). Mugglestone *et al.* (2000) apply these methods to sets of ecological and environmental data.

11.6   SPATIAL SAMPLING AND SPATIAL DESIGN

So far, all our discussions of spatial models and methods have implicitly assumed that the data to be analysed are already to hand and that we have not been able to play a role in their selection or collection. Sometimes, however, we can play a part in the data collection process, that is, in *sampling*.

11.6.1   Spatial sampling

Where sampling is required and before we can collect our data, we need to specify how this is to be done: we need to decide what precise *sampling scheme* or *sampling plan* would be appropriate for the problem in hand. We might also be able to intervene and impose certain 'treatments' on the chosen sites at which observations are to be taken, that is, to apply some *experimental design* for the purpose of investigating the effects on the spatial responses of imposed treatments.

Much has been written on *spatial sampling* and *spatial design*; see, for example, for different aspects of this theme, Myers and Montgomery (1995), McBratney *et al.* (1981), Cressie (1993, Section 5.6), Olea (1984), Fienberg and Tanur (1987), McArthur (1987), Yfantis *et al.* (1987) and Draper and Smith (1998). We can only briefly review this field by reference to some particular cases which merit special comment.

The process $Y(\mathbf{x})$ may be observable only at a discrete set of $\mathbf{x}$ values (i.e. a discrete set of locations). These might be on some **regular grid**, or **lattice**: a finite set of locations distributed over the region of interest in some regular way. However, what is meant by a grid or lattice does vary somewhat with different writers. Cressie (1993, Chapter 6) uses the term 'lattice' for any *countable set of locations* without any necessary regularity of spacing; Besag

(1974, 1975), in contrast, retains the regularity assumption. An environmental observation **network** may be so deployed over a regular grid or lattice (in intent, if not in reality; precise regularity may be inhibited by access or planning constraints). Network design and spatial prediction from network-based data have been considered among others by Bogardi *et al.* (1985), Caselton and Zidek (1984), Guttorp *et al.* (1993), Nychka *et al.* (1997) and Wu and Zidek (1992).

Inevitably, in practice, we will be restricting attention to what is happening over some finite geographic (or otherwise spatial) region. So we will be concerned with a finite population of possible responses $Y(\mathbf{x}_i), i = 1, 2, \ldots, N$. Thus we might consider employing the finite-population sampling methods developed in Chapter 4 above to select sites from this finite population. This takes us into the field of spatial sampling and the use of different sampling schemes or sampling plans. Cressie (1993, Section 5.6.1) briefly reviews this theme in the context of spatial studies.

In environmental investigations we often encounter restrictions on our freedom to observe a response at arbitrary locations, for reasons discussed at the beginning of Chapter 5, where we considered the concepts of encountered data and encounter sampling (Section 5.1). If such constraints do not arise, however, we will be free to use the finite sampling methods of Chapter 4. Simple random sampling, stratified sampling and cluster sampling may all be feasible and relevant, depending on prevailing circumstances. McArthur (1987) compares several finite sampling plans in a spatial context; Olea (1984) considers cluster sampling.

A special form of sampling was seen in Section 6.4 to be particularly appropriate for spatially distributed data. This is adaptive sampling, where we specifically direct sampling effort, successively as the sample is collected, to sites which are indicated from the current data to hand as likely to be most fruitful in inferring the properties of the population we are interested in studying.

In the finite-sampling context, suppose we observe a sample $y(\mathbf{x}_1), y(\mathbf{x}_2), \ldots, y(\mathbf{x}_n)$. If we wish to draw inferences about population characteristics such as the population mean $\overline{Y} = \sum_{i=1}^{N} Y(\mathbf{x}_i)$ or the population total $Y_T = N\overline{Y}$, it may thus sometimes be reasonable to suppress information about location and just use the basic finite-sampling methods of Chapter 4 applied to the resulting finite sample $y_1, y_2, \ldots, y_n$. Even if it is not reasonable to ignore the spatial element, we might still find the finite-sampling methods useful if we regard the $\mathbf{x}$ values as covariate components in the sense of ratio estimation or regression estimation (see Section 4.3.2). Covariates other than location (e.g. altitude in the climatological example in section 1.3.3, and Example 11.5 above) can also, of course, be incorporated in a similar way to inform our inferences.

These considerations apply whenever we obtain spatial data as a finite sample from a larger finite population of possible observations. The potential observation sites might, as we have remarked above, be restricted to lie on a finite set of regularly space locations – on a lattice.

But restriction to a lattice is a sampling plan in its own right even if we sample at every location on the lattice in the study region. Work has been done in the environmental context to compare the advantages and disadvantages of different forms of lattice (triangular, square, hexagonal, etc.) – principally for efficiency of kriging methods – for such full sampling, demonstrating some benefits in comparison with random sampling schemes. Olea (1984) compares these regular sampling approaches with simple and stratified random sampling, whilst McBratney *et al.* (1981) and Yfantis *et al.* (1987) concentrate more on the comparison of the different regular forms, with equilateral triangular lattices showing advantages. Zimmerman (2002) pursues the notion of *optimal spatial design* in the sense of 'the determination of the number, dimensions, and spatial arrangement of the sites that optimize ... the information content of the observations'.

We will now consider a more formal and classical notion of the 'design' of environmental experiments.

11.6.2   Spatial design

In some environmental studies, we may be able to influence more directly the very process by which the data are obtained. That is to say, it may be possible to set up an experiment under which the spatial locations are prespecified and specific treatments (combinations of further influencing factors, other than location) can be applied at such locations. We can then seek to examine the effects both of location and of the applied treatments on the values observed for the response variable. We now enter the fields of *design of experiments* and specifically of *spatial design*. The particular topics of *response surfaces* and *response designs* (with special emphasis on regression analysis of the effects of quantitative treatment variables on the response patterns) also relate to this interest – see, for example, Draper and Smith (1998) and Myers and Montgomery (1995).

Of course, we do not have space here to provide any extensive coverage of the field of **design of experiments** and **planning of experiments** so ably spelt out by others. The pioneering text by Cox (1958) on planning experiments remains the cornerstone of study of basic concepts on this topic. Hahn (1982) provides an informative categorized bibliography on design of experiments; among the many interesting and influential books it covers are: Cochran and Cox (1957), still a useful reference text; Davies (1956), which is practical and oriented to industrial applications; Fisher (1966), first published in 1935, and the 'classic in the field'; Box *et al.* (1978), a detailed intermediate-level methodological work; and John (1998), which is detailed and more advanced.

In the design of experiments we are concerned with applying to a selected and assumed 'representative' set of units (be they pieces of leather to be tanned in different ways, or pieces of sheet steel to be spray-painted in the car industry) a number of 'treatments' in a predetermined pattern of application to examine how the response variable (flexibility of the leather, number of blemishes in the paintwork) might vary with the treatments (e.g. pH level of the tanning fluid, paint drying method).

At this level of generality there is no specifically *environmental* or *spatial* component in the methodology. But there is no reason why the application might not be environmental; perhaps we are examining how nitrates are leached through agricultural land for different levels of application of a nitrogenous fertilizer to different varieties of winter wheat. The sampling units will now be the different plots of land on which the wheat is grown. The very term 'plot' is widespread in experimental design, reflecting the agricultural origins of the subject; see, for example, Barnett (1994), who traces the influence of the so-called 'classical experiments' set out at the Rothamsted agricultural station in the 1840s and 1850s on R.A. Fisher's development of the basic ideas of the design of experiments in the 1920s. So, in this example we see how spatial and environmental interest could be inevitable (notwithstanding the rather limited opportunities for the luxury of 'planning an experiment' in the environmental context; see Section 5.1 above).

To highlight a few of the interesting environmental and spatial components in a designed experiment, let us continue with the agro-ecological/environmental example of nitrate leaching. Thus we might have an agricultural land site which we divide into a 16 similar experimental plots on each which we will apply one of four (increasing) levels of nitrogenous fertilizer (A_1, A_2, A_3 and A_4) to wheat crops planted on the plots. We will be comparing the amounts of nitrates Y leached from the different plots by the end of the growing period with the intent of examining how level of leaching varies with the concentration of the fertilizer. It may be that different levels of wheat yield will also relate to the extent to which residual nitrates are leached.

Suppose the experimental site has rising wooded ground to the north and a river flowing across the southeast corner (Figure 11.7). The site is essentially a square/rectangular lattice of the 16 plot locations or sites. In terms of our earlier notation $Y(\mathbf{x})$, the location $\mathbf{x}$ represents the plot on which the observation is taken, but the response must also reflect the crucial 'treatment levels' of fertilizer concentration applied to the plot. So each response is essentially indexed by location $\mathbf{x}$ and treatment combination (A_i). In what way does the spatial variable $\mathbf{x}$ affect the response?

The minimal prospect is that plot location will be irrelevant to the amount of nitrate leached. If so, we could apply the classical *completely randomized design*

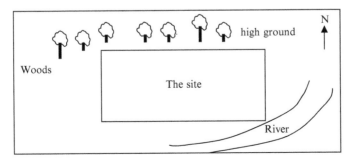

Figure 11.7   The nitrate leaching experimental site.

under which the four different possible treatment levels A_i, $i = 1, 2, 3, 4$, are each applied at random four times over the 16 experimental plots perhaps yielding, say, the layout in Figure 11.8. On the assumption of response being independent of location, a classical one-way analysis of variance (ANOVA) enables us to examine the treatment effects of fertilizer concentrations (subject to other necessary assumptions of linearity, independence and normality of errors). Inevitably no spatial dependence or effect is involved or possible in such an analysis.

But plot location might indeed be relevant. There could be systematic effects: growth, and hence yield, may be inhibited on the shady, higher, more northern plots compared with that on the sunnier, more irrigated, southeastern plots. This differential growth pattern and different drainage prospects on higher and lower ground could thus well affect extent of leachage over and above that attributable to fertilizer concentration.

If these systematic effects are indeed present, it is clear that the earlier completely randomized design might be far from ideal. As the treatments *fortuitously arranged themselves* (see Figure 11.8) in our layout we find all of the lowest fertilizer concentrations (A_1) along the higher northern fringe and all of the highest ones (A_4) on the more fertile south and eastern edges. Thus if results for levels A_1 and/or A_4 turn out to be anomalous in leaching effect we have no way of knowing if this is due to the low or high concentrations (A_1 or A_4) or is merely induced by ('confounded with') the differential fertility properties of the plots on which these treatments were applied.

More complex designs can be used to avoid such *confounding* and to ensure that observed effects are unambiguously attributable. Thus in a *randomized block* design we would 'block' the treatments to ensure that each of the levels of a treatment occurred equally often at each of the potentially different levels of systematic location effect. So if we want to protect against a north/south effect we would ensure that each of the varieties occurred equally often (namely, once) in each of the four east/west strips.

But we have more to protect against here: north/south differential drainage and terrain (height, shade) effects *and also* increasing fertility towards the east and south edges. Further complexity of design is thus needed, and Figure 11.9 shows a layout arising from a *Latin square* design, in which each treatment level A_i occurs equally often (namely, once) in each east/west strip and in each north/south strip. Clearly, we should by this design have avoided any prospect of confounding – any statistically inferred effects of fertilizer level on leachage

A_1	A_1	A_1	A_1
A_3	A_3	A_2	A_4
A_2	A_3	A_2	A_4
A_2	A_3	A_4	A_4

Figure 11.8 A completely randomized design.

A_1	A_2	A_3	A_4
A_2	A_3	A_4	A_1
A_3	A_4	A_1	A_2
A_4	A_1	A_2	A_3

Figure 11.9 A Latin square layout.

should be genuinely attributable to those treatments and not due to site location effects.

Another form of spatial effect in experimental design is to be found in what is called *residual effects* or *carry-over effects*. As the names suggest, the concern is that if we apply a treatment to one plot its effect may carry over to neighbouring plots; for example, the fertilizer which stimulates growth on the plot to which it is applied may also stimulate growth on an otherwise non-fertilized adjacent plot. One practical means of counteracting this effect is to intersperse 'guard strips' by leaving a gap between the plots on which treatments are to be applied. Interestingly, such guard strips do not need to be very wide in the agricultural context – typically just a metre or so. Alternatively, there are experimental designs which can be used specifically to protect against, or minimize, and carry-over effects; see, for example, Cochran and Cox (1957, Section 4.6a).

Aspects of the design of experiments which specifically consider the effects of spatial correlation are examined by Martin (1986).

Our brief study of experimental design considerations demonstrates that there is indeed a sense in which careful design can mitigate undesirable spatial effects in studying environmental phenomena.

11.7 SPATIAL-TEMPORAL MODELS AND METHODS

Of even greater interest and need than temporal or spatial models and methods separately for environmental studies are models and methods which conjointly encompass the effects of both time and space. This is a very difficult area of work, and progress on truly conjoint spatial-temporal models and methods is limited. There are, however, some emphases which have made useful inroads into this topic, and these include:

- covariance-structured methods (see, for example, Gneiting and Schlather, 2002);
- space-time Kalman filter approaches (see, for example, Cressie and Wikle, 2002);
- *ad hoc* case specific applications (see, for example, Blackwell, 1997);
- harmonic or spectral principles (extending the ideas of Renshaw, Mugglestone and others: see Section 11.5.2).

We will briefly examine some aspects of this work.

The general aim is to model and analyse the spatial/temporal random variable $Y_{\mathbf{s}}(t)$, commonly also denoted $Y(\mathbf{s}, t)$, indexed by spatial coordinates $\mathbf{s} \in \mathbb{R}^d$ and by time $t \in \mathbb{R}$. In most applications Y will be univariate, the spatial coordinate will be continuous and two-dimensional, and the time coordinate will be discrete and univariate.

A central feature of the study of spatial-temporal processes is the joint covariance structure: the pivotal *covariance function* is defined as

$$C(\mathbf{s}_1, \mathbf{s}_2; t_1, t_2) = \mathrm{Cov}[Y(\mathbf{s}_1, t_1) Y(\mathbf{s}_2, t_2)].$$

The process is said to be *covariance stationary* if the expectation of $Y(\mathbf{s}, t)$ is independent of $\mathbf{s}$ and t and if

$$C(\mathbf{s}_1, \mathbf{s}_2; t_1, t_2) = C(\mathbf{s}_1 - \mathbf{s}_2; t_1 - t_2) = C(\mathbf{d}; \tau),$$

where $\mathbf{d} = \mathbf{s}_1 - \mathbf{s}_2$ and $\tau = t_1 - t_2$. Most progress has been made for such stationary processes (or those readily tranformable to stationarity). Indeed, stationarity transformation is a major area of study for spatial/temporal processes.

Thus most approaches to space-time processes centre on the joint covariance function $C(\mathbf{d}; \tau)$. On the one hand, we may seek to estimate $C(\mathbf{d}; \tau)$ from the data. For an assumed Gaussian form of $Y(\mathbf{s}, t)$ this will lead to full specification and representation of the process. Alternatively, an initial assumption of a specific form for $Y(\mathbf{s}, t)$ yields an implied form for $C(\mathbf{d}; \tau)$ which can then be used for confirmative inference. The distinction is essentially between a data-driven and a (possibly physical) model-driven approach.

Of course, the stationarity assumption is a strong constraint and may not be justified or readily justifiable in a specific application – trends and periodicities are common and not always readily or usefully transformed away. Gneiting and Schlather (2002) briefly discuss the problems that can arise in seeking to achieve *stationarity* (and *isotropy*), in their interesting review of covariance-centred models for $Y(\mathbf{s}, t)$.

The simplest situation is one where we have stationarity but where $Y(\mathbf{s}, t)$ is also **separable** in the sense that

$$C(\mathbf{d}; \tau) = C_{\mathbf{s}}(\mathbf{d})C_t(\tau)$$

or

$$C(\mathbf{d}; \tau) = C_{\mathbf{s}}(\mathbf{d}) + C_t(\tau),$$

which can lead to distinct handling of the spatial and temporal components (e.g. using kriging and ARIMA-type processing, repectively).

Often, however, we need to deal with *non-separable* $Y(\mathbf{s}, t)$, in which case *harmonic analytic methods* or *direct situation-based models* may be needed. The latter ('physical') approach is exemplified in Jones and Zhang (1997). In the harmonic modelling approach for non-separable processes, Cressie and Huang (1999) consider the fundamental issue of characterizing $C(\mathbf{d}; \tau)$, showing that permissible forms admit of a Fourier transform representation.

Christakos and Hristopoulos (1997) develop spatial/temporal models and methods in environmental health problems. Specific other applications to environmental issues include studies of ozone (Guttorp *et al.*, 1994; Meiring *et al.*, 1998), rainfall (Cox and Isham, 1988; Rodríguez-Iturbe and Mejía, 1974), wind power (Haslett and Raftery, 1989) and general meteorological variables (Handcock and Willis, 1994).

A large body of work on spatial-temporal models and methods uses the *space-time Kalman filter (STKF)* – see Cressie and Wikle (2002). Over 40 years ago, Kalman (1960) proposed his new approach to linear filtering of time series in an engineering context and the Kalman filter has played a crucial role in this field (e.g. in *control theory*). In its application to spatial-temporal problems the STKF is typically directed (Wikle and Cressie, 1999) to a hierarchically expressed model representation of $Y(\mathbf{s}, t)$ where we assume that

$$Y(\mathbf{s}, t) = Z(\mathbf{s}, t) + \varepsilon(\mathbf{s}, t), \tag{11.18}$$

with

$$Z(\mathbf{s}, t) = \mu(\mathbf{s}, t) + v(\mathbf{s}, t) \tag{11.19}$$

and

$$\mu(\mathbf{s}, t) = \int \omega_{\mathbf{s}}(\mathbf{u})\mu(\mathbf{u}, t - 1)\mathrm{d}\mathbf{u} + \eta(\mathbf{s}, t). \tag{11.20}$$

Here $Y(\mathbf{s}, t)$ is expressed in terms of a basic (data) model $Z(\mathbf{s}, t)$ with a superimposed error component $\varepsilon(\mathbf{s}, t)$. A *process model* expresses $Z(\mathbf{s}, t)$ in terms of a temporally uprating form $\mu(\mathbf{s}, t)$ with further error components $v(\mathbf{s}, t)$ attached to $\mu(\mathbf{s}, t)$ itself and $\eta(\mathbf{s}, t)$ attached to its temporal uprating representation (11.20).

Usually $\mu(\mathbf{s}, t)$ is expressed, to ensure identifiability, in a finite separable form

$$\mu(\mathbf{s}, t) = \sum_{i=1}^{k} \alpha_i(t)\varphi_i(\mathbf{s}) \tag{11.21}$$

of products of deterministic spatial functions and finite time series (see Cressie and Wikle, 2002).

The hierarchical model (11.17)–(11.20), often adopted with assumed Gaussian error terms, constitutes a highly structured space-time model with many levels of assumptions. The expectation is that it is rich enough to accommodate any assumptions we are prepared to make for many a practical problem of interest.

The aim of the STKF method is to use data $y(\mathbf{s}_i, t_i), i = 1, 2, \ldots, n$, to predict the underlying process $Z(\mathbf{s}, t)$, seeking optimality in squared-error loss terms. A general review of the details of the method (along with references for deeper study) is given by Cressie and Wikle (2002).

Goodall and Mardia (1994) and Mardia *et al.* (1998) pursue the notion of space-time Kalman filtering, in the latter case for a somewhat simplified model, the 'kriged Kalman fitter', with no error term $v(\mathbf{s}, t)$ in (11.19). Huang and

Cressie (1996) use the approach for predicting snow-water levels. See also Berke (1998), Meiring *et al.* (1998) and Wikle and Cressie (1999).

An interesting specific development of spatial-temporal modelling is to be found in Blackwell (1997) who extends and applies earlier models (Dunn and Gipson, 1977; Worton, 1995) for representing animal movement in space and time with special reference to radio tracking. The particular subject of interest is the woodmouse. The models incorporate a finite (discrete) set of states relating to distinct behavioural (psychological) modes with different possible implications for the spatial-temporal movement process. These modes may reflect the subject feeding, travelling, hungry etc.

The models are constructed essentially as generalizations of a Ornstein–Uhlenbeck model for spatial diffusion in random temporal environments. The multivariate Ornstein–Uhlenbeck process takes the following form. If $\mathbf{X}_t = (X_{1t}, X_{2t}, \ldots, X_{pt})$ is the location (of an animal) in p-space at time t, then $\mathbf{X}_{s+t}$ (its location at a later time $s + t$) conditional on $\mathbf{X}_s = \mathbf{x}_s$ has a multivariate $N(\mathbf{\mu} + e^{\mathbf{B}t(\mathbf{X}_s - \mathbf{\mu})}, \mathbf{\Lambda} - e^{\mathbf{B}t}\mathbf{\Lambda}e^{\mathbf{B}t}t)$ distribution where $\mathbf{\mu}$ is a p-vector and $\mathbf{\Lambda}$ and $\mathbf{B}$ are $p \times p$ matrices. The matrix $\mathbf{B}$ 'controls the strength and form of the centralizing tendancy'. If $\mathbf{B}$ is stable in the sense that $e^{\mathbf{B}t} \to 0$ as $t \to \infty$, then the process will be stationary with mean vector $\mathbf{\mu}$ and covariance matrix $\mathbf{\Lambda}$; the limiting distribution will be $\mathbf{X}_t \sim N(\mathbf{\mu}, \mathbf{\Lambda})$.

Blackwell goes on to examine how reasonable such an Ornstein–Uhlenbeck process is for animal tracking studies and to explore generalizations ('mixed Ornstein–Uhlenbeck processes') to improve on data fit with particular reference to accommodating the different movement patterns in the different behavioural modes.

In conclusion, we might return to the theme of the spectral representation of spatial process discussed in Section 11.5.2. We noted in Section 10.6 how powerful harmonic analysis is for time series. Its extension as in Mugglestone and Renshaw (1998) to spatial processes is clearly informative and useful. It is tempting to consider whether the spectral approach might be worth pursuing as a basis for an *integrated joint representation of the spatial and temporal components in a space-time process*; merely by extending the index set in the spectrum from t or from $\mathbf{s}$ to $\mathbf{\nu} = (t, \mathbf{s})$. This has been suggested before but does not seem to have been followed up. It might be worth a try!

References

Abraham, B. and Box, G.E.P. (1979) Bayesian analysis of some outlier problems in time series. *Biometrika*, **66**, 229–236.

Adler, R.J. (1985) Markov random fields. In S. Kotz, N.L. Johnson and C.B. Read (eds), *Encyclopedia of Statistical Sciences*, Vol. 5. John Wiley & Sons, Inc., New York, pp. 270–273.

Agnew, M.D. and Palutikof, J.P. (1999) The impacts of climate on retailing in the UK with particular reference to the anomalously hot summer and mild winter of 1995. *International Journal of Climatology*, **19**, 1493–1507.

Anderson, C.W. (1984) Large deviations of extremes. In J. Tiago de Oliveira (ed.), *Statistical Extremes and Applications*. Reidel, Dordrecht, pp. 324–340.

Anderson, C.W.A., Barnett, V., Chatwin, P.C. and El-Shaarawi, A.H. (eds) (2002) *Quantitative Methods for Current Environmental Issues*. Springer, London.

Anderson, O.D. (ed.) (1982) *Time Series Analysis: Theory and Practice*. North-Holland, Amsterdam.

Anderson, T.W. (1994) *The Statistical Analysis of Time Series*. Wiley Interscience, New York.

Anscombe, F.J. (1960) Rejection of outliers. *Technometrics*, **2**, 123–147.

ANZECC and NHMRC (1992) *Australian and New Zealand Guidelines for the Assessment and Management of Contaminated Sites*. Australian and New Zealand Environment and Conservation Council and National Health and Medical Research Council.

ApSimon, H.M., Goddart, A.J.H. and Wrigley, J. (1985) Long-range transport of radio-isotopes. I. The MESOS model. *Atmospheric Environment*, **19**, 99–111.

Arnold, B.C. and Nagaraja, H.N. (1991) Lorentz ordering of exponential order statistics. *Statistics and Probability Letters*, **11**, 485–490.

Bailer, A.J. and Oris, J.T. (2002) Environmental toxicology. In A. El-Shaarawi and W.W. Piegorsch (eds), *Encyclopaedia of Environmetrics*. John Wiley & Sons, Ltd, Chichester, pp. 2212–2218.

Bailer, A.J. and Piegorsch, W.W. (2000) From quantal counts to mechanisms and systems: The past, present and future of biometrics in environmental toxicology. *Biometrics*, **56**, 327–336.

Baldock, F.C., Lyndalmurphy, M. and Pearse, B. (1990) An assessment of a composite sampling method for counting strongyle eggs in sheep faeces. *Australian Veterinary Journal*, **67**, 165–169.

Barnett, V. (1966) Order statistics estimators of the location of the Cauchy distribution. *Journal of the American Statistical Association*, **61**, 1205–1218.

Barnett, V. (1975) Probability plotting methods and order statistics. *Applied Statistics*, **24**, 95–108.

Barnett, V. (1976a) The ordering of multivariate data (with Discussion). *Journal of the Royal Statistical Society, Series A*, **139**, 318–354.

Barnett, V. (1976b) Convenient probability plotting positions for the normal distribution. *Applied Statistics*, **25**, 47–50.

Environmental Statistics   V. Barnett
© 2004 John Wiley & Sons, Ltd   ISBN: 0-471-48971-9 (HB)

Barnett, V. (1978a) The study of outliers: Purpose and model. *Applied Statistics*, **27**, 242–250.

Barnett, V. (1978b) Multivariate outliers: Wilks' test and distance measures. In *Bulletin of the International Statistical Institute: Proceedings of the 41st Session (New Delhi)*, Vol. 47, Book 4. ISI, Voorburg, Netherlands, pp. 37–40.

Barnett, V. (1979) Some outlier tests for multivariate samples. *South African Statistical Journal*, **13**, 29–52.

Barnett, V. (ed.) (1981) *Interpreting Multivariate Data*. John Wiley & Sons, Ltd, Chichester.

Barnett, V. (1983a) Marginal outliers in the bivariate normal distribution. In *Bulletin of the International Statistical Institute: Proceedings of the 44th Session (Madrid)*, Vol. 50, Book 4. ISI, Voorburg, Netherlands, pp. 579–583.

Barnett, V. (1983b) Principles and methods for handling outliers. In T. Wright (ed.), *Statistical Methods and the Improvement of Data Quality*. Academic Press, Orlando, FL, pp. 131–166

Barnett, V. (1988) Outliers and order statistics. *Communications in Statistics – Theory and Methods*, **17**, 2109–2118.

Barnett, V. (1994) Statistics and long-term experiments: Past achievements and future challenges. In R.A. Leigh and A.E. Johnston (eds), *Long-Term Experiments in Agricultural Ecological Sciences*. CAB International, Wallingford.

Barnett, V. (1997) Statistical analyses of pollution problems. In V. Barnett and K.F. Turkman (eds), *Statistics for the Environment 3: Pollution Assessment and Control*. John Wiley & Sons, Ltd, Chichester, pp. 3–41.

Barnett, V. (1999a) *Comparative Statistical Inference* (3rd edition). John Wiley & Sons, Ltd, Chichester.

Barnett, V. (1999b) Ranked set sample design for environmental investigations. *Environmental and Ecological Statistics*, **6**, 59–74.

Barnett,V. (2002a) *Sample Survey Principles and Methods* (3rd edition). Arnold, London.

Barnett, V. (2002b) Encounter sampling. In A. El-Shaarawi and W.W. Piegorsch (eds), *Encyclopaedia of Environmetrics*. John Wiley & Sons, Ltd, Chichester, pp. 667–668.

Barnett, V. (2002c) Sample ordering for effective statistical inference, with particular reference to environmental issues. In J. Panaretos (ed.), *Stochastic Musings: Perspectives from the Pioneers of the Late 20th Century*. Erlbaum, Mahwah, NJ.

Barnett, V. (2002d) Statistically evaluating risk in environmental issues. In P. Oliveira and E. Athayde (eds), *Um Olhar sobre a Estatística: Actas do VII Congresso Anual da Sociedade Portuguesa de Estatística*. Sociedade Portuguesa de Estatística, Lisbon.

Barnett, V. and Bown, M. (2002a) Standards, environmental: Statistical considerations. In A. El-Shaarawi and W.W. Piegorsch (eds), *Encyclopaedia of Environmetrics*. John Wiley & Sons, Ltd, Chichester, pp. 2113–2122.

Barnett, V. and Bown, M. (2002b) Setting environmental standards: a statistical approach. In C.W.A. Anderson, V. Barnett, P.C. Chatwin and A. El-Shaarawi (eds), *Quantitative Methods for Current Environmental Issues*. Springer-Verlag, London, pp. 99–109.

Barnett, V. and Bown, M. (2002c) Statistically meaningful standards for contaminated sites using composite sampling. *Environmetrics*, **13**, 1–13.

Barnett, V. and Bown, M. (2002d) Best linear quantile estimators for environmental standards. *Environmetrics*, **13**, 295–310.

Barnett, V. and Bown, M. (2002e) Setting consistent standards along the pollutant cause–effect chain. Mathematics and Statistics research report No.2/02, Nottingham Trent University, UK.

Barnett, V. and Kenward, M.G. (1996) Security systems and renewal processes. *Communications in Statistics – Theory and Methods*, **25**, 475–487.

Barnett, V. and Kenward, M.G. (1998) Testing a Poisson renewal process in the context of security alarm maintenance policies. *Communications in Statistics – Theory and Methods*, **27**, 3085–3094.

Barnett, V. and Lewis, T. (1967) A study of low-temperature probabilities in the context of an industrial problem (with Discussion). *Journal of the Royal Statistical Society, Series A*, **130**, 177–206.

Barnett, V. and Lewis, T. (1994) *Outliers in Statistical Data* (3rd edition). John Wiley & Sons, Ltd, Chichester.

Barnett, V. and Moore, K.L. (1997) Optimal estimates in ranked-set sampling with and without perfect ordering. *Journal of Applied Statistics*, **24**, 697–710.

Barnett,V. and O'Hagan, A. (1997) *Setting Environmental Standards*. Chapman & Hall, London.

Barnett, V, and Roberts, D. (1993) The problem of outlier tests in sample surveys. *Communications in Statistics – Theory and Methods*, **22**, 2703–2721.

Barnett, V. and Turkman, K.F. (eds) (1993) *Statistics for the Environment*. John Wiley & Sons, Ltd, Chichester.

Barnett, V. and Turkman, K.F. (eds) (1994) *Statistics for the Environment 2: Water-Related Issues*. John Wiley & Sons, Ltd, Chichester.

Barnett, V. and Turkman, K.F. (eds) (1997) *Statistics for the Environment 3: Pollution Assessment and Control*. John Wiley & Sons, Ltd, Chichester.

Barnett, V., Green, P.J. and Robinson, A. (1976) Concomitants and correlation estimates. *Biometrika*, **63**, 323–328.

Barnett, V., Landau, S., Colls, J.J., Craigon, J., Mitchell, R.A.C. and Payne, R.W. (1997) Predicting wheat yields: the search for valid and precise models. In G.R. Bock and J.A. Goode (eds), *Precision Engineering: Spatial and temporal Variability of Environmental Quality*, Novatis Foundation Symposium 210. John Wiley & Sons, Ltd, Chichester, pp. 79–92.

Barnett, V., Stein, A. and Turkman, K.F. (eds) (1999) *Statistics for the Environment 4: Statistical Aspects of Health and the Environment*. John Wiley & Sons, Ltd, Chichester.

Barreto, M.C.M. and Barnett, V. (1999) Best linear unbiased estimators for the simple linear regression model using ranked set sampling. *Journal of Environmental and Ecological Statistics*, **6**, 119–134.

Barreto, M.C.M. and Barnett, V. (2001) Estimators for a Poisson parameter using ranked set sampling. *Journal of Applied Statistics*, **28**, 929–941

Bartlett, M.S.(1955) *An Introduction to Stochastic Processes*. Cambridge University Press, Cambridge.

Bartlett, M.S. (1964) Spectral analysis of two-dimensional point processes. *Biometrika*, **51**, 299–311.

Bartlett, M.S. (1975) *The Statistical Analysis of Spatial Pattern*. Chapman & Hall, London.

Bates, D.M. and Watts, D.G. (1988) *Nonlinear Regression Analysis and Its Applications*. John Wiley & Sons, Inc., New York.

Bebbington, A.C. (1978) A method of bivariate trimming for robust estimation of the correlation coefficient. *Applied Statistics*, **27**, 221–226.

Berke, O. (1998) On spatiotemporal prediction for online monitoring data. *Communications in Statistics – Theory and Methods*, **27**, 2343–2369

Berthouex, P.M. and Brown, L.C. (1994) *Statistics for Environmental Engineers* Lewis, Boca Raton, FL.

Besag, J.E. (1974) Spatial interaction and the statistical analysis of lattice systems (with Discussion), *Journal of the Royal Statistical Society, Series B*, **36**, 192–236.

Besag, J.E. (1975) Statistical analysis of non-lattice data. *The Statistician*, **24**, 179–195.

Beveridge, Sir William H. (1921) Weather and harvest cycles. *Economic Journal*, **31**, 429–452.

Bickel, P.J. and Doksum, K.A. (1981) An analysis of transformation revisited. *Journal of the American Statistical Association*, **76**, 296–311.

Birch, J.B. and Myers, R.H. (1982) Robust analysis of covariance. *Biometrics*, **38**, 699–713.

Blackwell, P.G. (1997) Random diffusion models for animal movement. *Ecological Modelling*, **100**, 87–102

Blom, G. (1958) *Statistical Estimates and Transformed Beta Variables*. John Wiley & Sons, Inc., New York.

Bloomfield, P. (1976) *Fourier Analysis of Time Series: an Introduction*. John Wiley & Sons, Inc., New York.

Bloomfield, P. (2000) *Fourier Analysis of Time Series: an Introduction*. 2nd Ed. John Wiley & Sons, Inc., New York.

Bock, G.R. and Goode, J.A. (eds) (1999) *Environmental Statistics: Analysing Data for Environmental Policy*. Novatis Foundation Symposium 220, John Wiley & Sons, Ltd, Chichester.

Bock, R.D. (1975) *Multivariate Statistical Methods in Behavioural Research*. McGraw-Hill, New York.

Bogardi, I., Bardossy, A. and Duckstein, L. (1985) Multicriterion network design using geostatistics. *Water Resources Research*, **21**, 199–208.

Boswell, M.T. and Patil, G.P. (1987) A perspective of composite sampling. *Communications in Statistics – Theory and Methods*, **16**, 3069–3093.

Bown, M. (2002) A statistically meaningful approach to the setting of environmental standards. Ph.D. thesis, Nottingham Trent University, UK.

Box, G.E.P. and Cox, D.R. (1964) An analysis of transformations (with Discussion). *Journal of the Royal Statistical Society, Series B*, **26**, 211–252.

Box, G.E.P. and Draper, N.R. (1987) *Empirical Model-Building and Response Surfaces*. John Wiley & Sons, Inc., New York.

Box, G.E.P. and Tiao, G.C. (1968) A Bayesian approach to some outlier problems. *Biometrika*, **55**, 119–129.

Box, G.E.P., Hunter, W.G. and Hunter, J.S. (1978) *Statistics for Experimenters*. John Wiley & Sons, Inc., New York.

Box, G.E.P., Jenkins, G.M. and Reinsel, G.C. (1994) *Time Series Analysis, Forecasting and Control* (3rd edition). Prentice Hall, Englewood Cliffs, NJ.

Brain, P. and Marshall, E.J.P. (1999) Modelling cultivation effects using fast Fourier transforms. *Journal of Agricultural, Biological and Environmental Statistics*, **4**, 276–289.

Brillinger, D.R. (1975) *Time Series: Data Analysis and Theory*. Holden-Day, San Francisco.

Brockwell, P.J. and Davis, R.A. (1987) *Time Series, Theory and Methods*. Holden-Day, San Fransisco.

Brownie, C. and Robson, D.S. (1983) Estimation of time-specific survival rates from tag-resighting samples: A generalization of the Jolly–Seber model. *Biometrics*, **39**, 437–453.

Brumelle, S., Nemetz, P. and Casey, D. (1984) Estimating means and variances of the comparative efficiency of composite and grab samples. *Environmental Monitoring Assessment*, **4**, 81–84.

Buckland, S.T. (1987) On the variable circular plot method of estimating animal density. *Biometrics*, **43**, 363–384.

Buckland, S.T. and Tumock, B.J. (1992) A robust line transect method. *Biometrics*, **48**, 901–909.

Buckland, S.T., Anderson, D.R., Burnham, K.P. and Laake, J.L. (1993) *Distance Sampling: Estimating Abundance of Biological populations* Chapman & Hall, London.

Buckland, S.T., Thomas, L., Marques, F.F.C. *et al.* (2002) Distance sampling: Recent advances and future directions. In C.W.A. Anderson, V. Barnett, P.C. Chatwin and A. El-Shaarawi (eds), *Quantitative Methods for Current Environmental Issues.* Springer-Verlag, London, pp. 79–97.

Burnham, K.P. and Anderson, D.R. (1976) Mathematical models for nonparametric inferences from transect data. *Biometrics*, **32**, 325–336.

Burnham, K.P., Anderson, D.R. and Laake, J.L. (1980) *Estimation of Density from Line Transect Sampling of Biological Populations*, Wildlife Monographs 72. Bethesda, MD: The Wildlife Society.

Burnham, K.P., Anderson, D.R. and Laake, J.L. (1981) Line transect estimation of bird population density using a Fourier series. *Studies in Avian Biology*, **6**, 466–482.

Campbell, N.A. (1978) The influence function as an aid in outlier detection in discriminant analysis. *Applied Statistics*, **27**, 251–258.

Campbell, N.A. (1980) Robust procedures in multivariate analysis I: Robust covariance estimation. *Applied Statistics*, **29**, 231–237.

Campbell, N.A. (1982) Robust procedures in multivariate analysis II: Robust canonical variate analysis. *Applied Statistics*, **31**, 1–8.

Carn, S.A. (2002) Landform monitoring. In A. El-Shaarawi and W.W. Piegorsch (eds), *Encyclopaedia of Environmetrics.* John Wiley & Sons, Ltd, Chichester, pp. 1130–1132.

Carter, R.E. and Lowe, L.E. (1986) Lateral variability of forest floor properties under second-growth Douglas-fir stands and the usefelness pof composite sampling techniques. *Canadian Journal of Forest Research*, **16**, 1128–1132.

Caselton, W.F. and Zidek, J.V. (1984) Optimal monitoring network designs. *Statistics and Probability Letters*, **2**, 223–227.

Chapman, D.G. (1951) Some properties of the hypergeometric distribution with applications to zoological censuses. *University of California Publications in Statistics*, **1**, 131–160.

Chapman, D.G. (1955) Population estimation based on change of composition caused by selective removal. *Biometrika*, **42**, 279–290.

Chatfield, C. (1996) *The Analysis of Time Series: An Introduction* (5th edition). Chapman & Hall, London.

Chatterjee, S. and Price, B. (1991) *Regression Analysis by Example* (2nd edition). John Wiley & Sons, Inc., New York.

Chauvenet, W. (1863) Method of least squares. Appendix to *Manual of Spherical and Practical Astronomy*, Vol. 2, Lippincott, Philadelphia, pp. 469–566.

Chen, S.X. (1996) Studying school size effects in line transect sampling using the kernel method. *Biometrics*, **52**, 1283–1294

Christakos, G. and Hristopulos, D.T. (1997) Stochastic characterization of contaminated sites. *Journal of Applied Probability*, **34**, 988–1008.

Ciccone, G., Forastiere, F., Agabiti, N. *et al.* (1998) Road traffic and adverse respiratory effects in children. *Occupational and Environmental Medicine*, **55**, 771–778

Clayton, D. and Hills, M. (1993) *Statistical Models in Epidemiology*. Oxford University Press, Oxford.

Cochran, W.G. (1977) *Sampling Techniques* (3rd edition). John Wiley & Sons, Inc., New York.

Cochran, W.G. and Cox, G.M. (1957) *Design of Experiments*. John Wiley & Sons, Inc., New York.

Coles, S.G. and Walshaw, D. (1994) Directional modelling of extreme wind speeds. *Applied Statistics*, **43**, 139–157.

Coles, S.G. and Tawn, J.A. (1994) Statistical methods for multivariate extremes: An application to structural design. *Applied Statistics*, **43**, 1–48.

Cook, R.D. (1998) *Regression Graphics; Ideas for Studying Regressions through Graphics*. John Wiley & Sons, Inc., New York.

Cothern, C.R. and Ross, N.P. (1994) *Environmental Statistics, Asessment and Forecasting*. CRC, Boca Raton, FL.

Cox, D.R. (1955) Some statistical methods related with series of events (with Discussion). *Journal of the Royal Statistical Society, Series B*, **17**, 129–164.

Cox, D.R. (1958) *Planning of Experiments*. John Wiley & Sons, Ltd, London.

Cox, D.R. (1962) *Renewal Processes*. Methuen, London.

Cox, D.R. (1992) Causality – some statistical aspects. *Journal of the Royal Statistical Society, Series A*, **155**, 291–301.

Cox, D.R. and Isham, V. (1980) *Point Processes*. Chapman & Hall, London.

Cox, D.R. and Isham, V. (1988) A simple spatial-temporal model for rainfall. *Proceedings of the Royal Society (London), Series A*, **415**, 317–328

Cox, D.R. and Lewis, P.A.W. (1966) *The Statistical Analysis of Series of Events*. Chapman & Hall, London.

Cox, D.R. and Miller, H.D. (1965) *The Theory of Stochastic Processes*. Chapman & Hall, New York.

Cox, L.H., Guttorp, P., Sampson, P.D., Caccia, D.C. and Thompson, M.L. (1999) A preliminary statistical examination of the effects of uncertainty and variability on environmental regulatory criteria for ozone. In G.R. Bock and J.A. Goode (eds), *Environmental Statistics: Analysing Data for Environmental Policy*, Novatis Foundation Symposium 220. John Wiley & Sons, Ltd, Chichester, pp. 122–143.

Cressie, N.A.C. (1978) Removing nonadditivity from two-way tables with one observation per cell. *Biometrics*, **34**, 505–513.

Cressie, N. (1993) *Statistics for Spatial Data*. John Wiley & Sons, Inc., New York.

Cressie, N. and Huang, H.-C. (1999) Classes of non-separable, spatio-temporal stationary covariance functions. *Journal of the American Statistical Association*, **94**, 1330–1340.

Cressie, N. and Wikle, C.K. (2002) In A. El-Shaarawi and W.W. Piegorsch (eds), *Encyclopaedia of Environmetrics*. John Wiley & Sons, Ltd, Chichester, pp. 2045–2049.

Critchley, F. and Vitiello, C. (1991) The influence of observations on misclassification probability estimates in linear discriminant analysis. *Biometrika*, **78**, 677–690.

Crump, K. (2002) Benchmark analysis. In A. El-Shaarawi and W.W. Piegorsch (eds), *Encyclopaedia of Environmetrics*. John Wiley & Sons, Ltd, Chichester, pp. 163–170.

Darroch, J.N. (1958) The multiple-recapture census I: Estimation of a closed population. *Biometrika*, **45**, 343–359.

Darroch, J.N. (1959). The multiple-recapture census II: Estimation when there is immigration or death. *Biometrika*, **46**, 336–351.

David, H.A. (1981) *Order Statistics* (2nd edition). John Wiley & Sons, Inc., New York.

Davies, G. (1998). Extreme value theory. Undergraduate project, School of Mathematics, University of Nottingham, UK.

Davies, O.L. (ed.) (1956). *The Design and Analysis of Industrial Experiments* (2nd edition) Oliver and Boyd, Edinburgh.

Davis, A.C., Lovell, E.A., Smith, P.A. and Ferguson, M.A. (1998) The contribution of social noise to tinnitus in young people – a preliminary report *Noise and Health*, **1**, 40–46.

Davison, A.C. (1984) Modelling excesses over high thresholds, with an application. In J. Tiago de Olivei (ed.), *Statistical Extremes and Applications*. Riedel, Dordrecht.

Davison, A.C. and Smith, R.L. (1990) Models for exceedances over high thresholds (with Discussion). *Journal of the Royal Statistical Society, Series B*, **52**, 393–442.

Dell, T.R. and Clutter, J.L. (1972) Ranked set sampling theory with order statistics background. *Biometrics*, **28**, 545–555.

Department of the Environment, Transport and the Regions (2000) *The Air Quality Strategy for England, Scotland, Wales and Northern Ireland: A Consultation Document*. HMSO, London.

Devlin, S.J., Gnanadesikan, R. and Kettenring, J.R. (1975) Robust estimation of dispersion matrices and principal components. *Journal of the American Statistical Association*, **76**, 354–362.

Dey, D.K. (2002) Box–Cox transformation. In A. El-Shaarawi and W.W. Piegorsch (eds), *Encyclopaedia of Environmetrics*. John Wiley & Sons, Ltd, Chichester, pp. 220–223.

Diggle, P.J. (1975) Robust density estimation using distance methods. *Biometrika*, **62**, 39–48.

Diggle, P.J., Besag, J.E. and Gleaves, J.T. (1976) Statistical analysis of spatial point patterns by means of distance methods. *Biometrics*, **32**, 659–667.

Diggle P.J., Tawn, J.A. and Moyeed, R.A. (1998) Model-based geostatistics (with Discussion). *Applied Statistics*, **47**, 299–350.

Diggle, P.J., Liang, K.-Y. and Zeger, S.L. (1994) *The Analysis of Longitudinal Dta*. Oxford University Press, New York.

Dillman, D.A. (2000) *Mail and Telephone Surveys: The Total Design Method* (2nd edition). John Wiley & Sons, Inc., New York

Dillon, W.R. and Goldstein, M. (1984) *Multivariate Analysis*. John Wiley & Sons, Inc., New York.

Dixon, W.J. (1950) Analysis of extreme values. *Annals of Mathematical Statistics*, **21**, 488–506.

Dobson, A.J. (1990) *Introduction to Generalized Linear Models*. Chapman & Hall, London.

Dorfman, R. (1943) Detection of defective members of a large population. *Annals of Mathematical Statistics*, **14**, 436–488.

Downton, H.F. (1966) Linear estimates with polynomial coefficients. *Biometrika*, **53**, 129–141.

Draper, N.R. and Smith, H. (1998) *Applied Regression Analysis* (3rd edition). John Wiley & Sons, Inc., New York.

Drummer, T.D. and McDonald, L.L. (1987) Size bias in line transect sampling. *Biometrics*, **44**, 13–21.

DuMouchel, W. (1983) Estimating the stable index α in order to measure tail thickness. *Annals of Statistics*, **11**, 1019–1036.

Dunn, J.E. and Gipson, P. (1977) Analysis of radio-telemetry data in studies of home range. *Biometrics*, **33**, 85–101.

Eberhardt, L.L. (1968) A preliminary appraisal of line transects. *Journal of Wildlife Management*, **32**, 82–88

Elder, R.S., Thompson, W.O. and Myers, R.H. (1980) Properties of composite sampling procedures. *Technometrics*, **22**, 179–186.

Elliott, P., Westlake, A.J., Hills, M. *et al.* (1992) The Small Area Health Statistics Unit: A national facility for investigating health around point sources of environmental pollution in the United Kingdom. *Journal of Epidemiology and Community Health*, **46**, 345–349.

El-Shaarawi, A. and Piegorsch, W.W. (eds) (2002) *Encyclopaedia of Environmetrics* 4 vols. John Wiley & Sons, Ltd, Chichester.

Escoffier, B. and Le Roux, B. (1976) Factor's stability in correspondence analysis. How to control the influence of outlying data (in French). *Cahiers de l'Analyse des Données*, **1**, 297–318.

Feehrer, C.E. (2000) Dances with Wolf's: A short history of sunspot indices. www.aavso.org/observing/programs/ solar/dances.stm

Fellegi, I.P. (1975) Automatic editing and imputation of quantitative data (Summary). In *Bulletin of the International Statistical Institute: Proceedings of the 40th Session (Warsaw)*, Vol. 46, Book 3. ISI, Voorburg, Netherlands, pp. 249–253.

Feller, W. (1966) *An Introduction to Probability Theory and its Applications*. Vol. 2. John Wiley & Sons, Inc., New York.

Ferguson, T.S. (1961) On the rejection of outliers. In J. Neyman (ed.), *Proceedings of the Fourth Berkeley Symposium on Mathematical Statistics and Probability*, Vol. 1. University of California Press, Berkeley, pp. 253–287

Fienberg, S.E. and Tanur, J.M. (1987) Experimental and sampling structures: parallels diverging and maating. *International Statistical Review*, **55**, 75–96.

Finney, D.J (1971) *Probit Analysis* (3rd edition). Cambridge University Press, Cambridge.

Fisher, R.A. (1934) The effects of methods of ascertainment upon the estimation of frequencies. *Annals of Eugenics*, **6**, 13–25.

Fisher, R.A. (1966) *The Design of Experiments* (8th edition). Oliver and Boyd, Edinburgh.

Fisher, R.A. and Tippett, L.H.C. (1928) Limiting forms of the frequency distributions of the largest or smallest member of a sample. *Proceedings of the Cambridge Philosophical Society* **24**, 180–190.

Fuller, W.A. (1995) *Introduction to Statistical Time Series*. 2nd Ed. John Wiley & Sons, Inc., New York.

Galambos, J. (1987) *The Asymptotic Theory of Extreme Order Statistics*. 2nd Ed. John Wiley & Sons, Inc., New York.

Garner, F.C., Stapanian, M.A. and Williams, L.R. (1988) Composite sampling for environmental monitoring. In L.H. Keith (ed.), *Principles of Environmental Sampling*. American Chemical Society, Washington, DC, pp. 363–374.

Garthwaite, P.H., Jolliffe, I.T. and Jones, B. (1995) *Statistical Inference*. Prentice Hall, London.

Gates, C.E., Marshall, W.H. and Olson, D.P. (1968) Line transect method of estimating grouse population densities. *Biometrics*, **24**, 135–145.

Geladi, P. and Teugels, J. (1996) Extreme value theory and its potential role in chemometrics: An introduction. *Journal of Chemometrics*, **10**, 547–567.

Ganadesikan, R. (1977) *Methods of Statistical Data Analysis for Multivariate Observations*. John Wiley & Sons, Inc., New York.

Gnanadesikan, R. and Kettenring, J.R. (1972) Robust estimates, residuals and outlier detection with multiresponse data. *Biometrics*, **28**, 81–124.

Gneiting, T. and Schlather, M. (2002) Space-time covariance models. In A. El-Shaarawi and W.W. Piegorsch (eds), *Encyclopaedia of Environmetrics*. John Wiley & Sons, Ltd, Chichester, pp. 2041–2045.

Golub, G.H., Guttman, I. and Dutter, R. (1973) Examination of pseudo-residuals of outliers for detecting spuriosity in the general univariate linear model. In D.G. Kabe and P.R. Gupta (eds), *Multivariate Statistical Inference*. North-Holland, Amsterdam.

Gomes, M. I. (1993) On the estimation of parameters of rare events in environmental time series. In V. Barnett and K.F. Turkman (eds), *Statistics for the Environment*. John Wiley & Sons, Ltd, Chichester, pp. 225–241.

Gomes, M.I. (1994) Penultimate behaviour of extremes. In J. Galambos, J. Lechner and E. Simiu (eds), *Extreme Value Theory and Applications*. Kluwer, Dordrecht.

Goodall, C. and Mardia, K.V. (1994) Challenges in multivariate spatio-temporal modeling. In *Seventeenth International Biometrics Conference*. McMaster University, Hamilton, Ontario, Canada.

Goudey, R. and Laslett, G. (1999) Statistics and environmental policy: Case studies from long-term environmental monitoring data. In G.R. Bock and J.A. Goode (eds), *Environmental Statistics: Analysing Data for Environmental Policy*, Novatis Foundation Symposium 220. John Wiley & Sons, Ltd, Chichester, pp. 154–153.

Granger, C. (1966) The typical spectral shape of an economic variable. *Econometrica*, **34**, 150–161.

Green, E.J. and Strawderman, W.E. (1986) Reducing sample-size through the use of a composite estimator – an application to timber volume estimation. *Canadian Journal of Forest Research*, **16**, 1116–1118.

Groves, R.M. (1989) *Survey Errors and Survey Costs*. John Wiley & Sons, Inc., New York.

Grubbs, F.E. (1950) Sample criteria for testing outlying observations. *Annals of Mathematical Statistics*, **21**, 27–58.

Gumbel, E.J. (1958) *Statistics of Extremes*. Columbia University Press, New York.

Gupta, A.K. (1952) Estimation of the mean and standard deviation of a normal population from a censored sample. *Biometrika*, **39**, 260–273.

Guttman, I. (1973) Care and handling of univariate or multivatiate outliers in detecting spuriosity – a Bayesian approach. *Technometrics*, **15**, 723–738.

Guttorp, P. and Sampson, P.(1994) Methods for estimating heterogeneous spatial covariance functions with environmental applications. In G.P. Patil and C.R. Rao (eds), *Handbook of Statistics*, Vol. 12. Elsevier, Amsterdam, pp. 661–689.

Guttorp, P., Le, N.D., Sampson, P.D. and Zidek, J.V. (1993) Using entropy in the redesign of an environmental monitoring network. In G.P. Patil and C.R. Rao (eds), *Multivariate Environmental Statistics*, North-Holland, Amsterdam, pp. 175–202.

Guttorp, P., Meiring, M. and Sampson, P.D. (1994) A space-time analysis of groud-level ozone data. *Environmetrics*, **5**, 241–254.

Hahn, G.J. (1982) Design of experiments, an annotated bibliography. In S. Kotz, N.L. Johnson and C.B. Read (eds), *Encyclopedia of Statistical Sciences*, Vol. 2. John Wiley & Sons, Inc., New York, pp. 359–366.

Hall, W.J., and Wellner, J.A. (1981) Mean residual life. In M. Csörgő, D.A. Dawson, J.N.K. Rao and A.K.Md.E. Saleh (eds), *Proceedings of the International Symposium on Statistics and Related Topics*. North-Holland, Amsterdam, pp. 169–184.

Hampel, F.R. (1968) Contributions to the theory of robust estimation. Ph.D. thesis, University of California, Berkeley.

Hampel, F.R. (1971) A generalised qualitative definition of robustness. *Annals of Mathematical Statistics*, 42, 1887–1896.

Hampel. F.R. (1974) The influence curve and its role in robust estimation. *Journal of the American Statistical Association*, **69**, 383–393.

Hampel, F.R. Ronchetti, E.M., Rousseeuw, P.J. and Stahel, W.A. (1986) *Robust Statistics: The Approach Based on Influence Functions*. John Wiley & Sons, Inc., New York.

Handcock, M.S. and Wallis, J.R. (1994) An approach to statistical spatial-temporal modeling of meteorological fields (with Discussion). *Journal of the American Statistical Association*, **89**, 368–390.

Hannan, E.J. (1960) *Time Series Analysis*. Methuen, London.

Hannan, E.J. (1970) *Multiple Time Series*. John Wiley & Sons, Inc., New York

Harland, R., Pudsey, C. J., Howe, J.A. and FitzPatrick, M.E.J. (1998) Recent dinoflagellate cysts in a transect from the Falkland Trough to the Weddell Sea, Antarctica. *Palaeontology*, **41**, 1093–1131.

Harrison, P. (1992) *The Third Revolution*. Tauris, London.

Haslett, J. and Raftery, A.E. (1989) Space time modelling with long-memory dependence: Assessing Ireland's wind power resource (with Discussion). *Applied Statistics*, **38**, 1–50.

Hawkins, D.M. (1974) The detection of errors in multivariate data using principal components. *Journal of the American Statistical Association*, **69**, 340–344.

Hayne, D.W. (1949) An examination of the strip census method for estimating animal populations. *Journal of Wildlife Management*, **13**, 145–157.

Healy, M.R.J. (1968) Multivariate normal plotting. *Applied Statistics.*, **17**, 157–161.

Hedayat, A.S. and Sinha, B.K. (1991) *Design and Inference in Finite Population Sampling* John Wiley & Sons, Inc., New York.

Hewitt, C.N. (ed.) (1992) *Methods of Environmental Data Analysis*. Elsevier, London.

Hinkleman, K. and Kempthorne, O. (1994) *Design and Analysis of Experiments*, Vol. 1, John Wiley & Sons, Inc., New York.

Hocking, R.R. (1996) *Methods and Applications of Linear Models*. John Wiley & Sons, Inc., New York.

Hosking, J.R.M. (1985) Algorithm AS215: Maximum likelihood estimation of the parameters of the generalized extreme-value distribution. *Applied Statistics*, **34**, 301–310.

Hosking, J.R.M. and Wallis, J. R. (1987) Parameter and quantile estimation for the generalized Pareto distribution. *Technometrics*, **29**, 339–349.

Hosmer, D.W. and Lemeshow, S. (2000) *Applied Logistic Regression* (2nd edition). John Wiley & Sons, Inc., New York.

Huang, H.C. and Cressie, N. (1996) Spatio-temporal prediction of snow water equivalent using the Kalman filter. *Computational Statistics and Data Analysis*, **22**, 159–175.

Huber, P.J. (1981) *Robust Statistics*. John Wiley & Sons, Inc., New York

Hunter, J.S. (1994) Environmetrics: an emerging science. In G.P. Patil and C.R. Rao (eds), *Handbook of Statistics, Volume 12*. Elsevier, Amsterdam, pp. 1–7.

Hwang, F.K. (1972) A method for detecting all defective members in a population by group testing. *Journal of the American Statistical Association*, **67**, 605–608.

Jenkinson, D.S., Potts, J.M., Perry, J.N., Barnett, V., Coleman, K. and Johnston, A.E. (1994) Trends in herbage yields over the last century on the Rothamsted Long-term Continuous Hay Experiment. *Journal of Agricultural Science*, **122**, 365–374.

John, J.A. and Draper, N.R. (1980) An alternative family of transformations. *Applied Statistics*, **29**, 190–197.

John, P.W. (1998) *Statistical Design and Analysis of Experiments*. Society for Industrial and Applied Mathematics, Philadelphia.

Jolly, G.M. (1965) Explicit estimates from capture–recapture data with both death and immigration – stochastic model. *Biometrika*, **52**, 225–247.

Jones, R.H. and Zhang, Y. (1997) Models for continuous stationary space-time processes. In T.G. Gregoire, D.R. Brillinger, P.J. Diggle *et al.* (eds), *Modelling Longitudinal and Spatially Correlated Data: Methods, Applications, and Future Directions*, Lecture Notes in Statistics 122. Springer-Verlag, New York, pp. 289–298.

Jørgensen, B. (2002) Generalized linear models. In A. El-Shaarawi and W.W. Piegorsch (eds), *Encyclopaedia of Environmetrics*. John Wiley & Sons, Ltd, Chichester, pp. 873–880.

Kalman, R.E. (1960) A new approach to linear filtering and prediction problems. *Journal of Basic Engineering, Series D*, **82**, 32–45.

Kannan, N. and Raychaudhuri, A. (1994) Composite sampling: Effectiveness under a two stage scheme. *Communications in Statistics – Theory and Methods*, **23**, 2533–2540.

Karlin, S. and Taylor, H. (1981) *A First Course in Stochastic Processes*. Academic Press, New York.

Kaufmann, L. & Rousseeuw, P.J. (1990) *Finding Groups in Data*. John Wiley & Sons, Inc., New York.

Kaur, A., Patil, G.P. and Taillie, C. (1997). Unequal allocation models for ranked set sampling with skew distributions. *Biometrics*, **53**, 123–130.

Kenward, M.G. (2002) Repeated measures. In A. El-Shaarawi and W.W. Piegorsch (eds), *Encyclopaedia of Environmetrics*. John Wiley & Sons, Ltd, Chichester, pp. 1755–1757.

Kleiner, B., Martin, R.D. and Thompson, D.J. (1979). Robust estimation of power spectra (with Discussion). *Journal of the Royal Statistical Society, Series B*, **41**, 313–351.

Koch, G.G., Elasoff, J.D. and Amara, I.A. (1988) Repeated measurements – design and analysis. In S. Kotz, N.L. Johnson and C.B. Read (eds), *Encyclopedia of Statistical Sciences*, Vol. 8. John Wiley & Sons, Inc., New York, pp. 46–73.

Kotz, S., Johnson, N.L. and Read, C.B. (eds) (1982–1988) *Encyclopedia of Statistical Sciences*, 9 vols. John Wiley & Sons, Inc., New York.

Krige, D.G. (1966) Two-dimensional weighted moving average trend surfaces for ore-evaluation. *Journal of the South African Institute of Mining and Metallurgy*, **66**, 13–38.

Landau, S. and Barnett, V. (1996) A comparison of methods for climate data interpolation, in the context of yield predictions from winter wheat simulation models. In E.M. White, L.R. Benjamin, *et al.* (eds), *Aspects of Applied Biology: Modelling in Applied Biology. Vol. 46: Spatial Aspects*. Association of Applied Biologists, Wellesbourne, UK, pp 13–22.

Landau, S., Mitchell, R.A.C., Barnett, V., Colls, J.J., Craigon, J., Moore, K.L. and Payne, R.W. (1998) Testing winter wheat simulation models' predictions against observed UK grain yields. *Agricultural and Forest Meteorology*, **89**, 85–99.

Laplace, P.S. (1786) Sur les naissances, les marriages et les morts. *Mémoires de l'Académie Royale des Sciences, Année 1783*, 693–702.

Lawson, A. (1988) On tests for spatial trend in a nonhomogeneous Poisson process. *Journal of Applied Statistics*, **15**, 225–234.

Lawson, A.B. and Waller, L. (1996) A review of point pattern methods for spatial modelling of events around sources of pollution. *Environmetrics*, **7**, 471–488.

Leadbetter, M.R., Lindgren, G. and Rootzén, H. (1983) *Extremes and Related Properties of Random Sequences and Series*. Springer-Verlag, New York.

Lee, E.T. (1992) *Statistical Methods for Survival Data Analysis* 2nd Ed. John Wiley & Sons, Inc., New York.

Lehtonen, R. and Pahkinen, E.J. (2003) *Practical Methods for Design and Analysis of Complex Surveys* (2nd edition), John Wiley & Sons, Ltd, Chichester.

Levy, P.S. and Lemeshow, S. (1991) *Sampling of Populations: Methods and Applications* (3rd edition). John Wiley & Sons, Inc., New York.

Lieblein, J. (1954) Two early papers on the relation between extreme values and tensile strength. *Biometrika*, **41**, 559–560.

Lincoln, F.C. (1930) Calculating wildfowl abundance on the basis of banding returns. *US Department of Agriculture Circular* no. 118.

Linhart, H. and Zucchini, W. (1986) *Model Selection*. John Wiley & Sons, Inc., New York.

Lloyd, E.H. (1952) Least-squares estimation of location and scale parameters using order statistics. *Biometrika*, **34**, 41–67.

Loader, C. and Switzer, P. (1992). Spatial covariance estimation for monitoring data. In A. Walden and P. Guttorp (eds), *Statistics in Environmental and Earth Sciences*. Edward Arnold, London, pp. 52–70.

Lovison, G., Gore, S.D. and Patil, G.P. (1994) Design and analysis of composite sampling procedures: a review In G.P. Patil, and C.R. Rao, *Handbook of Statistics, Volume 12*. Elsevier, Amsterdam, pp 103–166.

Lund, R.B. and Seymour, L. (2002) Time series, periodic. In A. El-Shaarawi and W.W. Piegorsch (eds), *Encyclopaedia of Environmetrics*. John Wiley & Sons, Ltd, Chichester, pp. 2204–2210.

Lundberg, O. (1940) *On Random Processes and their Applications to Statistics and Accident Statistics*. Almqvist and Wiksells, Uppsala.

Macleod, A.J. (1989) AS R76 – a remark on algorithm AS 215: Maximum likelihood estimation of the parameters of the generalized extreme-value distribution. *Applied Statistics*, **38**, 198–199.

Maguire, B.A., Pearson, E.S. and Wynn, A.H.A. (1952) The time intervals between industrial accidents. *Biometrika*, **39**, 168–180.

Manly, B.F. (1976). Exponential data transformation. *The Statistician*, **25**, 37–42.

Manly, B.F.J. and McDonald, L.L. (1996) Sampling wildlife populations. *Chance*, **9**(2), 9–20.

Mann, N.R. & Schafer, R.E. & Singpurwalla, N.D. (1974) *Methods for Statistical Analysis of Reliability and Life Data*. John Wiley & Sons, Inc., New York.

Mardia, K.V. (1962) Multivariate Pareto distributions. *Annals of Statistics*, **33**, 1008–1015.

Mardia, K.V., Kent, J.T. and Bibby, J.M. (1995) *Multivariate Analysis*. Academic Press, New York.

Mardia, K.V., Goodall, C.R., Redfern, E. and Alonso, F.J. (1998). The kriged Kalman filter (with Discussion). *Test*, **7**, 217–285.

Maronna, R.A. (1976) Robust M-estimators of multivariate location and scatter. *Annals of Statistics*, **4**, 51–67.

Martin, R.J. (1986) On the design of experiments under spatial correlation. *Biometrika*, **73**, 247–277.

Matérn, B. (1960) Spatial variation: Stochastic models and their applications to problems in forest surveys and other sampling investigations. *Meddelanden från Statens Skogsforskningsinstitut*, **49**, 1–144.

Matheron, G. (1962) *Traité de Geostatistique Appliquée*, Vol. 1. Editions Technip, Paris.

Matheron, G. (1963) Principles of geoststistics. *Economic Geology*, **58**, 1246–1266.

Matheron, G. (1965) *Les Variables Regionalisées et Leur Estimation*. Masson, Paris.

Matheron, G. (1969) *Le Krigeage Universel*. Cahiers du Centre de Morphologie Mathematique de Fontainebleau, No. 1. Ecole Nationale Supérieure des Mines, Paris.

McArthur, R.D. (1987) An evaluation of sample designs for estimating a locally concentrated pollutant. *Communications in Statistics – Theory and Methods*, **16**, 739–759.

McBratney, A.B. and Webster, R. (1983) Optimal interpolation and isarithmic mapping of soil properties. V. Coregionalization and multiple sampling strategy. *Journal of Soil Science*, **34**, 137–162.

McBratney, A.B., Webster, R. and Burgess, T.M. (1981) The design of optimal sampling schemes for local estimation and mapping of regionalized variables. *Computers and Geosciences*, **7**, 331–334.

McCullagh, P. and Nelder, J.A. (1989) *Generalized Linear Models* (2nd edition). Chapman & Hall, London

McCulloch, C.E. and Searle, S.R. (2001) *Generalized, Linear, and Mixed Models*. John Wiley & Sons, Inc., New York.

McIntyre, G.A. (1952). A method of unbiased selective sampling, using ranked sets. *Aust. J. Agric. Res.*, **3**, 385–390.

Meiring, W., Monestiez, P., Sampson, P. D. and Guttorp, P. (1997) Developments in the modelling of nonstationary spatial convariance structure from space-time monitoring data. In Baafi, E. Y. and Schofield, N. (eds), *Geostatistics Wollongong '96*. Kluwer Academic, Dordrecht.

Meiring, W., Guttorp, P. and Sampson, P.D. (1998) Space-time estimation of grid-cell hourly ozone levels for assessment of a deterministic model. *Environmental and Ecological Statistics*, **5**, 197–222

Millard, S.P. and Neerchal, N.K. (2000) *Environmental Statistics with S-PLUS*. CRC Press, Boca Raton, FL.

Montgomery, D.C. and Peck, E.A. (1992) *Introduction to Linear regression Analysis* (2nd edition). John Wiley & Sons, Inc., New York.

Mood, A.M., Graybill, F.A. and Boes, D.C. (1974) *Introduction to the Theory of Statistics* (3rd edition). McGraw-Hill Kogakusha, Tokyo.

Morgan, B.J.T. (1992) *Analysis of Quantal Response Data*. Chapman & Hall, New York.

Morgan, B.J.T. (ed.) (1996) *Statistics in Toxicology*. Oxford University Press, Oxford.

Moser,C.A. and Kalton, G. (1999) *Survey Methods in Social Investigation (2nd edition)*. Heinemann, London.

Mugglestone, M.A. and Renshaw, E. (1996a) A practical guide to the spectral analysis of spatial point processes. *Computational Statistics and Data Analysis*, **21**, 43–65.

Mugglestone, M.A. and Renshaw, E. (1996b) The exploratory analysis of bivariate spatial point patterns using cross-spectra. *Environmetrics*, **7**, 361–377.

Mugglestone, M.A. and Renshaw, E. (1998) Detection of geological lineations on aerial photographs using two-dimensional spectral analysis. *Computers and Geosciences*, **24**, 771–784.

Mugglestone, M.A. and Renshaw, E. (2001) Spectral tests of randomness for spatial point patterns. *Environmental and Ecological Statistics*, **8**, 237–251.

Mugglestone, M.A., Barnett, V., Nirel, R. and Murray, D.A. (2000) Modelling and analysing outliers in spatial lattice data. *Mathematical and Computer Modelling*, **32**, 1–10.

Muttlak, H.A. and McDonald, L.L. (1990) Ranked set sampling with size-biased probability of selection. *Biometrics*, **46**, 435–445.

Myers, R.M. and Montgomery, D.C. (1995) *Response Surface Methodology* (2nd edition). John Wiley & Sons, Inc., New York.

Nakai, S., Itoh, T. and Morimoto, T. (1999). Deaths from heat-stroke in Japan: 1968–1994. *International Journal of Biometeorology*, **16**, 983–1004.

National Environmental Research Council (1975) *Flood Studies Report, Volume IV: Hydrological Data*. HMSO, London. Reprinted 1993.

Nelder, J.A.N. and Wedderburn, R.W.M. (1972). Generalized linear models. *Journal of the Royal Statistical Society Series A*, **135**, 370–384.

Nemetz, P.N. and Dreschler, H.D. (1978) The role of effluent monitoring in environmental control. *Water, Air and Soil Pollution*, **10**, 477–497.

Newman, S.C. (2001) *Biostatistical Methods in Epidemiology*. John Wiley & Sons, Inc., New York.

Neyman, J. (1939) On a new class of 'contagious' distributions, applicable in entomology and bacteriology. *Annals of Mathematical Statistics*, **10**, 35–57.

Neyman, J. and Scott, E.L. (1958). Statistical approach to problems of cosmology (with Discussion). *Journal of the Royal Statistical Society, Series B*, **20**, 1–43.

Nirel, R., Mugglestone, M.A. and Barnett, V. (1998) Outlier-robust spectral estimation for spatial lattice processes. *Communications in Statistics – Theory and Methods*, **27**, 3095–3111.

North, M. (1980) Time-dependent stochastic models of floods. *Journal of the Hydrology Division, American Society of Civil Engineers*, **106**, 649–655.

Nychka, D., Yang, Q. and Royle, J.A. (1997) Constructing spatial designs for monitoring air pollution using subset regression. In V. Barnett and K.F. turkman (eds), *Statistics for the Environment 3: Pollution Assessment and Control*. John Wiley & Sons, Ltd, Chichester, pp. 129–154.

Nychka, D., Wikle, C. and Royle, J.A. (2000) Large spatial prediction problems and nonstationary random fields. *Research report*, Geophysical Statistics Project, National Center for Atmospheric Research, Bolder, CO.*http://www.cgd.ucar.edu/stats/manuscripts/krig5.ps*.

Nychka, D, Piegorsch, W.W. and Cox, L.H. (eds) (1998) *Case Studies in Environmental Statistics*. Springer-Verlag, New York.

Oehlert, G.W. (1993) Regional trends in sulfate wet deposition. *Journal of the American Statistical Association*, **88**, 390–399.

Olea, R.A. (1984) Sampling design optimization for spatial functions. *Mathematical Geology*, **16**, 369–392.

Otis, D.L., Burnham, K.P., White, G.C. and Anderson, D.R. (1978) *Statistical Inference for Capture Data on Closed Animal Populations*, Wildlife Monographs, **62**. Bethesda, MD: The Wildlife Society.

Otis, D.L., McDonald, L.L. and Evans, M.A. (1993) Parameter estimation in encounter sampling surveys. *Journal of Wildlife Management*, **57**, 543–548.

Ott, W.R. (1995) *Environmental Statistics and Data Analysis*. Lewis, Boca-Raton, FL.

Otto, M.C. and Pollock, K.H. (1990) Size bias in line transect sampling: a field test. *Biometrics*, **46**, 239–245.

Parzen, E. (1962) *Stochastic Processes*. Holden-Day.

Patil, G.P. (1991) Encountered data, statistical ecology, environmental statistics and weighted distribution methods. *Environmetrics*, **2**, 377–433.

Patil, G.P. (2002) Composite sampling. In A. El-Shaarawi and W.W. Piegorsch (eds), *Encyclopaedia of Environmetrics*. John Wiley & Sons, Ltd, Chichester, pp. 387–391.

Patil, G.P. and Rao, C.R. (1977) Weighted distributions and a survey of their applications. In P.R. Krishnaiah (ed.), *Applications of Statistics*. North-Holland, Amsterdam.

Patil, G.P. and Taillie, C. (1989) Probing encountered data, meta analysis and weighted distribution methods. In Y. Dodge (ed.), *Statistical Data Analysis and Inference*. North- Holland, Amsterdam.

Patil, G.P., Rao, C.R. and Zelen, M. (1988) Weighted distributions. In S. Kotz, N.L. Johnson and C.B. Read (eds), *Encyclopedia of Statistical Sciences*, Vol. 9. John Wiley & Sons, Inc., New York, pp. 565–571.

Patil, G.P., Boswell, M.T., Gore, S.D. and Lovison G. (1996) Annotated bibliography of composite sampling: Part A: 1936–1992. *Environmental and Ecological Statistics*, **3**(1), 1–50.

Pearson, E.S. and Hartley, H.O. (1976) *Biometrika Tables for Statisticians*, Vol. 2. Griffin, London.

Peirce, B. (1852) Criterion for the rejection of doubtful observations. *Astronomical Journal*, **2**, 161–163.

Pentecost, A. (1999) *Analysing Environmental Data*. Longman, Harlow.

Phatarfod, R.M and Sudbury, A. (1994) The use of a square array scheme in blood testing. *Statistics in Medicine*, **13**, 2337–2343.

Pickands, J. (1975) Statistical inference using extreme order statistics. *Annals of Statistics*, **3**, 119–131.

Piegorsch, W.W. and Bailer, A.J. (1997) *Statistics for Environmental Biology and Toxicology*. Chapman & Hall, London.

Pielou, E.C. (1974) *Population and Community Ecology*. Gordon and Breach, New York.

Pollock, K.H. (1975) A *K*-sample tag-recapture model allowing for unequal survival and catchability. *Biometrika*, **62**, 577–583.

Pollock, K.H. (1978). A family of density estimators for line transect sampling *Biometrics*, **34**, 475–487

Pollock, K.H. (1981). Capture–recapture models allowing for age-dependent survival and capture rates. *Biometrics*, **37**, 521–529.

Pollock, K.H. (1991) Modelling capture, recapture and removal statistics for estimation of demographic parameters for fish and wildlife populations: Past, present and future. *Journal of the American Statistical Association*, **86**, 225–238.

Prescott, P. and Walden, A.T. (1980) Maximum likelihood estimators of the parameters of the generalized extreme value distribution. *Biometrika*, **67**, 723–724.

Prescott, P. and Walden, A.T. (1983) Maximum likelihood estimation of the parameters of the three-parameter generalized extreme-value distribution from censored samples. *Journal of Statistical Computation and Simulation*, **16**, 241–250.

Priestley, M.B. (1981) *Spectral Analysis and Time Series, Vol. 1: Univariate Series*. Academic Press, New York.

Quang, P.X. (1991) A nonparametric approach to size-biased line transect sampling. *Biometrics*, **47**, 269–279.

Quang, P.X. (1993) Nonparametric estimators for variable circular plot surveys. *Biometrics*, **49**, 837–852.

Ramsey, F.L. and Scott, J.M. (1979) Estimating population densities from variable circular plot surveys. In R.M. Cormack, G.P. Patil, and D.S. Robson (eds), *Sampling Biological Populations*. International Cooperative Publishing House, Burtonsville, MD, pp. 155–181.

Ramsey, F.L., Gates, C.E., Patil, G.P. and Taillie, C. (1988) On transect sampling to assess wildlife populations and marine resources. In P.R. Krishnaiah and C.R. Rao, (eds), *Handbook of Statistics, Volume 6: Sampling*. Elsevier, Amsterdam.

Ramsey, F.L., Wildman, V. and Engbring, J. (1987) Covariate adjustments to effective area in variable-area wildlife surveys. *Biometrics*, **43**, 1–11.

Rao, C.R. (1965). On discrete distributions arising out of methods of ascertainment. In G.P. Patil (ed.), *Classical and Contagious Discrete Distributions*. Pergamon, Calcutta

Rao, C.R. (1973) *Linear Statistical Inference and its Applications*. 2nd Ed. John Wiley & Sons, Inc., New York.

Renshaw, E. and Ford, E.D. (1983) The interpretation of process from pattern using two-dimensional spectral analysis: Methods and problems of interpretation. *Applied Statistics*, **32**, 51–63.

Renshaw, E. and Ford, E.D. (1984) The description of spatial pattern using two-dimensional spectral analysis. *Vegetatio*, **56**, 75–85.

Resnick, S.I. (1987) *Extreme Values, Regular Variation and Point Processes*. Springer-Verlag, New York.

Ridout, M.S. and Cobby, J.M. (1987) Ranked set sampling with non-random selection of sets and errors in ranking. *Applied Statistics*, **36**, 145–152.

Ripley, B.D. (1981) *Spatial Statistics*. John Wiley & Sons, Inc., New York

Ripley, B.D. and Kelley, F.P. (1977) Markov point processes. *Journal of the London Mathematical Society*, **15**, 188–192.

Robinette, W.L., Loveless, C.M. and Jones, D.A. (1974) Field tests of strip census methods. *Journal of Wildlife Management*, **38**, 81–96.

Robson, D.S. (1969) Mark-recapture methods of population estimation. In N.L. Johnson, and H. Smith, (eds) (1969) *New Developments in Survey Sampling*. John Wiley & Sons, Inc., New York

Rodríguez-Iturbe, I. and Mejía, J.M. (1974) The design of rainfall networks in time and space. *Water Resources Research*, **10**, 713–728.

Rohde, C.A (1976) Composite sampling. *Biometrics*, **32**, 273–282.

Rohlf, F.J. (1975) Generalisation of the gap test for the detection of multivariate outliers. *Biometrics*, **31**, 93–101.

Ross, G.J.S. (1995) Extended tables of variances of normal order statistics. Personal correspondence.

Rousseeuw, P.J. and Leroy, A.M. (1987) *Robust Regression and Outlier Detection*. John Wiley & Sons, Inc., New York

Rousseeuw, P.J. and van Zomeren, B.C. (1990) Unmasking multivariate outliers and leverage points. *Journal of the American Statistical Association*, **85**, 633–639.

Royal Commission on Environmental Pollution (1998) *Setting Environmental Standards*. The Stationary Office, London.

Sampson, P.D. (2002) Spatial covariance. In A. El-Shaarawi and W.W. Piegorsch (eds), *Encyclopaedia of Environmetrics*. John Wiley & Sons, Ltd, Chichester, pp. 2058–2067.

Sampson, P.D. and Guttorp, P. (1999) Operational evaluation of air quality models. In G.R. Bock and J.A. Goode (eds), *Environmental Statistics: Analysing Data for Environmental Policy*, Novatis Foundation Symposium 220. John Wiley & Sons, Ltd, Chichester, pp. 33–45.

Sarhan, A.E. and Greenberg, B.G. (eds) (1962) *Contributions to Order Statistics*. John Wiley & Sons, Inc., New York

Scarf, P. and Laycock, J. (1996) Estimation of extremes in corrosion engineering. *Journal of Applied Statistics*, **23**, 621–643.

Schnabel, Z.E. (1938) The estimation of the total fish population of a lake. *American Mathematical Monthly*, **45**, 348–352.

Seber, G.F. (1965). A note on the multiple recapture census. *Biometrika*, **52**, 330–349.

Seber, G.A.F. (1982) *The Estimation of Animal Abundance and Related Parameters* (2nd edition). Griffin. London

Seber, G.A.F. (1986) A review of estimating animal abundance. *Biometrics*, **42**, 267–292.

Seber, G.A.F. (1992) A review of estimating animal abundance II. *International Statistical Review*, **60**, 129–166.

Seber, G.A.F. and Manly, B.F.J. (1985) Approximately unbiased variance estimation for the Jolly–Seber mark–recapture model: Population size. In B.J.T. Morgan and P.M. North (eds), *Statistics in Ornithology*, Lecture Notes in Statistics 29. Springer-Verlag, Berlin, pp. 363–373.

Sen, A.R., Tourigny, J. and Smith, G.E.H. (1974) On the line transect sampling method. *Biometrics*, **30**, 329–340.

Silbergleit, V.M. (1998) On the statistics of maximum sunspot numbers. *Journal of Atmospheric and Solar-Terrestrial Physics*, **60**, 1707–1710.

Sinha, B.K., Sinha, B.K. and Purkayastha, S. (1996) On some aspects of ranked set sampling for estimation of normal and exponential parameters *Statistical Decisions*, **14**, 223–240.

Skinner, C.J., Holt, D. and Smith, T.M.F. (eds) (1989) *Analysis of Complex Surveys* John Wiley & Sons, Inc., New York

Silverman, BW. (1986) *Density Estimation*. Chapman & Hall, London

Skalski, J.R. and Robson, D.S. (1982). A mark and removal field procedure in estimating population abundance. *Journal of Wildlife Management*, **46**, 741–751.

Smith, R.L. (1982). Uniform rates of convergence in extreme-value theory. *Advances in Applied Probability*, **14**, 600–622.

Smith, R.L. (1985) Maximum likelihood estimation in a class of nonregular cases. *Biometrika*, **72**, 67–92.

Smith, R.L. (1987) Approximations in extreme value theory. Preprint, University of North Carolina.

Smith, R.L. (1999) Air pollution statistics in policy applications. In G.R. Bock and J.A. Goode (eds), *Environmental Statistics: Analysing Data for Environmental Policy*, Novatis Foundation Symposium 220. John Wiley & Sons, Ltd, Chichester, pp. 227–239.

Sobel, M. and Groll, P.A. (1966) Group testing to estimate efficiently all defectives in a binomial sample. *Bell System Technical Journal*, **38**, 1179–1252.

Solow, A.R. (2002) Time series, ecological. In A. El-Shaarawi and W.W. Piegorsch (eds), *Encyclopaedia of Environmetrics*. John Wiley & Sons, Ltd, Chichester, pp 2202–2204.

Speed, T. (1993) Modelling and managing a salmon population. In V. Barnett and K.F. Turkman (eds), *Statistics for the Environment*. John Wiley & Sons, Ltd, Chichester, pp. 267–292.

Spence, I. and Lewandowski, S. (1989) Robust multidimensional scaling. *Psychometrika*, **54**, 501–513.

Standards Australia (1997) *Guide to the Sampling and Investigation of Potentially Contaminated Soil. Part 1: Non-volatile and Semi-volatile Compounds*. AS 4482.1-1997. Standards Australia, Sydney.

Stapleton, J.H. (1995) *Linear Statistical Models*. John Wiley & Sons, Inc., New York.

Stehman, S.V. and Overton, W.S. (1994) Environmental sampling and monitoring. In G.P. Patil and C.R. Rao, *Handbook of Statistics, Volume 12: Environmental Statistics*. Elsevier, Amsterdam, pp. 263–306.

Sterrett, A. (1957). On the detection of defective members of a large population. *Annals of Mathematical Statistics*, **28**, 1033–1036.

Stokes, S.L. (1977) Ranked set sampling with concomitant variables. *Communications in Statistics – Theory and Methods*, **6**, 1207–1211.

Stokes, S.L. (1980) Estimation of variance using judgement ordered ranked-set samples. *Biometrics*, **36**, 35–42.

Stokes, S.L. (1995) Parametric ranked set sampling. *Annals of the Institute of Statistical Mathematics*, **47**, 465–482

Stone, R. (1993) The assumption on which causal inferences rest. *Journal of the Royal Statistical Society, Series B*, **55**, 455–466.

Stoyan, D. (1988) Thinnings of point processes and their use in the statistical analysis of a settlement pattern with deserted villages. *Statistics*, **19**, 45–56.

Stuart, A., Ord, J.K. and Arnold, S. (1999) *Kendall's Advanced Theory of Statistics, Vol. 2a. Classical Inference and Relationship*. Arnold, London.

Subba Rao, T. (1979) Discussion of Kleiner *et al.* (1979).

Takahasi, K. and Wakimoto, K. (1968) On unbiased estimates of the population mean based on a sample stratified by means of ordering. *Annals of the Institute of Statistical Mathematics*, **20**, 1–31.

Tawn, J.A. (1993) Extreme sea levels. In V. Barnett and K.F. Turkman (eds), *Statistics for the Environment*. John Wiley & Sons, Ltd, Chichester, pp. 243–263.

Tawn, J.A. and Vassie, J.M. (1989) Extreme sea levels: The joint probabilities method revisited and revised. *Proceedings of the Institution of Civil Engineers, Part 2*, **87**, 429–442.

Thompson, M.L., Cox, L.H., Sampson, P.D. and Caccia, D.C. (2002) Statistical hypothesis testing formulations for U.S. environmental regulatory standards for ozone. *Environmental and Ecological Statistics*, **9**, 321–339

Thompson, S.K. (2002) *Sampling* (2nd edition). John Wiley & Sons, Inc., New York.

Thompson, S.K. and Seber, G.A.F. (1996) *Adaptive Sampling*. John Wiley & Sons, Inc., New York.

Titterington, D.M. (1978) Estimation of correlation coefficients by ellipsoidal trimming. *Applied Statistics*, **27**, 227–234

Todorovic, P. and Rousselle, J. (1971) Some problems of flood analysis. *Water Resources Research*, **7**, 1144–1150.

Todorovic, P. and Zelenhasic, E. (1970) A stochastic model for flood analysis. *Water Resources Research* **6**, 1641–1648.

Townend, J. (2002) *Practical Statistics for Environmental and Biological Scientists*. John Wiley & Sons, Ltd, Chichester.

Tsutakawa, R.K. (1982) Bioassay, statistical methods. In S. Kotz, N.L. Johnson and C.B. Read (eds), *Encyclopedia of Statistical Sciences*, Vol. 1. John Wiley & Sons, Inc., New York, pp. 236–243.

Tukey, J.W. (1957) On the comparative anatomy of transformations. *Annals of Mathematical Statistics*, **28**, 602–632

Tukey, J.W. (1977) *Exploratory Data Analysis*, Vol 1. Addison-Wesley, Reading, MA.

Tukey, P.A. and Tukey, J.W. (1981) Data-driven view selection; agglomeration and sharpening. In V. Barnett (ed.), *Interpreting Multivariate Data*. John Wiley & Sons, Ltd, Chichester, pp. 215–243.

Upton, G.J.G. and Fingleton, B. (1985) *Spatial Data Analysis by Example, Vol. I: Point Pattern and Quantitative Data*. John Wiley & Sons, Ltd, Chichester.

Upton, G.J.G. and Fingleton, B. (1989) *Spatial Data Analysis by Example, Vol. II: Categorical and Directional Data*. John Wiley & Sons, Ltd, Chichester.

Urquhart, N.S., Overton, W.S. and Birkes, D.S. (1993) Comparing sampling designs for monitoring ecological status and trends: Impact of temporal patterns. In V. Barnett and K.F. turkman (eds), *Statistics for the Environment*. John Wiley & Sons, Ltd, Chichester, pp. 71–85.

USEPA (1995). *EPA Observational Economy Series Volume 1: Composite Sampling Guidelines*, EPA-230-R95-005. USEPA, Washington DC.

USEPA (1997). *National Ambient Air Quality Standards for Ozone; Final Rule*. At: www.epa.gov/fedrgstr/ EPA-AIR/1997/July/ Day-18/a18580.htm

van Belle, G. , Griffith, W.C. and Edland, S.D. (2001) Contributions to composite sampling. *Environmental and Ecological Statistics*, **8**, 171–180.

Velleman, P.F. and Hoaglin, D.C. (1981) *Applications, Basics, and Computing of Exploratory Data Analysis*, Duxbury Press.

Voltz, M. and R. Webster. (1990). A comparison of kriging, cubic splines and classification for predicting soil properties from sample information. *Journal of Soil Science*, **41**, 473–490.

Waldemeier, M. (1961). *The Sunspot-Activity in the Years 1610–1690*. Schulthess, Zurich.

Watson, G.H. (1936) A study of the group screening method *Technometrics*, **3**, 371–388.

Webster, R. and Oliver, M. (1990) Statistical methods in soil and land resource survey. New York: Oxford University Press.

Webster, R. and Oliver, M. (2001) *Geostatistics for Environmental Scientists*. John Wiley & Sons, Ltd, Chichester.

Wheater, C.P. and Cook, P.A. (2000) *Using Statistics to Understand the Environment*. Routledge, London.

White, G.C., Anderson, D.R., Burnham, K.P. and Otis, D.L. (1982). *Capture–Recapture and Removal Methods for Sampling Closed Populations*. Los Alamos National Laboratory, Los Alamos, NM.

Wikle, C.K. and Cressie, N. (1999) A dimension-reduced approach to space-time Kalman filtering. *Biometrika*, **86**, 815–829.

Wilks, S.S. (1963) Multivariate statistical outliers. *Sankhyā, A*, **25**, 407–426

Wilson, D.A., Butcher, D.P. and Labadz, J.C. (1997) Prediction of travel times and dispersion of pollutant spillages in non-tidal waters. In *Proceedings of Sixth National Hydrology Symposium*. Institute of Hydrology, Wallingford, pp. 4.13–4.19.

Wilson, B.J. and Brain, P. (1991) Long term stability of *Alopecurus myosuroides Huds.* within cereal fields. *Weed Research*, **31**, 367–373

Wise, M.E. (1982) Epidemiological statistics. In S. Kotz, N.L. Johnson and C.B. Read (eds), *Encyclopedia of Statistical Sciences*, Vol. 2. John Wiley & Sons, Inc., New York, pp. 523–537.

Worton, B.J. (1995) Modelling radio-tracking data. *Environmental and Ecological Statistics*, **2**, 15–23.

Worsley, K.J., Cao, J., Paus, T., Petrides, M. and Evans, A.C. (1998) Applications of random field theory to functional connectivity. *Human Brain Mapping*, **6**, 364–367.

Wu, S. and Zidek, J.V. (1992) An entropy-based analysis of data from selected NADP/NTN network sites for 1983–1986. *Atmospheric Environment*, **26A**, 2089–2103.

Yang, G.L. (1978) Estimation of a biometric function. *Annals of Statistics*, **6**, 112–116.

Yfantis, E.A., Flatmann, G.T. and Behar, J.V. (1987) Efficiency of kriging estimation for square, triangular and hexagonal grids. *Mathematical Geology*, **19**, 183–205.

Zidek, J., Sun, W. and Le, N.D. (2000) Designing and integrating composite networks for monitored multivariate Gaussian pollution fields. *Applied Statistics*, **49**, 63–79.

Zimmerman, D.L. (2002). Spatial design. Optimal. In A. El-Shaarawi and W.W. Piegorsch (eds), *Encyclopaedia of Environmetrics*. John Wiley & Sons, Ltd, Chichester, pp. 2067–2071.

Index

Note: *italic page numbers* indicate Figures and Tables

Environmental Statistics   V. Barnett
© 2004 John Wiley & Sons, Ltd   ISBN: 0-471-48971-9 (HB)

Index compiled by Paul Nash

WILEY SERIES IN PROBABILITY AND STATISTICS
ESTABLISHED BY WALTER A. SHEWHART AND SAMUEL S. WILKS

Editors: *David J. Balding, Peter Bloomfield, Noel A. C. Cressie, Nicholas I. Fisher, Iain M. Johnstone, J. B. Kadane, Louise M. Ryan, David W. Scott, Adrian F. M. Smith, Jozef L. Teugels*
Editors Emeriti: *Vic Barnett, J. Stuart Hunter, David G. Kendall*

The *Wiley Series in Probability and Statistics* is well established and authoritative. It covers many topics of current research interest in both pure and applied statistics and probability theory. Written by leading statisticians and institutions, the titles span both state-of-the-art developments in the field and classical methods.

Reflecting the wide range of current research in statistics, the series encompasses applied, methodological and theoretical statistics, ranging from applications and new techniques made possible by advances in computerized practice to rigorous treatment of theoretical approaches.

This series provides essential and invaluable reading for all statisticians, whether in academia, industry, government, or research.

ABRAHAM and LEDOLTER · Statistical Methods for Forecasting
AGRESTI · Analysis of Ordinal Categorical Data
AGRESTI · An Introduction to Categorical Data Analysis
AGRESTI · Categorical Data Analysis, *Second Edition*
ALTMAN, GILL, and McDONALD · Numerical Issues is Statistical Computing for the Social Scientist
AMARATUNGA and CABRERA · Exploration and Analysis of DNA Microarray and Protein Array Data
ANDĚL · Mathematics of Chance
ANDERSON · An Introduction to Multivariate Statistical Analysis, *Second Edition*
*ANDERSON · The Statistical Analysis of Time Series
ANDERSON, AUQUIER, HAUCK, OAKES, VANDAELE and WEISBERG · Statistical Methods for Comparative Studies
ANDERSON and LOYNES · The Teaching of Practical Statistics
ARMITAGE and DAVID (editors) · Advances in Biometry
ARNOLD, BALAKRISHNAN and NAGARAJA · Records
*ARTHANARI and DODGE · Mathematical Programming in Statistics
*BAILEY · The Elements of Stochastic Processes with Applications to the Natural Sciences
BALAKRISHNAN and KOUTRAS · Runs and Scans with Applications
BARNETT · Comparative Statistical Inference, *Third Edition*
BARNETT · Environmental Statistics: Methods and Applications
BARNETT and LEWIS · Outliers in Statistical Data, *Third Edition*
BARTOSZYNSKI and NIEWIADOMSKA-BUGAJ · Probability and Statistical Inference
BASILEVSKY · Statistical Factor Analysis and Related Methods: Theory and Applications
BASU and RIGDON · Statistical Methods for the Reliability of Repairable Systems

---

*Now available in a lower-priced paperback edition in the Wiley Classics Library.

BATES and WATTS · Nonlinear Regression Analysis and Its Applications

BECHHOFER, SANTNER and GOLDSMAN · Design and Analysis of Experiments for Statistical Selection, Screening, and Multiple Comparisons

BELSLEY · Conditioning Diagnostics: Collinearity and Weak Data in Regression

BELSLEY, KUH and WELSCH · Regression Diagnostics: Identifying Influential Data and Sources of Collinearity

BENDAT and PIERSOL · Random Data: Analysis and Measurement Procedures, *Third Edition*

BERRY, CHALONER and GEWEKE · Bayesian Analysis in Statistics and Econometrics: Essays in Honour of Arnold Zellner

BERNARDO and SMITH · Bayesian Theory

BHAT and MILLER · Elements of Applied Stochastic Processes, *Third Edition*

BHATTACHARYA and JOHNSON · Statistical Concepts and Methods

BHATTACHARYA and WAYMIRE · Stochastic Processes with Applications

BILLINGSLEY · Convergence of Probability Measures, *Second Edition*

BILLINGSLEY · Probability and Measure, *Third Edition*

BIRKES and DODGE · Alternative Methods of Regression

BLISCHKE and MURTHY (editors) · Case Studies in Reliability and Maintenance

BLISCHKE AND MURTHY · Reliability: Modeling, Prediction, and Optimization

BLOOMFIELD · Fourier Analysis of Time Series: An Introduction, *Second Edition*

BOLLEN · Structural Equations with Latent Variables

BOROVKOV · Ergodicity and Stability of Stochastic Processes

BOULEAU · Numerical Methods for Stochastic Processes

BOX · Bayesian Inference in Statistical Analysis

BOX · R. A. Fisher, the Life of a Scientist

BOX and DRAPER · Empirical Model-Building and Response Surfaces

*BOX and DRAPER · Evolutionary Operation: A Statistical Method for Process Improvement

BOX, HUNTER and HUNTER · Statistics for Experimenters: An Introduction to Design, Data Analysis, and Model Building

BOX and LUCENO · Statistical Control by Monitoring and Feedback Adjustment

BRANDIMARTE · Numerical Methods in Finance: A MATLAB-Based Introduction

BROWN and HOLLANDER · Statistics: A Biomedical Introduction

BRUNNER, DOMHOF and LANGER · Nonparametric Analysis of Longitudinal Data in Factorial Experiments

BUCKLEW · Large Deviation Techniques in Decision, Simulation, and Estimation

CAIROLI and DALANG · Sequential Stochastic Optimization

CHAN · Time Series: Applications to Finance

CHATTERJEE and HADI · Sensitivity Analysis in Linear Regression

CHATTERJEE and PRICE · Regression Analysis by Example, *Third Edition*

CHERNICK · Bootstrap Methods: A Practitioner's Guide

CHERNICK and FRIIS · Introductory Biostatistics for the Health Sciences

CHILÈS and DELFINER · Geostatistics: Modeling Spatial Uncertainty

CHOW and LIU · Design and Analysis of Clinical Trials: Concepts and Methodologies

CLARKE and DISNEY · Probability and Random Processes: A First Course with Applications, *Second Edition*

*COCHRAN and COX · Experimental Designs, *Second Edition*

CONGDON · Applied Bayesian Modelling

CONGDON · Bayesian Statistical Modelling

CONOVER · Practical Nonparametric Statistics, *Second Edition*

COOK · Regression Graphics

COOK and WEISBERG · Applied Regression Including Computing and Graphics

COOK and WEISBERG · An Introduction to Regression Graphics

---

*Now available in a lower priced paperback edition in the Wiley Classics Library.

GNANADESIKAN · Methods for Statistical Data Analysis of Multivariate Observations, *Second Edition*

GOLDSTEIN and LEWIS · Assessment: Problems, Development, and Statistical Issues

GREENWOOD and NIKULIN · A Guide to Chi-Squared Testing

GROSS and HARRIS · Fundamentals of Queueing Theory, *Third Edition*

*HAHN · Statistical Models in Engineering

HAHN and MEEKER · Statistical Intervals: A Guide for Practitioners

HALD · A History of Probability and Statistics and Their Applications Before 1750

HALD · A History of Mathematical Statistics from 1750 to 1930

HAMPEL · Robust Statistics: The Approach Based on Influence Functions

HANNAN and DEISTLER · The Statistical Theory of Linear Systems

HEIBERGER · Computation for the Analysis of Designed Experiments

HEDAYAT and SINHA · Design and Inference in Finite Population Sampling

HELLER · MACSYMA for Statisticians

HINKELMAN and KEMPTHORNE: · Design and Analysis of Experiments, Volume 1: Introduction to Experimental Design

HOAGLIN, MOSTELLER and TUKEY · Exploratory Approach to Analysis of Variance

HOAGLIN, MOSTELLER and TUKEY · Exploring Data Tables, Trends and Shapes

*HOAGLIN, MOSTELLER and TUKEY · Understanding Robust and Exploratory Data Analysis

HOCHBERG and TAMHANE · Multiple Comparison Procedures

HOCKING · Methods and Applications of Linear Models: Regression and the Analysis of Variables

HOEL · Introduction to Mathematical Statistics, *Fifth Edition*

HOGG and KLUGMAN · Loss Distributions

HOLLANDER and WOLFE · Nonparametric Statistical Methods, *Second Edition*

HOSMER and LEMESHOW · Applied Logistic Regression, *Second Edition*

HOSMER and LEMESHOW · Applied Survival Analysis: Regression Modeling of Time to Event Data

HØYLAND and RAUSAND · System Reliability Theory: Models and Statistical Methods

HUBER · Robust Statistics

HUBERTY · Applied Discriminant Analysis

HUNT and KENNEDY · Financial Derivatives in Theory and Practice

HUSKOVA, BERAN and DUPAC · Collected Works of Jaroslav Hajek–with Commentary

IMAN and CONOVER · A Modern Approach to Statistics

JACKSON · A User's Guide to Principle Components

JOHN · Statistical Methods in Engineering and Quality Assurance

JOHNSON · Multivariate Statistical Simulation

JOHNSON and BALAKRISHNAN · Advances in the Theory and Practice of Statistics: A Volume in Honor of Samuel Kotz

JUDGE, GRIFFITHS, HILL, LÜTKEPOHL and LEE · The Theory and Practice of Econometrics, *Second Edition*

JOHNSON and KOTZ · Distributions in Statistics

JOHNSON and KOTZ (editors) · Leading Personalities in Statistical Sciences: From the Seventeenth Century to the Present

JOHNSON, KOTZ and BALAKRISHNAN · Continuous Univariate Distributions, Volume 1, *Second Edition*

JOHNSON, KOTZ and BALAKRISHNAN · Continuous Univariate Distributions, Volume 2, *Second Edition*

---

*Now available in a lower priced paperback edition in the Wiley Classics Library.

---

*Now available in a lower priced paperback edition in the Wiley Classics Library.

McLACHLAN · Discriminant Analysis and Statistical Pattern Recognition

McLACHLAN and KRISHNAN · The EM Algorithm and Extensions

McLACHLAN and PEEL · Finite Mixture Models

McNEIL · Epidemiological Research Methods

MAGNUS and NEUDECKER · Matrix Differential Calculus with Applications in Statistics and Econometrics, *Revised Edition*

MALLER and ZHOU · Survival Analysis with Long Term Survivors

MALLOWS · Design, Data, and Analysis by Some Friends of Cuthbert Daniel

MANN, SCHAFER and SINGPURWALLA · Methods for Statistical Analysis of Reliability and Life Data

MANTON, WOODBURY and TOLLEY · Statistical Applications Using Fuzzy Sets

MARDIA and JUPP · Directional Statistics

MASON, GUNST and HESS · Statistical Design and Analysis of Experiments with Applications to Engineering and Science

MEEKER and ESCOBAR · Statistical Methods for Reliability Data

MEERSCHAERT and SCHEFFLER · Limit Distributions for Sums of Independent Random Vectors: Heavy Tails in Theory and Practice

*MILLER · Survival Analysis, *Second Edition*

MONTGOMERY, PECK and VINING · Introduction to Linear Regression Analysis, *Third Edition*

MORGENTHALER and TUKEY · Configural Polysampling: A Route to Practical Robustness

MUIRHEAD · Aspects of Multivariate Statistical Theory

MÜLLER and STOYAN · Comparison Methods for Stochastic Models and Risks

MURRAY · X-STAT 2.0 Statistical Experimentation, Design Data Analysis, and Nonlinear Optimization

MURTHY, XIE, and JIANG · Weibull Models

MYERS and MONTGOMERY · Response Surface Methodology: Process and Product Optimization Using Designed Experiments, *Second Edition*

MYERS, MONTGOMERY and VINING · Generalized Linear Models. With Applications in Engineering and the Sciences

NELSON · Accelerated Testing, Statistical Models, Test Plans, and Data Analyses

NELSON · Applied Life Data Analysis

NEWMAN · Biostatistical Methods in Epidemiology

OCHI · Applied Probability and Stochastic Processes in Engineering and Physical Sciences

OKABE, BOOTS, SUGIHARA and CHIU · Spatial Tesselations: Concepts and Applications of Voronoi Diagrams, *Second Edition*

OLIVER and SMITH · Influence Diagrams, Belief Nets and Decision Analysis

PALTA · Quantitative Methods in Population Health: Extensions of Ordinary Regressions

PANKRATZ · Forecasting with Dynamic Regression Models

PANKRATZ · Forecasting with Univariate Box-Jenkins Models: Concepts and Cases

*PARZEN · Modern Probability Theory and Its Applications

PEÑA, TIAO and TSAY · A Course in Time Series Analysis

PIANTADOSI · Clinical Trials: A Methodologic Perspective

PORT · Theoretical Probability for Applications

POURAHMADI · Foundations of Time Series Analysis and Prediction Theory

PRESS · Bayesian Statistics: Principles, Models, and Applications

PRESS · Subjective and Objective Bayesian Statistics, *Second Edition*

PRESS and TANUR · The Subjectivity of Scientists and the Bayesian Approach

PUKELSHEIM · Optimal Experimental Design

---

*Now available in a lower priced paperback edition in the Wiley Classics Library.

---

*Now available in a lower priced paperback edition in the Wiley Classics Library.

TANAKA · Time Series Analysis: Nonstationary and Noninvertible Distribution Theory

THOMPSON · Empirical Model Building

THOMPSON · Sampling, *Second Edition*

THOMPSON · Simulation: A Modeler's Approach

THOMPSON and SEBER · Adaptive Sampling

THOMPSON, WILLIAMS and FINDLAY · Models for Investors in Real World Markets

TIAO, BISGAARD, HILL, PEÑA and STIGLER (editors) · Box on Quality and Discovery: with Design, Control, and Robustness

TIERNEY · LISP-STAT: An Object-Oriented Environment for Statistical Computing and Dynamic Graphics

TSAY · Analysis of Financial Time Series

UPTON and FINGLETON · Spatial Data Analysis by Example, Volume II: Categorical and Directional Data

VAN BELLE · Statistical Rules of Thumb

VESTRUP · The Theory of Measures and Integration

VIDAKOVIC · Statistical Modeling by Wavelets

WEISBERG · Applied Linear Regression, *Second Edition*

WELSH · Aspects of Statistical Inference

WESTFALL and YOUNG · Resampling-Based Multiple Testing: Examples and Methods for p-Value Adjustment

WHITTAKER · Graphical Models in Applied Multivariate Statistics

WINKER · Optimization Heuristics in Economics: Applications of Threshold Accepting

WONNACOTT and WONNACOTT · Econometrics, *Second Edition*

WOODING · Planning Pharmaceutical Clinical Trials: Basic Statistical Principles

WOOLSON and CLARKE · Statistical Methods for the Analysis of Biomedical Data, *Second Edition*

WU and HAMADA · Experiments: Planning, Analysis, and Parameter Design Optimization

YANG · The Construction Theory of Denumerable Markov Processes

*ZELLNER · An Introduction to Bayesian Inference in Econometrics

ZHOU, OBUCHOWSKI and McCLISH · Statistical Methods in Diagnostic Medicine

---

*Now available in a lower priced paperback edition in the Wiley Classics Library.